Frontiers of Physics: 1900–1911

By the same author:

Imagery in Scientific Thought: Creating 20th-Century
   Physics, 1984
Albert Einstein's Special Theory of Relativity:
   Emergence (1905) and Early Interpretation (1905–1911),
   1981

Frontiers of Physics: 1900–1911

Selected Essays

With an Original Prologue and Postscript

Arthur I. Miller

A *Pro Scientia Viva* Title

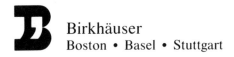
Birkhäuser
Boston • Basel • Stuttgart

Arthur I. Miller

Departments of Philosophy and History
University of Lowell
Lowell, Massachusetts 01854

Department of Physics
Harvard University
Cambridge, Massachusetts 02138

Library of Congress Cataloging in Publication Data
Miller, Arthur I.
  Frontiers of physics, 1900–1911.
  "A Pro scientia viva title."
  1. Physics—History. 2. Electrodynamics—History.
3. Special relativity (Physics)—History. I. Title.
QC7.M55   1986   530'.09'041   85-7488

CIP-Kurztitelaufnahme der Deutschen Bibliothek
Miller, Arthur I.:
Frontiers of physics: 1900–1911: selected
essays / Arthur I. Miller—Boston ; Basel ;
Stuttgart : Birkhäuser, 1986.
  (A Pro Scientia Viva title)
  ISBN 3-7643-3203-4 (Basel)
  ISBN 0-8176-3203-4 (Boston)

ISBN 0-8176-3203-4
     3-7643-3203-4

9 8 7 6 5 4 3 2 1

Printed in the USA

To My Mother

Contents

Prologue ix
Author's Notes xxiv
Acknowledgments xxv

Part I. On Electrodynamics Without Relativity: 1892–1911
  Essay 1: On Some Other Approaches to Electrodynamics in
      1905 2
  Essay 2: A Study of Henri Poincaré's "Sur la Dynamique de
      l'Électron" 29

Part II. A Technological Interlude
  Essay 3: Unipolar Induction: a Case Study of the Interaction
      between Science and Technology 153

Part III. Special Relativity
  Essay 4: On Einstein's Invention of Special Relativity 191

Part IV. On Some Contemporary Approaches to Special Relativity
  Essay 5: On Lorentz's Methodology 219
  Postscript: Further Comments on what Einstein, Lorentz, Planck,
      and Poincaré Did—circa 1905 237
  Essay 6: P.W. Bridgman and the Special Theory of Relativity 253

Prologue

Physicists in 1905 were after bigger game than a theory of space and time rooted in kinematics. Much as today, they sought a unified description of the then-known forces: electromagnetism and gravitation. They were optimistic about achieving this goal, principally for the following reasons. The proper theory was available: H.A. Lorentz's electromagnetic theory. The particle essential for grand unification had been discovered: the electron. The method for achieving the goal of this electromagnetic world-picture had been agreed upon: patient examination of the rich empirical data. Many of the best physicists of the first decade of the twentieth century were hard at work on this frontier of physics. Chief among them were France's greatest living mathematician and philosopher–scientist *par excellence*, Henri Poincaré; Holland's master of theoretical physics, Hendrik Antoon Lorentz; Germany's doyen of physics, Max Planck; Göttingen's brash young theorist, Max Abraham; Europe's first high-velocity elementary-particle experimentalist, Walter Kaufmann; and France's up-and-coming physicist, Paul Langevin. Their combined research efforts were the high watermark of classical electromagnetic theory. The fundamental problems with which they grappled remained to be dealt with in the quantum theory of the electron. Although they never attained their goal of unification, their struggles, methods, and intent are in every way exemplars of theoretical physics at its highest level.

Today, however, the elegance and depth of their work is overshadowed by a development in 1905 that was at first appreciated largely for the wrong reasons, if at all — namely, what its originator, Albert Einstein, Patent Clerk Third Class, Berne, Switzerland, referred to as the "so-called theory of relativity" (1909). In 1907 Einstein wrote that he chose the kinematical route to resolve what he took to be the basic problems in the physics of 1905 mainly because he realized that neither electromagnetic theory nor mechanics, nor thermodynamics, could serve as the basis for all science (Einstein, 1907). In 1907, however, no major physicist shared this view.

The essays brought together in this volume explore the rich tradition in electrodynamics, electrical engineering, and mathematics from which emerged the special theory of relativity. Although Einstein was out of the academic mainstream from 1902 to 1909, he moved within the scientific and philosophical currents of his day. Among those problems analyzed herein are: What was the situation in electrodynamics in 1905 that engendered in physicists the confidence that unification could be achieved? What was the interplay between developments in electromagnetic theory and electrical engineering from which Einstein would draw his unique view of physics? What was the effect of fundamental problems in the electromagnetic world-picture on subsequent theories of elementary particles? The essays in this book trace these prerelativistic developments in electrodynamics and electrical engineering from their beginnings into the present.

There are many reasons why it is important for the student, teacher, and researcher in the sciences to be reminded of the past. As with any other field of intellectual endeavor, to have a knowledge of the scientific tradition adds to one's effectiveness as a teacher and, sometimes, as a researcher. There is much good physics in theories that have fallen short of their goals. Looking back on failed theories, we see that in 1900 Henri Poincaré wrote in a chapter entitled, "Theories of Modern Physics," for his classic reprint volume *Science and Hypothesis*, that the "ruins may be still good for something." His point is that certain aspects of fallen theories survive to find their rightful place in their successors. Thus it is that from the failed theory of elementary particles of circa 1905 (i.e., the classical theory of the electron), the following survived: the Hamilton-Lagrange formulation of the electromagnetic field; group-theoretical methods applied to electrodynamics; radiation-reaction forces; the notion of Lorentz covariance and four-vector methods, among the other contributions that have been absorbed into standard physics courses and the physicist's repertoire. The techniques were developed by scientists who for the first time were faced with problems that concerned a domain beyond sense perceptions: the submicroscopic world of the electron. Their struggle to seek a fundamental theory in this new domain involved philosophical considerations as well. Throughout the history of science seminal advances in theoretical physics have been made by scientists with a philosophical bent. By this I do not mean scientists who try to spin new philosophical systems. Rather, I mean scientists such as Poincaré, Einstein, Niels Bohr, and Werner Heisenberg, who found it necessary to their scientific research to probe the foundations of science by assessing anew the structure of scientific theories, the origins of scientific concepts like space, time, causality, and substance, and then thinking itself. In this process philosophical currents of their day had to be borne in mind, and many of these contemporary ideas became transformed.

Another problem is how science progresses — on which the knowledge of the state of physics of the period 1900–1911 bears strongly — for it is essential to keep in mind that only in retrospect is the year 1905 an *Annus Mirabilis*. Not until 1909 did Einstein leave the Patent Office to take a position as Associate Professor at the University of Zurich, and this appointment was a result of his research on the quantum theory of solids, not on relativity. There was no scientific revolution in 1905; quite the contrary, in fact. Owing to the mathematical identity of certain formulae in Einstein's first paper on electrodynamics — which subsequently became known as the relativity paper — with those in Lorentz's 1904 theory of the electron, and the ease with which Einstein obtained them, there emerged a Lorentz-Einstein theory of the electron. The Patent Clerk did not protest having his name linked with one of the world's foremost physicists. It was not until 1911, owing mainly to Paul Ehrenfest's fundamental investigations on the contraction of moving bodies, that the views of Lorentz and Einstein were separated. Then, almost simultaneously, special relativity theory was declared to be no longer on the frontier of physics. With confidence we can even specify when the pronouncement was made that something new and very important had occurred in 1905. In September of 1911 Arnold Sommerfeld declared to the physicists gathered in Karlsruhe for the prestigious annual Meeting of German Natural Scientists and Physicians that the special relativity theory was in the "safe possession of physicists." He went on to define the new frontier as the meaning of Planck's energy quantum and the light quantum that Einstein had proposed in the "same memorable year of 1905, even before presenting the principle of relativity" (Sommerfeld, 1911). The Einstein legend would not break upon the public scene until November 1919, when a war-weary world learned of Arthur Stanley Eddington's empirical verification of general relativity's prediction for the bending of starlight near the sun. The years 1905 and 1915 then were declared *anni mirabili*. This, however, is a whole other story, and beyond the scope of this book.

The plan of this book is as follows:

In 1976 I discovered in Paris some of the correspondence and manuscripts of Henri Poincaré that had been missing since his death in 1912. Elsewhere in Europe I located additional letters of Poincaré. Four of these letters between Poincaré and Lorentz illustrate the fascinating process of science in the making, and are analyzed in Essay 1. In these letters Poincaré's acumen as a critic and philosopher whose point of view is in jeopardy, is played out against that of Lorentz, whose style was to push ahead and leave fundamental matters and problems of consistency to others. The issues that concerned these correspondents were the validity of the principle of action and reaction in Lorentz's electromagnetic theory, consistency problems in his 1904 theory of the electron, and how this theory conflicted with the empirical data of the time.

These newly discovered correspondences further illuminate a more detailed analysis of Poincaré's classic paper, "Sur la dynamique de l'électron," which is analyzed in Essay 2. There we see that Poincaré's results, obtained simultaneously and independently of Einstein's, are mathematically identical to those in the special theory of relativity. A goal of this essay is to distinguish between Poincaré's and Einstein's intents. Thus in Essay 2 I trace the roots and ramifications of Poincaré's paper, and then bring them up to date with developments in present-day quantum electrodynamics. The effect of Poincaré's complex philosophical view on his practice of physics — particularly concerning electromagnetic theory — is surveyed in order to illustrate that the interplay between Poincaré's philosophy and science was predominantly one way — from philosophy to science. Clearing up the state of physics in 1905 required the more balanced interplay between philosophy and science that Einstein introduced.

Essay 3 broadens the net of the inquiry through use of a case study in a style of mental imagery that I found influential in the developments in technology and science as those disciplines were practiced in the German cultural milieu. In this chapter an instance is discussed in which that style of imagery was counterproductive in German electrical engineering — namely, the research and development of the unipolar dynamo. This result is highlighted by comparing the German electrical industry with those in England and the United States. Certain fundamental problems concerning electromagnetic induction in systems undergoing relative rotatory motion were important to Einstein's invention of the special theory of relativity. (Einstein had attended the Swiss Polytechnic Institute in Zurich, where applied electricity was emphasized, and then had worked in a patent office from 1902 to 1909 assessing, among other things, patents on electrical dynamos.) In the special relativity paper of 1905 Einstein brusquely dismissed these fundamental problems as "meaningless" because he could solve them in principle. Their detailed resolution has yet to be accomplished. Essay 3 shows that even their resolution in principle is not straightforward and requires careful thought before the assertion can be made, based on the mental imagery of magnetic lines of force, that "everything is relative."

Essays 1–3 set the stage for Essay 4, which puts forth a scenario for Einstein's invention of the special theory of relativity. The proposed scenario takes into account Einstein's own recollections whenever they can be supported with primary sources, as well as the small amount of extant archival material pertaining to his route to special relativity. Essays 5 and 6 discuss certain interpretations of developments during the first decade of the twentieth century.

Essay 5 analyzes an attempt to interpret the relation between Lorentz's theory of the electron and Einstein's view of physics within the guidelines of Imre Lakatos' "Methodology of Scientific Research Programs" (MSRP). The progenitor of this brand of philosophy of science is Thomas S. Kuhn, although its roots go further back through Hans Reichenbach and Sir Karl Popper. For reasons that will soon become clear, I hasten to add that Kuhn's intent is to base case studies on as firm an historical ground as possible. The great interest in Kuhn's (1962) notion of scientific progress is in no small part due to the viable alternative it offered to the by-then sterile analytic approach to the analysis of the structure of a scientific theory. Kuhn emphasized such extrascientific components as the dynamics of a scientific community and the importance of taking into account cognitive psychological analysis toward studying what he then judged to be the essentially unquantifiable act of scientific discovery. Roughly, in Kuhn's view, scientific progress occurs in abrupt, that is, revolutionary, movement from one paradigm to another. In turn there were (and still are) reactions to Kuhn's view. Among the sharpest is Lakatos' "Methodology of Scientific Research Programs" (MSRP). The MSRP offers scenarios in which science advances in a totally rational manner (i.e., logical according to the MSRP), in contrast to Kuhn's scenario in which scientific change is a "mystical conversion...a kind of religious change" (Lakatos, 1970). Lakatos' student Elie Zahar applied the MSRP to the case of Lorentz and Einstein. Essay 5 is a critique of Zahar's results and of the MSRP generally.[1] This essay also demonstrates what a stumbling block the emergence of special relativity theory is for any methodology without a strong historical basis, especially one that intentionally excludes the intuitive or irrational dimension of creative thinking (see also Note 4 in Essay 4). The Postscript is my reply to Zahar's attempt in 1978 to counter the criticisms originally made in Essay 5.

Essay 6 employs previously unpublished archival materials to shed new light on the development of Percy W. Bridgman's operational view. In particular, the previously unpublished manuscripts of lectures and of early versions of Bridgman's monograph *A Sophisticate's Primer of Relativity* lend insight into his lifelong struggle to understand within the operational framework the connections between special relativity, the quantity we call light, the isotropy of space, clock synchronization, and the notion of the conventionality of distant simultaneity proposed by Hans Reichenbach and then elaborated on by Adolf Grünbaum. Discussion of attempts by Bridgman's self-acknowledged philosophical antecedents to understand these problems further illuminates physics during the first decade of the twentieth century.

Background

In order to set the stage for the analyses to follow I should like to develop the following topics: (1) the state of the discipline of physics circa 1900 so that we can better appreciate the importance to the physicists of that era of the quest for unification; and (2) the notion of scientific change because the problem of whether there are revolutions in science arises with a development like special relativity theory. (Essay 5 discusses this problem from another point of view.)

(1) Physics Circa 1900

The quest for an electromagnetic world-picture was in no small part due to the new esteem in which science held itself in 1900.[2] In turn this esteem resulted, on the one hand, from exciting new empirical data that had been discovered in the late 1890s and, on the other hand, from the outgrowth of the positive approach to science that had been advocated since the 1870s by, among others, Ernst Mach and Gustav Kirchhoff. The positive approach, or positivism, in turn had been a response to both the malaise into which science had been dropping and the then-current antiscience sentiment. This malaise resulted from the unfulfilled promises of the bygone period that a mechanical foundation would underlie all knowledge, as had been advocated by Newton and his followers, and which had served as the stimulus for the Age of Enlightenment. Giambattista Vico's cycles of history hold for science as well. The period that began in the 1870s and that ran into the first decade of the twentieth century was one characterized by a predominantly anti-science bias.

It was in 1872 that the German scientist Emil du Bois-Reymond declared that understanding a phenomenon meant describing it completely in mechanical terms. Even du Bois-Reymond acknowledged that certain phenomena, despite their importance, would remain forever a mystery because apparently they were not reducible to the laws of mechanics. Among these phenomena are the workings of the human brain — that is, the nature of thinking itself. About these great problems du Bois-Reymond wrote, "We shall never know," and so we should say *"ignorabimus"* instead of *"ignoramus"* (we do not know). The word *ignorabimus* (we shall not know) became the slogan of the antiscience movement and of defeatism in science itself. Shortly thereafter *ignorabimus* was replaced by the pungent slogan, "Bankruptcy of science." After all, science offered no solutions to the truly deep and pressing questions of life. Nor could science respond to the needs of society; rather, it often helped to increase society's miseries. The rise of technology seemed only to result in overcrowded cities, polluted air, and better instruments for destruction.

The problem of what is the proper domain for science was settled by the late 1880s. The second problem — that of the rise of technology — of course, is still with us, but by the late 1890s it, too, would be dealt with to some satisfaction.

Ernst Mach's hard-hitting criticism of the idle metaphysical baggage of science led the way toward resurrecting science from the blight of *ignorabimus*. Mach defined the proper domain for science to be the study of data of the senses and of the laboratory. He developed the view that the goal of science was merely to describe data economically: "Science itself therefore should be regarded as a minimal problem, consisting of the completest possible presentation of facts with the least possible expenditure of thought" (1883). Some of Mach's early attacks on the doctrine of *ignorabimus* were in his influential *Science of Mechanics* (1883), and from a biological standpoint in the *Analysis of Sensations* (1886). There was great resistance at first, as Mach recalled: "My first publications were, very naturally, received extremely coolly, and negatively by both physicists and philosophers." (1910) By the late 1880s he found allies. Among them was Gustav Kirchhoff, who argued in his 1876 *Lectures on Mechanics* that the goal of mechanics should be to describe phenomena in the simplest possible manner. Perhaps, though, it was Heinrich Hertz who put it most succinctly in his 1894 book *Principles of Mechanics* in which he declared that certain questions are "illegitimate" because they impede progress in science. Thus questions like "What is the nature of force, of velocity, of the mind?" are illegitimate, and the serious scientist does not concern himself with such problems.

Although in his mechanics book Hertz expressed his intellectual debt to Mach, Hertz was among the new wave of philospher-scientists who used positivism as a springboard to philosophical views that emphasized the role of mathematics and the primacy of the imagination. Mach had been led by his own empirical emphasis to conclude that the laws of mechanics, and of every branch of physics, can never be exact; therefore, mechanics could not be the basis of the sciences. But Hertz wrote that his goal was "Tracing the phenomena of nature back to the principal laws of mechanics." In contrast to Mach, Hertz advocated a mechanical world-picture based on mechanisms that are beyond sense-perceptions.

Another advocate of a different sort of positivism was Henri Poincaré. Poincaré's view was positivist to the extent that it emphasized sense-perceptions and empirical data, although he went on to weave it around organizing principles that he assumed to be innate. In other words, Poincaré's view was neo-Kantian (see Essay 2).

Fundamental analyses such as those of Mach, Kirchhoff, and Hertz in Germany, and Poincaré in France, were considered to have cleansed mechanics of metaphysical notions and to have clarified its goals. The mechanical world-picture supported by Hertz and others in Germany and by most British physicists, however, fed into the *fin-de-siècle* mood concerning science. According to the antiscience factions, not only had science and technology not lived up to their much-vaunted promises of improving the quality of life, but mechanism, with its materialistic basis, seemed to have as one of its goals the stripping away of all remaining vestiges of a spiritual quality. Thus, all phenomena would be reduced to matter in motion, and the completely deterministic clockwork universe of Newton would come into being.

But in 1896 the stock of science began to rise dramatically, and by 1900 it shattered the *fin-de-siècle* mood. In late 1895 Konrad Röntgen discovered X-rays, and his first X-ray photographs were nothing short of astounding. This basic scientific discovery was developed quickly for use in industry and medicine; for example, by the spring of 1896 the British Expeditionary Force took an X-ray machine to the Sudan. Then in 1897 J.J. Thomson announced his discovery of what he took to be the primordial element: the electron. In 1898 the Curies found radium, whose seemingly inexhaustible supply of heat appeared to violate basic laws of physics. At the same time the noble gases were discovered by Lord Rayleigh and William Ramsey.

As the prejudice against physics waned, so too, to some extent, did the negative attitude toward technology. Particularly in Germany, in contrast to the United States, close ties were forged between physics and electro-technology — that is, electrical engineering and, more specifically, the design and manufacture of electrical dynamos (Essay 3). More than anyone else, the "Edison" of Europe, Werner von Siemens, fostered these ties. Von Siemens was an inventor, scientist, entrepreneur, and patron of science and technology. In a ceremony in 1899 at the Berlin Technische Hochschule, for which Siemens had provided funds, the Kaiser unveiled monuments to those heroes to whom Germany owed her technological prominence: Siemens and Krupp. Among Siemens's other philanthropic endeavors was providing the funds in 1886 to establish a physical-technical institute like the one in Berlin at the Swiss Polytechnic Institute in Zurich. Just as at the Berlin institute, there would be close ties between scientists and engineers, a point of note when we consider that during the years 1896–1900 a student named Albert Einstein had studied at the institute in Zurich.

By 1900 the mechanical world-picture was sinking under the weight of its ponderous mechanical models of the ether. Lorentz's electromagnetic theory — in which the equations of electromagnetism were assumed to be axiomatic, requiring no mechanical models of the ether — was immensely successful, however. Thus in 1900 Wilhelm Wien suggested research toward an "electromagnetic basis for mechanics," or an "electromagnetic world-picture" as it would soon be called. According to this research program first mechanics and then all of physical theory would be reduced to Lorentz's electromagnetic theory. The electron's mass was assumed to originate in its self-electromagnetic field. Thus, it was predicted to be velocity-dependent, for which Walter Kaufmann in 1901 provided the empirical evidence. Flushed with excitement, Kaufmann boldly went further and asked, "Is it then possible to regard all masses as only apparent?" (1901) In 1902, Kaufmann's colleague at Göttingen, Max Abraham, formulated the first field theory of an elementary particle. Even Mach, who had opposed any effort to reduce science to the laws of one discipline, was willing to take a wait-and-see attitude. But by 1905 the Patent Clerk alone knew otherwise (Essay 4).

(2) On what is "sehr revolutionär"?

To remark that Einstein was busy during the spring of 1905 is an understatement. Besides spending eight hours per day, five days a week, at his job at the Patent Office in Berne, and tending to family duties, he sent off to the prestigious *Annalen der Physik,* at intervals of approximately eight weeks, three papers that were published in its Volume 17.[3] Any one of these papers would have won him a Nobel Prize — eventually, that is. They solved problems that were of varying degrees of concern to everyone and that had been set out, in fact, in Poincaré's classic *Science and Hypothesis,* which Einstein had read sometime during the period 1902–1904. But Einstein alone realized that the thread connecting them was the nature and constitution of light. The main problem, solved in the first paper, was why certain metals emit electrons only when ultraviolet light is shone on them (i.e., the photoelectric effect) — although Einstein's intent transcended this problem (1905a). (Incidentally, in 1921 Einstein was awarded the Nobel Prize "for his services to Theoretical Physics, and especially for his discovery of the law of the photoelectric effect.") The second paper in the *Annalen* deals with the erratic dance of dust particles in the air and pollen grains in solution, that is, Brownian motion (1905b). The triad is completed with Einstein's maiden effort at electrodynamics, which was applauded as having solved what Einstein considered a problem of secondary consequence — namely, exact equations for the motion of an electron (1905c). The physics community at first disregarded and then did its best to resist the first paper, with its proposal of a granular structure for light; the well-known Polish physicist Marian von Smoluchowski's 1906 solution to the problem of Brownian motion was found more comprehensible than Einstein's, although its methods were less deep; and the electrodynamics paper was interpreted as a nice generalization of a theory that Einstein was not even familiar with in any detail: Lorentz's theory of the electron.

Let us see how Einstein, in that glorious spring of 1905, described these three papers in a letter to his friend Conrad Habicht (in Seelig, 1954):

> ...the first...deals with radiation and energy characteristics of light and is very revolutionary [*sehr revolutionär*]...[another concerns] "Brown's molecular motion" [and then there is one that] exists in first draft and is an electrodynamics of moving bodies employing a modification of the doctrine of space and time; the purely kinematical part of this work will certainly interest you.

We notice that Einstein considered the paper on light quanta as *sehr revolutionär,* and not the relativity paper. This brings us to the much-debated question in the history of science of whether there are scientific revolutions. Einstein thought not. For example, in a manuscript of February 1947 he wrote (in Klein, 1975): "The reader gets the impression that every five minutes there is a revolution in science, somewhat like the *coups d'état* in some of the smaller unstable republics." Rather, in Einstein's view, science is transformed gradually by a "process of development to which the best brains of successive generations add by untiring labor [a process which] slowly leads to a deeper conception of the láws of nature." Then why did he refer to his paper as *sehr revolutionär*? The reason I offer below permits further understanding of his own style of research.

Let us look at the opening passages of the *sehr revolutionär* paper.

Concerning an Heuristic Point of View toward the Emission and Transformation of Light

By A. Einstein

A profound formal distinction exists between the theoretical concepts which physicists have formed regarding gases and other ponderable bodies and the Maxwellian theory of electromagnetic processes in so-called empty space.... According to the Maxwellian theory, energy is to be considered a continuous spatial function in the case of all purely electromagnetic phenomena including light, while the energy of a ponderable object should, according to the present conceptions of physicists, be represented as a sum carried over the atoms and electrons.... It seems to me that the observations associated with "blackbody radiation," fluorescence, the production of cathode rays by ultraviolet light, and other related phenomena connected with the emission or transformation of light are more readily understood if one assumes that the energy of light is discontinuously distributed in space.

Thus Einstein went far beyond Planck's original proposal to quantize only the energy exchanged between the radiation in a cavity and the charged oscillators that constitute the cavity's walls. Einstein quantized the radiation field itself, whether or not it was in a cavity. At first glance we might assume that this was what Einstein had in mind when he referred to the light-quantum paper as *sehr revolutionär*. But a close analysis, based in part on Einstein's first published papers on the wave-particle duality in 1909 and his correspondence, leads me to the following conjecture: What Einstein meant by *sehr revolutionär* is that the light quantum runs counter to our intuitions that we have constructed from observing phenomena in the world of sense-perceptions. Thus, for example, how can light, or anything for that matter, be both localized and extended at the same time? In fact, the light quantum was strongly resisted until 1927 when it found its place in the new atomic theory; by that time Einstein had long since abandoned the light quantum as not even having a heuristic significance. Until 1927 one clever hypothesis after another was proposed in the effort to avoid the light quantum and to deal only with the time-honored wave mode of light, with its intuitive pictures abstracted from such phenomena as water waves.[4]

There actually *is* something revolutionary about Einstein's first paper on light quanta, namely, his combination of symmetry principles and aesthetics in a scientific paper. Einstein's motivating argument for light quanta omits mention of empirical data to be explained, such as the photoelectric effect, as was *de rigueur* in the physics of 1905 (see Essay 4). There is a need, writes Einstein, to remove the "profound formal distinction between theoretical ideas which physicists have formed concerning gases and other ponderable bodies and the Maxwell theory of electromagnetic processes in so-called empty space." What he is getting at here is that it is asymmetrical or unaesthetic to have, side-by-side in the same theory, discrete atoms and continuous waves. This was the case in Lorentz's electromagnetic theory, where a discrete source of light (an electron) emits continuous waves of radiation, like the situation of throwing a stone into a pond, which results in forming spherical waves. Einstein offers a heuristic proposal or theoretical expedient for the purpose of removing the tension between the discontinuous and the continuous in favor of the discontinuous — that is, to consider discrete atoms and discrete light. Then he goes on to give a mathematical argument for light quanta and to discuss empirical data.

Around 1905 symmetry considerations had been employed in physics by Poincaré to deal with mathematical symmetries in the Lorentz transformation (Essays 1 and 2); by Lorentz in 1904 to deal with the approximate symmetry between phenomena that originate in moving dielectric and in magnetic matter (Essay 3); and by Abraham to investigate the symmetry properties of the electron's electromagnetic fields under spatial inversion in order to determine the conditions for an electron to undergo inertial motion (1903). Einstein was aware of Abraham's work and probably of Lorentz's, but not of Poincaré's (Essay 4). Einstein was probably also aware of Abraham's pronouncement, based on the initial success of his theory of the rigid sphere electron: "The electron's inertia is produced exclusively by its electromagnetic field. Therefore: atomistic structure of electricity, but continuous space distribution of the ether! Let that be our solution." (1902) Thus, according to Abraham's sense of aesthetics, unity of foundations took precedence over unity of physical forms. Particularly after the general relativity theory was completed in 1915, the theme of unity of foundations and of physical forms would surface in Einstein's quest for the unified field theory.

Einstein's use of symmetry arguments in the relativity paper went beyond Abraham, Lorentz, Poincaré, and even beyond his own methods in the light-quantum paper. In the relativity paper Einstein combined the notion of symmetry with the thought experiment for the purpose of supplementing the "known facts," as he referred in 1946 to the much-emphasized empirical data of 1905 (Essay 4). He began the relativity paper as follows: "That Maxwell's electrodynamics — in the way that it is usually understood — when applied to moving bodies, leads to asymmetries which do not appear to be inherent in the phenomena is well known. Consider, for example, the reciprocal electrodynamic action of a magnet and conductor." Although the phenomenon of electromagnetic induction was considered to be not yet entirely understood in 1905, all fundamental problems were assumed to reside in the workings of an apparatus with relative rotatory motion between magnet and conductor, and basically with the source of the moving magnet's electric field, that is, with the magnet's constitutive electrons (Essay 3). In the electrodynamics paper Einstein developed the simplest possible case of electromagnetic induction (magnet and conducting loop in relative inertial motion) in order to show that contemporary electromagnetic theory "contains asymmetries which do not appear to be inherent" (1905c) in this phenomenon. After deftly linking this blemish on the contemporary treatment of electromagnetic induction with problems in the optics of moving bodies, Einstein declared that fundamental problems in electromagnetism and optics resided in "insufficient consideration" of kinematics. Left unsaid, as was so much else in this scientific-literary masterpiece, is that for this reason the electromagnetic world-picture could not succeed.

Yet even during the paper's germination Einstein had taken pains to emphasize that his new theory required changes that were gradual, that were a "modification of the doctrine of space and time" (letter to Habicht in Spring 1905). In 1917 he wrote that the "special theory of relativity does not depart from classical mechanics through the postulate of relativity, but through the postulate of the constancy of light." In 1947, as we have seen, he was critical of the notion of scientific revolutions. Then in a letter of 1952 to his biographer Carl Seelig (1954) he wrote: "With respect to the theory of relativity it is not at all a question of a revolutionary act, but of a natural development of a line which can be pursued through centuries."

Thus in 1905 there was no forceful overthrow of the *ancien régime*. Its demise happened gradually until 1911, by which time it was displaced for most serious physicists. In each of the seminal developments in early twentieth-century physics — special relativity (1905), general relativity (1915), and quantum mechanics (1925) — the stated desire by the principal physicists initially was to salvage notions based on intuitions constructed from the world of perceptions, and then gradually to transform them in such a way that the new ones were linked in a well-defined manner to the familiar linguistic-perceptual anchors to the world in which we live. Subsequent research extended the quantum theory to quantum field theory (1929), nuclear theory (1932), and renormalization theory (1948), which required further transformation of basic concepts. But here, too, historical research shows that the transformations were gradual and the emphasis was on correspondence limit rules.

In summary, the notion of scientific revolutions describes at best only the gross structures of scientific change. In the fine structure, where change is gradual, resides such fascinating problems as the nature of creative scientific thinking.[5]

Notes

1. In its intent the MSRP is more ambitious than Kuhn's viewpoint. Lakatos hoped that the MSRP would be used as the set of guidelines to assess research proposals by funding organizations such as the National Science Foundation. Happily this did not come to pass. Lakatos will be remembered for his contributions to logic.
2. This section is based, in part, on material in Brush (1967), Frank (1947), Heilbron (1982), Holton (1981) and Miller (1981).
3. In addition, Einstein completed a Ph.D. thesis (1905d) and published a fourth paper on a result that he had overlooked from the electrodynamics paper — the equivalence of mass and energy (1905e).
4. This point is developed further in my (1984).
5. This theme is explored in my (1984).

Bibliography

Abraham, Max (1902): Prinzipien der Dynamik des Elektrons, *Phys. Z.*, *4*, 57-63.

(1903): Prinzipien der Dynamik des Elektrons, *Ann. Phys.*, *10*, 105-179.

Brush, Stephen G. (1967): Thermodynamics and History, *The Graduate Journal*, *7*, 477-565.

Einstein, Albert (1905a): Über einen die Erzeugung und Verwandlung des Lichtes betreffenden heuristischen Gesichtspunkt, *Ann. Phys.*, *17*, 132-148 (1905), translated by A.B. Arons and M.B. Peppard, *Am. J. Phys.*, *33*, 367-374 (1965).

(1905b): Die von der molekularkinetischen Theorie der Wärme geforderte Bewegung von in ruhenden Flüssigkeiten suspendierten Teilchen, *Ann. Phys.*, *17*, 549-560 (1905). Reprinted in *A. Einstein: Investigations on the Theory of the Brownian Movement* (New York: Dover, 1956), translated by A.D. Cowper, with notes by R. Furth.

(1905c): *Zur Elektrodynamik bewegter Körper, Ann. Phys.*, *17*, 891-921 (1905). Reprinted in Miller (1981), pp. 392-415.

(1905d): Eine neue Bestimmung der Moleküldimensionen, Inaugural-Dissertation, Zürich Universität.

(1905e): Ist die Trägheit eines Körpers von seinem Energieinhalt abhängig?, *Ann. Phys.*, *18*, 639-641 (1905). Reprinted in *H. A. Lorentz, A. Einstein, H. Minkowski, Das Relativitätsprinzip, eine Sammlung von*

Abhandlungen (Leipzig: Teubner, 1st ed., 1913; 2nd and 3rd enlarged eds., 1919, 1923), translated from the edition of 1923 by W. Perrett and G.B. Jeffery as *The Principle of Relativity: A Collection of Original Memoirs on the Special and General Theories of Relativity by H.A. Lorentz, A. Einstein, H. Minkowski and H. Weyl* (London: Methuen, 1923); the Methuen version was reprinted (New York: Dover, n.d.), pp. 69-71, where the volume number is stated incorrectly and the title misspelled.

(1907): Die vom Relativitätsprinzip geforderte Trägheit der Energie, *Ann. Phys., 23,* 371-384.

(1909): Entwicklung unserer Anschauungen über das Wesen und die Konstitution der Strahlung, *Phys. Z., 10,* 817-825.

(1917): *Relativity, the Special and the General Theory* (Braunschweig: Vieweg, 1917; New York: Holt, 1920), translated from the fifth German Edition by R.W. Lawson.

Frank, Philipp (1947): *Einstein: Sein Leben und seine Zeit* (New York: Knopf, 1947), translated by G. Rosen, and edited and revised by S. Kitsaka (New York: Knopf, 1953).

Heilbron, John L. (1982): *Fin-de Siècle Physics, in Science, Technology and Society in the Time of Alfred Nobel,* C.F. Bernhard, E. Crawford and P. Sörbom (eds.) (New York: Pergamon Press).

Hertz, Heinrich (1894): *The Principles of Mechanics* (Leipzig: Teubner, 1894; London: Macmillan, 1899; New York: Dover, 1956), translated by D.E. Jones and J.T. Walley, with a Preface by Hermann von Helmholtz. The Dover edition contains an Introduction by R.S. Cohen.

Holton, Gerald (1981): Einstein's Search for the *Weltbild, Proceedings of the American Philosophical Society, 125,* 1-15.

Kaufmann, Walter (1901): Die Entwicklung des Elektronenbegriffs, *Phys. Z., 3,* 9-15 (1901), translated in *Electrician, 48,* 94-97 (1901).

Klein, Martin J. (1975): Einstein on Scientific Revolutions, *Vistas in Astronomy, 17,* 113-120.

Kuhn, Thomas S. (1962): *The Structure of Scientific Revolutions* (Chicago: University of Chicago Press, 1962; enlarged ed., 1970).

Lakatos, Imre (1970): "Falsification and the Methodology of Scientific Research Programmes," in I. Lakatos and A.E. Musgrave (eds.): *Criticism and the Growth of Knowledge* (Cambridge University Press, Cambridge), pp. 91-195.

Mach, Ernst (1883): *Die Mechanik in ihrer Entwicklung historisch-kritisch dargestellt* (Leipzig: F.A. Brockhaus, 1883, 1889, 1897, 1901, 1904, 1908, 1912, 1921, 1933), translated in 1893 by T.J. McCormack from the second German edition and revised in 1942 to include additions and alterations up to the ninth German edition as *The Science of Mechanics: A Critical and Historical Account of Its Development* (La Salle, IL: Open Court, 1960).

(1886): *Analysis of Sensations,* (New York: Dover, 1959) translated by C.M. Williams from the 1st German edition of 1880 with Revisions and Supplemented from the 5th German edition.

(1910): Die Leitgedanken meiner naturwissenschaftlichen Erkenntnislehre und ihre Aufnahme durch die Zeitgenossen, *Phys. Z., 11,* 599-606.

Miller, Arthur I. (1981): *Albert Einstein's Special Theory of Relativity: Emergence (1905) and Early Interpretation (1905-1911).* (Reading, MA: Addison-Wesley, Advanced Book Program).

(1984): *Imagery in Scientific Thought: Creating 20th–Century Physics* (Cambridge, MA: Birkhäuser Boston, Inc., 1984; Cambridge, MA: MIT Press, 1986).

Seelig, Carl (1954): *Albert Einstein: Eine dokumentarische Biographie* (Zürich: Europa-Verlag).

Sommerfeld, Arnold (1911): Das Plancksche Wirkungsquantum und seine allgemeine Bedeutung für die Molekularphysik, *Phys. Z., 12,* 1057-1069.

Author's Notes

The essays are reproduced here as they were originally printed. Further development of Essay 2 is in my (1984). For results of my more recent research on the electron deflection experiments, particularly of Kaufmann, see my (1981). Occasionally, my (1984) is referred to as *On the Nature of Scientific Discovery*.

The overlap between essays here works to advantage in that the essays thereby are complementary.

Acknowledgments

For permission to reproduce the previously published essays I gratefully acknowledge the following publishers:

Essay 1: Reprinted from H. Woolf (ed.), *Some Strangeness in the Proportion* (Reading, MA: Addison-Wesley, Advanced Book Program, © 1980).

Essay 2: Reprinted from *Archive for History of the Exact Sciences, 10,* 207–328 (1973), © Springer-Verlag Publishers.

Essay 3: Reprinted from *Annals of Science, 38,* 155–189 (1981), © Taylor & Francis Ltd.

Essay 4: Reprinted from the *PSA Proceedings,* 1983, © Philosophy of Science Association.

Essay 5: Reprinted from *The British Journal for the Philosophy of Science, 25,* 29–45 (1974), © British Society for the Philosophy of Science.

Essay 6: Reprinted from my "Introduction" to P.W. Bridgman's *Sophisticate's Primer of Relativity* (2nd ed.; Middletown, CT: Wesleyan University Press, 1983) [Introduction © 1983, A.I. Miller.]

In each essay there is proper acknowledgment to the scholars with whom I discussed my work and to the funding agencies that made possible my research in the history of science. For permission to reprint material from their archives I thank the Estates of Albert Einstein and Henri Poincaré, and the Center for History of Physics, American Institute of Physics.

Part I

On Electrodynamics Without Relativity: 1892–1911

Essay 1

5. ON SOME OTHER APPROACHES TO ELECTRODYNAMICS IN 1905

The thrust of fundamental research in 1905 was, as it is today, toward the unification of physics within a field-theoretical framework. During the first decade of the twentieth century many physicists believed that this goal was imminent, and their researches possessed an elegance and significance that seventy-four years of relativity have tended to obscure. Albert Einstein's paper "On the Electrodynamics of Moving Bodies," the relativity paper, appeared September 26, 1905, but until 1911 many physicists referred to a "Lorentz–Einstein theory." Einstein's first paper on electrodynamics was considered to have enriched current research. This essay discusses some different approaches to those problems in electrodynamics that we now think of as posing the same puzzle — that is, relativity. The approaches to be analyzed are the theories of the electron that were proposed by Max Abraham, Paul Langevin, and H. A. Lorentz; these theories will be compared with each other and with Einstein's work. Recently discovered correspondence of Lorentz and Henri Poincaré will enable us to catch a glimpse of the drama that is science in the making.[1]

It is natural to begin with H. A. Lorentz's electromagnetic theory that he first proposed in his paper of 1892, "La théorie électromagnétique de Maxwell et son application aux corps mouvants,"[2] because his theory was central to developments in what was deemed to be basic research at the beginning of the twentieth century. Lorentz's theory was the successful result of over two decades of elaborations and purifications of James Clerk Maxwell's electromagnetic theory. One of Lorentz's fundamental premises was that the sources of the electromagnetic field were undiscovered electrons that moved about in an all-pervasive absolutely resting ether. Lorentz described the state of the ether with five fundamental equations that he considered axiomatic:

$$\nabla \times \mathbf{E} = -\frac{1}{c}\frac{\partial \mathbf{B}}{\partial t} \tag{5.1}$$

$$\nabla \times \mathbf{B} = \frac{1}{c}\frac{\partial \mathbf{E}}{\partial t} + \frac{4\pi}{c}\rho\mathbf{v} \qquad \text{Maxwell–Lorentz equations} \tag{5.2}$$

$$\nabla \cdot \mathbf{E} = 4\pi\rho \tag{5.3}$$

$$\nabla \cdot \mathbf{B} = 0$$

Harry Woolf (ed.), Some Strangeness in the Proportion: A Centennial Symposium to Celebrate the Achievements of Albert Einstein   ISBN 0-201-09924-1
Copyright © 1980 by Addison-Wesley Publishing Company, Inc., Advanced Book Program.

$$\mathbf{F} = \rho\mathbf{E} + \rho\,\frac{\mathbf{v}}{c} \times \mathbf{B}, \qquad\qquad \text{Lorentz force equation} \qquad (5.5)$$

where $\mathbf{E}$ and $\mathbf{B}$ are the electromagnetic fields, ρ is the electron's charge density, and the equations are written relative to a reference system fixed in the ether that I shall call S^3; c is the velocity of light relative to S; and $\mathbf{v}$ is the electron's velocity relative to S. Lorentz's equations are consistent with the basic requirement of a wave theory of light: namely, that relative to the ether the velocity of light is independent of the source's motion and is always c. But measuring the velocity of light in moving reference systems may give a different result. In such a case, the measured velocity of light should be direction-dependent in a manner given by Newton's addition law of velocities. However, as is so well known, despite much effort, no measurable effects of the moving earth on optical or electromagnetic phenomena had been found — that is, the velocity of light measured on the earth turned out to be c.

In his 1895 monograph entitled *Treatise on a Theory of Electrical and Optical Phenomena in Moving Bodies*,[19] Lorentz was able to explain systematically the failure of ether-drift experiments accurate to first order in the quantity v/c, where v was the velocity of neutral, nonmagnetic, non-dielectric matter relative to the ether. Using Galilean spatial transformations and a mathematical "local-time coordinate" t_L, with certain "new" vectors for the electromagnetic fields, Lorentz demonstrated that the electromagnetic field equations have the same form in an inertial reference system S_r as in S (see Fig. 5.1). Thus, to order v/c, neither optical nor electromagnetic experiments could reveal the motion of S_r. Lorentz called this stunning and desirable result the "theorem of corresponding states," because if a state of optics is described in S by $\mathbf{E}$ and $\mathbf{B}$ as functions of x, y,

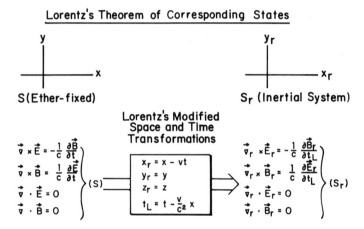

Fig. 5.1 Lorentz's modified space and time transformations contain the "local-time coordinate" t_L, and Lorentz referred to the electromagnetic field quantities $\mathbf{E}_r = \mathbf{E} + \mathbf{v}/c \times \mathbf{B}$ and $\mathbf{B}_r = \mathbf{B} - (\mathbf{v}/c) \times \mathbf{E}$ as "new" vectors.

z, and t, there is a corresponding state in S_r characterized by the "new" electromagnetic-field quantities as functions of x_r, y_r, z_r, and t_L. Needless to say, the physical time was taken to be identical with the absolute time because there was no reason to consider that time depended on the motion of a reference system.

By 1900 the electron Lorentz had hoped for had been discovered and the theory was able to explain systematically most extant empirical data. So successful had Lorentz's theory become that Henri Poincaré's assessments of it had gone from the "least defective" (in 1895) to the "most satisfactory" (1900).[4] However, Poincaré was disturbed by two blemishes on Lorentz's theory that Lorentz had been willing to overlook: 1) Lorentz's ad hoc 1892 contraction hypothesis for explaining the 1887 Michelson and Morley experiment involving an effect in second order in v/c; Poincaré scathingly criticized the contraction hypothesis; "hypotheses are what we lack least."[5] 2) But even worse, in Poincaré's opinion, Lorentz's theory violated Newton's principle of action and reaction. Lorentz had built this violation into his theory since the ether acted on bodies, but not vice versa. In his publication of 1895 Lorentz had dodged this issue by asserting that Newton's principle of action and reaction need not be universally valid. Newton's principle of action and reaction, however, was on the highest level of Poincaré's hierarchical view of a scientific theory because its generality precluded its experimental disconfirmation.[6]

The 1900 *Lorentz-Festschrift*, celebrating the twenty-fifth anniversary of Lorentz's doctorate from Leiden, included Poincaré's paper, "The Theory of Lorentz and the Principle of Reaction,"[7] in which he demonstrated the lengths to which he was prepared to go in order to save a principle. In his work, using the Maxwell–Lorentz equations, Poincaré reexpressed the Lorentz force on a charged body enclosed in a volume V, owing to charges external to V, as

$$\mathbf{F} = \nabla \cdot \overset{\leftrightarrow}{\mathbf{T}} - \frac{d}{dt} \frac{\mathbf{E} \times \mathbf{B}}{4\pi c} \tag{5.6}$$

where $\overset{\leftrightarrow}{\mathbf{T}}$ is Maxwell's stress tensor. He calculated the net force $\mathbf{F}_N$ on the charged body by integrating over the volume V, which he extended out to infinity in order to include the sources of the external electromagnetic fields. Poincaré's result is

$$\mathbf{F}_N = - \frac{d}{dt} \int \frac{\mathbf{E} \times \mathbf{B}}{4\pi c} dV, \tag{5.7}$$

since the contribution from the Maxwell stress tensor vanishes. Consequently, in Lorentz's theory a charged body cannot attain equilibrium. Poincaré traced this result to the absence of isolated systems in Lorentz's theory in the sense that this term was used in mechanics. Setting

$$\mathbf{F}_N = \frac{d}{dt} m_o \mathbf{v} \tag{5.8}$$

where m_o is the charged body's inertial mass, Poincaré rewrote Eq. (5.7) as

$$\frac{d}{dt} \left[m_o \mathbf{v} + \int \frac{\mathbf{E} \times \mathbf{B}}{4\pi c} dV \right] = \mathbf{0}. \tag{5.9}$$

4

The second term in Eq. (5.9) violated the principle of action and reaction as this principle was understood in Newton's mechanics. Poincaré next took the short but bold step that his philosophic view demanded: in order to rescue Newton's principle of action and reaction he compared the electromagnetic field to a "fictitious fluid" with a mass and "momentum"

$$\mathbf{G} = \frac{1}{4\pi c} \int \mathbf{E} \times \mathbf{B} \, dV. \tag{5.10}$$

Thus, Poincaré's electromagnetic momentum permitted him to simulate isolated closed systems in Lorentz's theory. In these sytems the net force $\mathbf{F}_N$ was cancelled by compensatory mechanisms in the ether arising from the electromagnetic momentum's temporal variation. But further hypotheses were necessary. Suppose, continued Poincaré, that an emitter of unidirectional radiation and an absorber were in relative inertial motion. Using Lorentz's local-time coordinate Poincaré demonstrated the insufficiency of the electromagnetic momentum for saving the principle of action and reaction separately for emitter and absorber. For this purpose he postulated an "apparent complementary force." In order to emphasize the necessity for this desperate step to save a principle, Poincaré reiterated that in conservative mechanical systems the principle of action and reaction could be considered as a consequence of the principles of energy conservation and of relative motion. According to Poincaré's view of mechanics, what he called the "principle of relative motion" could never be overthrown because it embodied the covariance of Newton's laws in inertial reference systems *and* Newton's second law;[8] furthermore, the principle of relative motion asserted the meaningfulness of only the relative motion between ponderable bodies, which is information drawn from the world of our perceptions; and so, wrote Poincaré, the "contrary hypothesis is singularly repugnant to the mind."

In a collection of Poincaré documents in Paris there is a letter of January 20, 1901 from Lorentz to Poincaré (Fig. 5.2). In this letter Lorentz expressed his admiration for Poincaré's contribution to the *Festschrift*, and then launched into an eight-page rebuttal. Lorentz's opinion of Poincaré's valiant attempt at saving the principle of action and reaction was, in his words: "But must we, in truth, worry ourselves about it?" Lorentz added in his forthright way: "I must claim to you that it is impossible for me to modify the theory in such a way that the difficulty that you cited disappears." Lorentz went on to emphasize several times that his ether acted on bodies but that there was no reaction on the ether. He explained that the "phenomena of aberration," that is, first-order effects, had "forced him" to assume a motionless ether; in fact, in Lorentz's seminal paper on electromagnetic theory,[2] second-order effects were nowhere mentioned.[10] Lorentz continued in his letter: "I deny therefore the principle of reaction in these elementary actions." In mechanics, Lorentz continued, action and reaction were instantaneous because disturbances were not mediated by an ether; however, in electromagnetic theory the reaction of an emitter of radiation was not compensated simultaneously by the action on the absorber. Poincaré had avoided this problem by attempting to satisfy the principle of reaction separately by emitter and absorber. Consistent with his desire to maintain an absolutely immobile ether, Lorentz protested Poincaré's naming the quantity in Eq. (5.10), which Lorentz compared to Poynting's vector, to be an electromagnetic momentum. To Lorentz the term momentum, of course, connoted motion. Lorentz was willing to concede only that Poincaré's electromagnetic momentum was formally "'equivalent' to a momentum." Thus, in the 1901 letter, Lorentz informed the critical Poincaré of his own sensitivity toward adding further hypotheses to an already overburdened theory, especially hypo-

5

Leiden, le 20 janvier 1901

Monsieur et très honoré collègue,

Permettez moi de vous remercier bien sincèrement de la part que vous avez bien voulu prendre au recueil de travaux que m'a été offert à l'occasion du 25eme anniversaire de mon doctorat. J'ai été profondiment touché de ce que tant d'illustres savants aux choisi ce jour pour me témoigner leur sympathie et l'intérêt qu'ils peuvent à mes études, malgré l'imperfection des resultats auxquels elles m'ont conduit. Cette imperfection est telle que je n'ose presque pas regarder comme un signe d'approbation le livre qu'on m'a dédié; j'y verrai plutôt un encouragement qui m'est précieuse.

Comme votre jugement a, à mes yeux une tres grande importance, vous m'avez particulière-ment obligé par le choise de votre sujet et par les paroles qui précèdent votre article. J'ai suivi vos raisonnements avec toute l'attention qu'ils demandent et je sens toute la force de vos remarques. Je dois vous avouer qu'il m'est impossible de modifier la théorie de telle façon que la difficulté que vous signalez disparaitre. Il me semble même guère probable qu'on puisse y réussir; je crois plutôt — et c'est aussi le résultat auquel tendent vos remarques — que la violation du principe de réaction est nécessaire dans toutes les théories que peuvent expliquer l'expérience de Fizeau. Mais faut-il en vérité que nous nous en inquiétions. Il y a un certain rapport entre vos considérations et une ques-tion qui a été soulevée, comme vous savez, par Helmholtz dans un de ces derniers mémoires. En ef-fet, vos formules démontrent que l'éther contenu dans une surface fermée ne sera pas en équilibre sous l'influence des pressions de Maxwell exercées à cette surface, des que le vecteur de Poynting est une fonction du temps. De ceci, Helmholtz tire la conclusion que l'éther sera mis en mouve-ment dans un tel cas, et il cherche à établir les équations qui déterminent ce mouvement.

J'ai préféré une autre manière de voir. Ayant toujours en vue les phénomènes de l'aberration, j'ai admis que l'éther est absolument immobile — je veux dire que ses éléments de volume ne se dé-placement pas, bien qu'ils puissent être le siège de certains mouvements internes. Or, si un corps ne se dèplace jamais, il n'y a aucune raison pour laquelle on parlerait de forces exercées sur ce corps. C'est ainsi que j'ai été amené à ne plus parler de forces qui agissent sur l'éther.

Je dis que l'éther agit sur les électrons, mais je ne dis pas qu'il éprouve de leur côté une réac-tion; je nie donc le principe de la réaction dans ces actions élémentaires. Dans cet ordre d'idées je ne puis pas non plus parler d'une force exercée par une partie de l'éther sur l'autre; les pressions de Maxwell n'ont plus d'existence réelle et ne sont que des fictions mathématiques qui servent à calculer d'une manière simple la force qui agit sur un corps pondérable. Evidemment, je n'ai plus à me soucier de ce que les pressions qui agisseront à la surface d'une portion le partie de l'éther ne seraient pas en équilibre.

Quant au principe de la réaction, il ne me semble pas qu'il doive être regardé comme un prin-cipe fondamental de la physique. Il est vrai que dans tous les cas où un corps acquiert une certaine quantité de mouvement a, notre esprit ne sera pas satisfait tout que nous ne puissons indiquer un changement simultané dans quelque autre corps, et que dans tous les phénomènes dans lesquels l'éther n'internent, ce changement consiste dans l'acquisition d'une quantité de mouvement $-a$. Mais je crois qu'on pourrait etre également satisfait si ce changement simultané ne fût pas lui même la production d'un mouvement. Vous avez déduit la belle formule

$$\Sigma M v_x + \int d\tau(\gamma g - \beta h) = \text{Const.}$$

6

Il me semble qu'on pourrait se borner à considerer $\int d\tau(\gamma g - \beta h)$,

$$\int d\tau(\alpha h - \gamma j), \int d\tau(\beta j - \alpha g)$$

comme des quantites dépendantes de l'état de l'éther qui sont pour ainsi dire "équivalentes" à une quantité de mouvement. Votre théoreme nous donne pour toute modification de la quantité de mouvement de la matière pondérable une modification simultanée de cette quantité équivalente; je crois qu'on pourrait bien se contenter de cela.

Je ne veux pas prétendre que cette manière de voir soit aussi simple qu'on pourrait le désirer; aussi n'aurais-je pas été conduit a cette théorie si les phénomènes de l'aberration ne m'y eussent pas force. Du reste, il va sans dire que la théorie ne doit être considerée que comme provisionare. Ce que je viens d'appeler "équivalence" pourra bien un jour nous apparaître comme une "identité"; cela pourrait arriver si nous parvenons à considérer la matière ponderable comme une modification de l'éther lui-même.

Il est presque inutile de dire qu'on pourrait aussi se tirer d'embarras en attribuant à l'éther une masse infiniment (ou très) grande. Alors les électrons pourraient réagir sur l'éther sans que ce milieu se mit en mouvement. Mais cette issue me semble assez artificielle.

Je désirerais bien vous faire encore quelques remarques au sujet de la compensation des termes en v^2, mais cette lettre devrait trop longue. J'espère donc que vous me permettrez de revenir sur cette question une autre fois. Il y a là encore bien des difficultes; vous pourriez peut être parvenir à les surmonter.

Veuillez agréer, Monsieur et tres honoré collègue, l'assurance de ma sincère considération. Votre bien devoué.

H. A. Lorentz

Fig. 5.2 Typed copy of a handwritten letter of H. A. Lorentz to H. Poincaré, 1901.

theses invented solely to save a principle whose violation permitted the theory's formulation in the first place. Lorentz concluded the 1901 letter by emphasizing that at this time, his chief concern was folding the contraction hypothesis into his electromagnetic theory.

Possibly as a result of this letter, one of Poincaré's reasons in papers of 1904 and 1908[11] for rejecting the principle of action and reaction was Lorentz's argument that the compensation between action and reaction could not be simultaneous.

To summarize: Poincaré's 1900 attempt to save the principle of action and reaction is a fine example of his application of his immense powers of mathematics, and his philosophic view, to criticizing, clarifying, and extending previously proposed physical theories. Lorentz's 1901 letter to Poincaré is typical of Lorentz's style at this point in his research on electromagnetic theory — that is, pushing ahead and leaving problems of foundations and of philosophy to others. It turned out that the road to a foundationally sound electrodynamics could be achieved only through an analysis of fundamental problems within a philosophic framework that was, to some extent, liberated from empirical data. In this way, Einstein could formulate a basis for investigating what it meant for action and reaction not to be simultaneous.[12]

7

We next turn to Wilhelm Wien's contribution to the *Lorentz-Festschrift* that served to change the direction of physical theory in a way that rendered superfluous Poincaré's attempts to rescue the principle of reaction, and demonstrated the conservatism of Lorentz's opinion of electromagnetic momentum. Wien was so impressed with the successes of Lorentz's electromagnetic theory and with its possibilities for unifying electromagnetism and gravitation that he suggested research toward an "electromagnetic basis for mechanics," that is, an electromagnetic world-picture.[13] In this scheme mechanics and then all of physical theory would have been deduced from Lorentz's electromagnetic theory. The oppositely directed research effort, a mechanical world-picture, had produced only increasingly complicated mechanical models in order to simulate the contiguous actions of an ether. One implication of an electromagnetic world-picture was that the electron's mass originated in its self-electromagnetic fields as a self-induction effect. Consequently, the electron's mass was predicted to be a velocity-dependent quantity.

Using high-velocity electrons from radium salts, Walter Kaufmann at Göttingen gave data for the velocity-dependence of the electron's mass.[14] Kaufmann's colleague, Max Abraham formulated the first field-theoretical description of an elementary particle.[15] Putting aside problems concerning the motion of Lorentz's ether, Abraham interpreted Poincaré's electromagnetic momentum as the electron's momentum as a result of its self-fields. Then, with a restriction on the acceleration of a rigid-sphere electron, and with new techniques of Hamilton-Lagrange mechanics applied to electrodynamics, Abraham deduced Newton's second law from Lorentz's force.[16] His principal result was that the electron's mass was a two-component quantity, which depended on whether the electron was acted on by external forces transverse or parallel to its trajectory. Abraham's transverse (m_T) and longitudinal (m_L) masses are where $\beta = v/c$, $m_o^e = e^2/2Rc^2$ is the electron's electrostatic mass due to its self-electrostatic field, and R is the electron's radius. Kaufmann's data agreed with Abraham's equation for the transverse mass m_T. Abraham's theory agreed with every first-order experiment since it was predicated on Lorentz's electromagnetic theory, but at this point it explained no second-order data other than

$$m_T = \frac{m_o^e}{2\beta^3}\left[(1 + \beta^2)\ln\left(\frac{1 + \beta}{1 - \beta}\right) - 2\beta\right], \tag{5.11}$$

$$m_L = \frac{m_o^e}{\beta^3}\left[\frac{2\beta}{1 - \beta^2} - \ln\left(\frac{1 + \beta}{1 - \beta}\right)\right] \tag{5.12}$$

Kaufmann's.[17] Contrary to postrelativity myths, the first-order experiments were of primary importance to everyone.[18] The Michelson–Morley experiment was, of course, much discussed, but, like Abraham, many physicists tried assiduously to avoid the unpalatable contraction hypothesis.[19]

Therefore, by the end of 1903, it seemed that a great deal of progress had been made toward a unified description of nature that was based on the most agreeable aspects of Lorentz's electromagnetic theory. The hallmark of the new world picture, proclaimed Abraham, was "atomistic structure of electricity, but continuous spatial distribution of the ether."[20] And the electromagnetic world-picture squared with Abraham's philosophic view: since the electron's dynamics could be deduced from a single quantity — its Lagrangian, then the electromagnetic world-picture, as Abraham wrote in 1903, has an "economical meaning," a term that had been introduced by the widely read philosopher–scientist Ernst Mach.[21]

In response to Poincaré's criticisms, Kaufmann's data, the new second-order ether-drift experiments of Trouton and Noble, and those of Rayleigh and Brace, in 1904 Lorentz extended his electromagnetic theory to a theory of the electron.[22] Whereas Abraham's electron was a rigid sphere, Lorentz's was deformable and he postulated that when at rest it was a sphere, but it contracted when moving. Lorentz's electron can be likened to a balloon uniformly smeared with charge. More than ten interlocking postulates enabled Lorentz to extend the 1895 theorem of corresponding states to apply to any order of accuracy. Then, using the same approximation as Abraham, Lorentz derived transverse (m_T) and longitudinal (m_L) masses for his electron:

$$m_T = \frac{4}{3} m_o^e / \sqrt{1 - \beta^2}$$
(5.13)

$$m_L = \frac{4}{3} m_o^e / (1 - \beta^2)^{3/2}$$
(5.14)

where $\beta = v/c$, $m_o^e = e^2/2Rc^2$ is the electron's electrostatic mass due to its self-electrostatic field, and R is the electron's radius. Lorentz showed that his m_T also agreed with Kaufmann's data. This time Poincaré found no faults with the foundations of Lorentz's theory, but he quickly caught several technical errors, and his keen sense of aesthetics revealed to him certain symmetries that Lorentz's original mathematical formulation had obscured. Poincaré's completed work appeared in his classic paper of 1906, "On the Dynamics of the Electron"; a short version of this paper had appeared in June 1905.[23] The development of certain key results that survived the eventual demise of Lorentz's theory of the electron are in three letters that Poincaré wrote to Lorentz, which I discovered at the Algemeen Rijksarchief, The Hague. In order to set the stage for describing these three letters, I shall review the interpretation of Lorentz's space and time transformations of 1904, which Poincaré dubbed the "Lorentz transformations":

$$x' = klx_r$$
(5.15)

$$y' = ly_r$$
(5.16)

$$z' = lz_r$$
(5.17)

$$t' = \frac{l}{k} t_r - kl \frac{v}{c^2} x_r$$
(5.18)

$$(k = 1 / \sqrt{1 - \beta^2})$$

where the coordinates (x', y', z', t') refer to an auxiliary system Σ' in which the Maxwell–Lorentz equations have the same form as they do relative to an ether-fixed reference system; the coordinates (x_r, y_r, z_r, t_r) refer to an inertial reference system S_r; and $x_r = x - vt$, $y_r = y$, $z_r = z$, $t_r = t$, where (x, y, z, t) refer to the ether-fixed system Σ; l admits the possibility of any sort of deformation, where $l = 1$ yields Lorentz's contraction and also is the only value of l that is consistent with Lorentz's theorem of corresponding states that applies to any order in v/c. Lorentz's goal was to transform the fundamental equations of his electromagnetic theory from the inertial system S_r to the fictitious reference system Σ' in which he could perform any necessary calculations, and then transform back to S_r. For example, it was particularly useful to take Σ' to be the electron's instantaneous rest system in order to reduce problems of the electrodynamics and optics of moving bodies to problems concerning bodies at rest.

In summary (see Fig. 5.3) there were three reference systems in Lorentz's theory — S, S_r, Σ' — where Σ' was an auxiliary system possessing no physical interpretation. Consequently, there were two sorts of electrons: the "real" electron of the reference systems S and S_r, and the "imaginary" electron in Σ'. Taking Σ' to be the electron's instantaneous rest system meant that in order for Σ' to have the same properties as the ether-fixed system S, the imaginary electron had to be a sphere in Σ'. But from the spatial portion of the Lorentz transformations [Eqs. (5.15)–(5.17)], the electron in Σ' is deformed into a shape with axes (kl, l, l). Therefore in order to maintain the theorem of corresponding states, the moving electron in S_r must be deformed in a manner that compensates for the deformation in Σ' — that is, by an amount $(1/kl, 1/l, 1/l)$. Lorentz's supporting argument for this postulate of deformation included inverting the spatial part of the Lorentz transformations. In order to render the postulate of contraction further "plausible," to use Abraham's word from his subsequent criticism[24] of Lorentz's theory of the electron, Lorentz proposed further interlocking postulates; among them was a postulate that the as yet unknown intermolecular forces transformed like the electromagnetic force in order that macroscopic matter contracted when in motion. Thus, to Lorentz and Poincaré, covariance was a mathematical property that was achieved through postulated transformations, which did not possess an unambiguous physical interpretation, because the transformations contained unknown velocities relative to the ether. The reason is that in electrodynamics and optics of 1905, the ether-fixed reference system S was cavalierly taken to be the laboratory system. The tacit assumption was that the ratio of the laboratory's velocity relative to the ether with that of light was small enough to be neglected. In other words, the notion of at rest relative to the laboratory was not well defined.

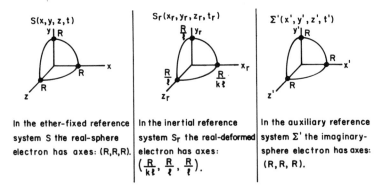

$S(x, y, z, t)$	$S_r(x_r, y_r, z_r, t_r)$	$\Sigma'(x', y', z', t')$
In the ether-fixed reference system S the real-sphere electron has axes: (R, R, R).	In the inertial reference system S_r the real-deformed electron has axes: $\left(\dfrac{R}{kl}, \dfrac{R}{l}, \dfrac{R}{l}\right)$.	In the auxiliary reference system Σ' the imaginary-sphere electron has axes: (R, R, R).

Fig. 5.3 The three different reference sytems employed in the electromagnetic world-picture for theories of a deformable electron.

With this background, we turn to a letter (Fig. 5.4) by Poincaré written to Lorentz shortly after Poincaré had returned from the September 1904 Congress of Arts and Science at St. Louis, Missouri. Although Poincaré's letters are undated, as was customary in France, this and the next two can be set easily into the period between late 1904 and mid-1905. At St. Louis, Poincaré had discussed elatedly the main results of Lorentz's 1904 theory of the electron, and, using a term well known from fundamental studies in geometry, he had renamed Lorentz's new theorem of corresponding states the "principle of relativity," according to which "the laws of physical phenomena must be the same for a stationary observer as for an observer carried along in a uniform motion of translation; so that we have not and can not have any means of discerning whether or not we are carried along in such a motion."[25]

Poincaré began the first of this set of three letters in a familiar way, that is, by noting certain errors in Lorentz's 1904 paper. Lorentz's use of three different reference systems had led him to deduce incorrect equations for the electron's velocity, charge density, and Lorentz force as these quantities were related between S_r and Σ'. Lorentz had been aware of these errors, and he had ensured their not affecting his final results by having taken Σ' as the electron's instantaneous rest system, thereby causing offending terms to vanish. Using only S and Σ', Poincaré obtained the correct transformations for the general case where Σ' was not the electron's instantaneous rest system.

Poincaré continued by pointing out where the fundamental difficulty with Lorentz's theory lay, although he did not know yet how to resolve it: two of the quantities characterizing the moving electron (its momentum and longitudinal mass) differed when calculated from the Lagrangian and from the energy.[26] Since Lorentz had calculated the electron's mass directly from its momentum, he had missed this problem. For Poincaré, investigating the foundations of Lorentz's theory meant reformulating it within the framework of that most elegant and powerful formulation of physical theory — the mechanics of Hamilton and Lagrange. Consistent with his style of treating every problem in the most general possible manner, Poincaré's reformulation was based on a generalized version of the dynamics of the electron in which he could study every possible model of the electron, that is, all connections between k and l. It turned out as Poincaré wrote in this letter, that only the 1904 model of the electron by Poincaré's former student, Paul Langevin,[27] offered consistency regarding the electron's momentum. According to Langevin the real electron in S_r was deformed in such a way that the volumes of the real and imaginary electrons were equal; that is, $l^3 k = 1$.[28]

In the conclusion of this letter, Poincaré expressed his dissatisfaction with Lorentz's method of proof for $l = 1$. Lorentz's proof[29] employed formulae for forces and accelerations valid only in the electron's instantaneous rest system, and so the proof was not general enough.

But in the next letter (Fig. 5.5), Poincaré elegantly resolved this insufficiency. By eliminating the intermediate inertial reference system, and thereby relating directly the fictitious reference system Σ' with the unobservable reference system S, Poincaré was able to write the Lorentz transformations in a form possessing a high degree of mathematical symmetry. Considering two successive Lorentz transformations along the same direction, it was easy for him to prove that the Lorentz transformations form a group, and that l must be equal to one.[30] As a bonus, Poincaré also obtained the new addition law for velocities that was independent of l.

Mon cher Collègue,

J'ai énormément regretté les circonstances qui m'ont empêché d'abord d'entendre votre conférence et ensuite de causer avec vous pendant votre séjour à Paris.

Depuis quelque temps j'ai étudié plus en détail votre mémoire électromagnetic phenomena in a system moving with any velocity smaller than that of Light, mémoire dont l'importance est extrême et dont j'ai déjà cité les principaux résultats dans ma conférence de St Louis.

Je suis d'accord avec vous sur tous les points essentiels; cependant il y a quelques divergences de détail.

Ainsi page 813, au lieu de poser:

$$\frac{1}{k\ell^3}\rho = \rho' \; ; \; k^2 u_x = u'_x, \; k^2 u_y = u'_y$$

il me semble qu'on doit poser:

$$\frac{1}{k\ell^3}\rho(1+\varepsilon v_x) = \rho' \quad \frac{1}{k\ell^3}\rho(v_x+\varepsilon) = \rho' u'_x$$

où $\varepsilon = -\frac{w}{c}$ ou $\varepsilon = -w$ si nous choisissons les unités de telle façon que $c = 1$.

Cette modification me semble s'imposer si l'on veut que la charge apparente de l'électron se conserve.

Les formules (10) page 813 se trouvent

alors modifiées, et je trouve *alors* pour le
dernier terme au lieu de

$$\ell^2 \frac{w}{c^2}\left(u'_y d'_y + u'_z d'_z\right), \quad -\frac{\ell^2}{k}\frac{w}{c^2}\, u'_x d'_y, \quad -\frac{\ell^2}{k}\frac{w}{c^2}\, u'_x d'_z$$

je trouve

$$\ell^2 \frac{w}{c^2}\left(u'_x d'_x + u'_y d'_y + u'_z d'_z\right), \quad 0, \quad 0$$

C'est la force de Liénard, que vous
trouvez aussi *mais* pas d'une de différences.

Et alors la question se pose de savoir
si cette force est ou non compensée.

Ceci montre qu'entre les forces réelles
X, Y, Z et les forces apparentes X', Y', Z'
il y a les relations

$$X' = A\left(X + \varepsilon X v_x\right), \quad Y' \stackrel{.}{=} BY, \quad Z' \stackrel{.}{=} BZ$$

A et B étant des coïff. et $A \varepsilon \sum X v_x$
représentant la force de Liénard.

Si toutes les forces sont d'origine électrique
les conditions d'équilibre (ou du principe de
d'Alembert modifié) donnent

$$X = Y = Z = 0$$

d'où
$$X' = Y' = Z' = 0$$

Si toutes les forces ne sont pas d'origine
électrique, il y aura encore compensation
pourvu qu'elles se comportent toutes
comme si elles étaient d'origine électrique.

Mais il y a autre chose.

Vous supposez $\ell = 1$

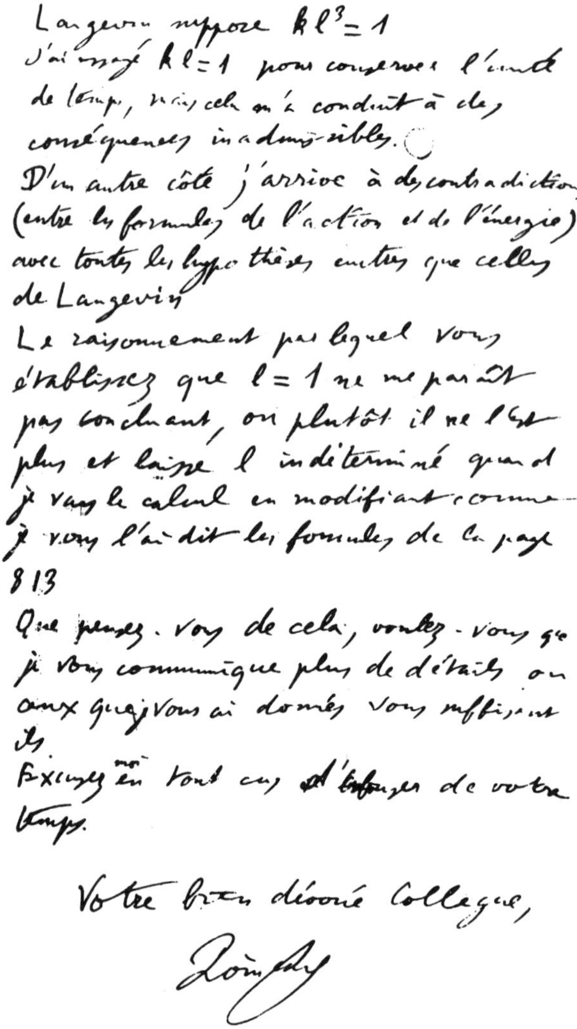

Langevin suppose $k\ell^3 = 1$
J'ai essayé $k\ell = 1$ pour conserver l'unité
de temps, mais cela m'a conduit à des
conséquences inadmissibles.
D'un autre côté j'arrive à des contradictions
(entre les formules de l'action et de l'énergie)
avec toutes les hypothèses autres que celles
de Langevin
Le raisonnement par lequel vous
établissez que $\ell = 1$ ne me paraît
pas concluant, ou plutôt il ne l'est
plus et laisse ℓ indéterminé quand
je vais le calcul en modifiant comme
je vous l'ai dit les formules de la page
813
Que pensez-vous de cela, voulez-vous que
je vous communique plus de détails ou
ceux que je vous ai donnés vous suffisent
ils.
Excusez moi en tout cas et n'abusez de votre
temps.

Votre bien dévoué Collègue,

Poincaré

Fig. 5.4 Letter from H. Poincare to H. A. Lorentz, 1904.

Mon cher Collègue,

Merci de votre aimable lettre.
Depuis que je vous ai écrit mes idées
se sont modifiées sur quelques points.
Je trouve comme vous, $l = 1$ par une
autre voie.
Soit $-\varepsilon$ la vitesse de la translation celle de
la lumière étant prise pour unité.

$$k = \left(1 - \varepsilon^2\right)^{-\frac{1}{2}}$$

On a la transformation

$$x' = kl(x + \varepsilon t), \quad t' = kl(t + \varepsilon x)$$

$$y' = ly, \quad z' = lz.$$

Cette transformation forment un groupe.
soient deux transformations composantes
correspondant à

$$k, \quad l, \quad \varepsilon$$

et

$$k', \quad l', \quad \varepsilon'$$

leur résultante correspondra à

$$k'', \quad l'', \quad \varepsilon''$$

où:

$$k'' = \left(1 - \varepsilon''^2\right)^{-\frac{1}{2}}, \quad l'' = ll', \quad \varepsilon'' = \frac{\varepsilon + \varepsilon'}{1 + \varepsilon \varepsilon'}$$

15

Si nous poulons maintenant prendre

$$\ell = \left(1 - \varepsilon^2\right)^m, \quad \ell' = \left(1 - \varepsilon'^2\right)^m$$

nous n'aurons:

$$\ell'' = \left(1 - \varepsilon''^2\right)^m$$

que pour $m = 0$

D'un autre côté je ne trouve d'accord entre le calcul des masses par le moyen des quantités de mouvement électromagnétique et par le moyen de la moindre action, et par le moyen de l'énergie que dans l'hypothèse de Langevin.

J'espère tirer bientôt au clair cette contradiction, je vous tiendrai au courant de mes efforts.

Votre très dévoué Collègue,

Poincaré

Fig. 5.5 Letter from H. Poincaré to H. A. Lorentz, written between late 1904 and mid-1905.

In his next letter (Fig. 5.6), Poincaré's key result was that only Lorentz's 1904 theory of the electron offered "perfect compensation (which prevents the experimental determination of absolute motion)." Poincaré linked this result to solving the problem concerning the longitudinal mass of Lorentz's electron. Poincaré's solution was to add a term to the electron's Lagrangian that preserved the Lorentz invariance of the principle of least action. Thus, he was led to interpret the additional term as the energy due to the internal pressure that served to prevent the deformable electron from exploding in its rest system.[31] Now the theory was freed of all removable blemishes. We can well believe that Poincaré was, as he put it, in "parfait accord" with the "beaux travaux" of Lorentz. Lorentz's 1904 theory of the electron was really the culmination of over a decade of painstaking work and interaction by Lorentz and Poincaré.

Mon cher Collègue,

J'ai continué les recherches dont je
vous avais parlé. Mes résultats confirment
pleinement les vôtres en ce sens que
la compensation parfaite (qui empêche
la détermination expérimentale du
mouvement absolu) ne peut se faire
complètement que dans l'hypothèse $l = 1$.
Seulement pour que cette hypothèse soit
admissible, il faut admettre que chaque
électron est soumis à des forces complémentaires,
dont le travail est proportionnel aux
variations de son volume.

Ou si vous aimez mieux, que chaque
électron se comporte comme s'il était
une capacité creuse soumise à une
pression interne constante (d'ailleurs,

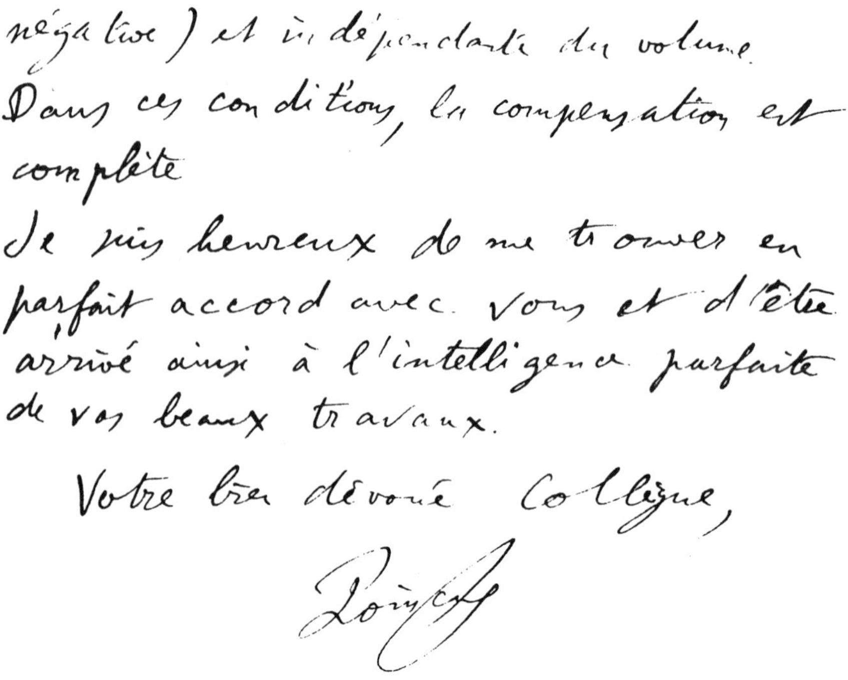

négative) et indépendante du volume

Dans ces conditions, la compensation est complète

Je suis heureux de me trouver en parfait accord avec vous et d'être arrivé ainsi à l'intelligence parfaite de vos beaux travaux.

Votre bien dévoué Collègue,

Poincaré

Fig. 5.6   Letter from H. Poincaré to H. A. Lorentz, written between late 1904 and mid-1905.

Of the existing electron models, only Lorentz's was really adequate. It could explain optical data accurate to second order in v/c. It explained effects such as the dependence of mass on velocity, the contraction of moving objects, and the isotropy of the velocity of light in inertial reference systems, all to have been *caused* by the interaction between the electrons constituting matter and the ether. Lorentz's theory of the electron appeared to many physicists to be the most likely candidate for the cornerstone of a unified field-theoretical description of nature. As he showed in his 1905-6 publications, Poincaré had cast the theory into the Hamilton-Lagrange formalism, replete with group theory and four-dimensional spaces, and then extended it to a Lorentz-covariant theory of gravity.

From the relativity theory we have learned that it is more fruitful to consider as axiomatic that the space of inertial reference systems is isotropic for light propagation, rather than to attempt to explain this property. Then, as Einstein wrote in 1907,[19] certain postulates of Lorentz's theory become "secondary consequences," — for example, the contraction of moving bodies.[32]

Let us turn next to Kaufmann's reception of Lorentz's theory of the electron. On July 10, 1904, Kaufmann wrote to Lorentz that Lorentz's m_T agreed better with Kaufmann's most recent data than even Lorentz had imagined.[33] In order to decide the issue between the electron theories of Abraham and Lorentz, Kaufmann promised to increase his accuracy in the next series of experiments. He was as good as his word, but the results were not to please Lorentz.

In a short paper of November 30, 1905, and then in a longer one of early 1906, Kaufmann reported that his new measurements agreed best with Abraham's theory — a close second was Bucherer's — but as he put it, the "measurement results are not compatible with the Lorentz–Einstein fundamental assumption."[34] As Kaufmann wrote in 1905, "a recent publication by Mr. A. Einstein on the theory of electrodynamics . . . leads to results which are formally identical with those of Lorentz's theory" — that is, the "formal equivalence" of Einstein's 1905 relativity transformations for space and time and his prediction for the electron's transverse mass with these quantities in Lorentz's theory led to the names "Lorentz–Einstein theory" and "Lorentz–Einstein principle of relativity."[35] Just how devastated Lorentz was by Kaufmann's results is clear from his unpublished letter to Poincaré, dated March 8, 1906 (Fig. 5.7). It is a very revealing moment in the

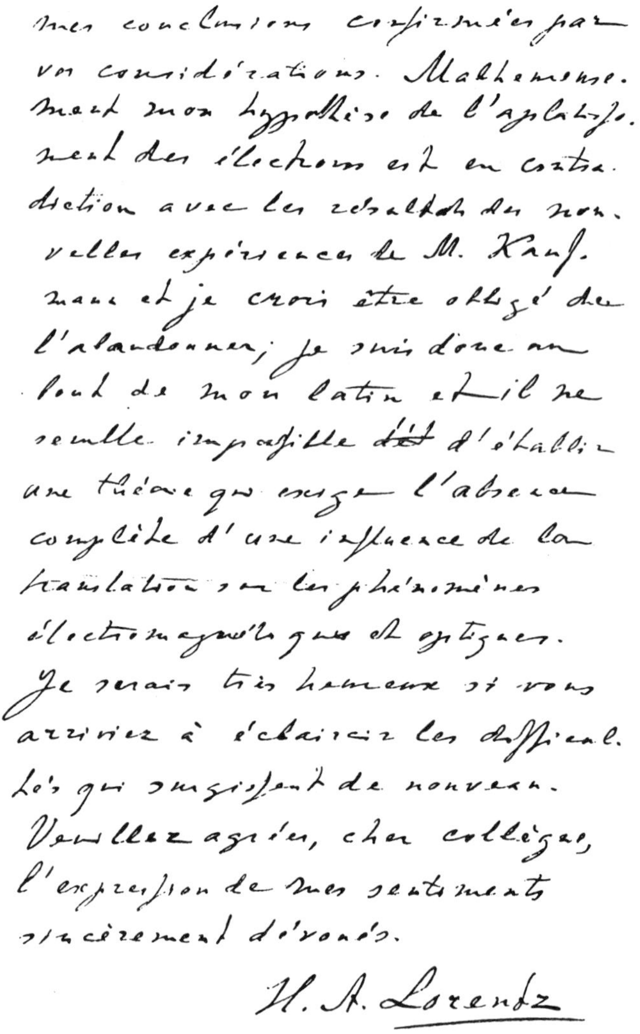

mes conclusions confirmées par
vos considérations. Malheureuse.
ment mon hypothèse de l'aplatisse.
ment des électrons est en contra.
diction avec les résultats des nou.
velles expériences de M. Kauf.
mann et je crois être obligé de
l'abandonner; je suis donc au
bout de mon latin et il ne
semble impossible d'établir
une théorie qui exige l'absence
complète d'une influence de la
translation sur les phénomènes
électromagnétiques et optiques.
Je serais très heureux si vous
arriviez à éclaircir les difficul.
tés qui surgissent de nouveau.
Veuillez agréer, cher collègue,
l'expression de mes sentiments
sincèrement dévoués.

H. A. Lorentz

Fig. 5.7   Letter from H. A. Lorentz to H. Poincaré, 1906.

history of relativity theory. After congratulating Poincaré for the essay "On the Dynamics of the Electron," Lorentz wrote that, nevertheless all their work may have been for nothing:

> Unfortunately my hypothesis of the flattening of electrons is in contradiction with Kaufmann's new results, and I must abandon it. I am, therefore, at the end of my Latin. It seems to me impossible to establish a theory that demands the complete absence of an influence of translation on the phenomena of electricity and optics. I would be very happy if you would succeed in clarifying the difficulties which arise again.

What a remarkable confession, and what a clear-cut example of falsification. After all those years of work Lorentz was willing to abandon his theory because of the report of a single experiment. Lorentz repeated this message in his lectures at Columbia University in March and April 1906 that appeared in 1909 as *Theory of Electrons*: "But, so far as we can judge at present, the facts are against our hypothesis." For his part, Poincaré, too, was strongly affected by Kaufmann's new results; but his reaction was not as radical as Lorentz's. In the introductory paragraphs to the 1905 version of "On the Dynamics of the Electron," Poincaré had written that the theories of Abraham, Langevin, and Lorentz "agreed with Kaufmann's experiments." Of course he meant those based on data prior to 1905. But in the last paragraph of the introduction to the 1906 version, Poincaré wrote, concerning Lorentz's theory, that at "this moment the entire theory may well be threatened" by Kaufmann's new 1905 data.[36] Poincaré was still willing for the sake of argument to consider the principle of relativity for the moment as valid in order "to see what consequences follow from it." He called for someone to repeat Kaufmann's experiment.[37] Whereas for Lorentz, Kaufmann's results threatened a theory; for Poincaré, a philosophic view was also at stake, one that emphasized a principle of relative motion.

It turned out that Kaufmann's data were incorrect. In a letter of September 7, 1908, A. H. Bucherer in Bonn wrote Einstein in Berne of his recent experimental results concerning the mass of high-velocity electrons: "by means of careful experiments, I have elevated the validity of the principle of relativity beyond any doubt."[38] On September 23, 1908 Bucherer presented his results to the Meeting of German Scientists and Physicians at Cologne; his communication was entitled "Measurements on Becquerel-rays. The Experimental Confirmation of the Lorentz–Einstein Theory."[44] Poincaré and Lorentz welcomed Bucherer's results, and did not call for someone to repeat his experiments.[39] although, incidentally, thirty years later it turned out that Bucherer's data were inconclusive.[40] As Poincaré wrote in 1912, further weight was added to Bucherer's data because they disconfirmed Bucherer's own 1904 theory of the electron that was identical with Langevin's.[41]

The principle of relativity of Bucherer, Lorentz, and Poincaré resulted from careful study of a large number of experiments, and it was the basis of a theory in which empirical data could be explained to have been *caused* by electrons interacting with an ether. Einstein's principle of relativity excluded the ether of electromagnetic theory and did not explain anything. For example, in Einstein's relativity theory, the negative results of the ether-drift experiments were a foregone conclusion. As Lorentz wrote in his 1909 book, *Theory of Electrons*, "Einstein simply postulates what we have deduced."

In a 1907 paper[42] reviewing the status of the principle of relativity, Einstein discussed Kaufmann's new data from late 1905, those same data that, in 1906, had driven Lorentz to the "end of [his] Latin." Einstein acutely emphasized that the "systematic deviation" between the Lorentz–Einstein predictions and the data could indicate a hitherto "unnoticed source of error"; consequently, Einstein called for further experiments. Then he dismissed Kaufmann's data because in his "opinion" they supported theories that did not "embrace a greater complex of phenomena" — that is, the electron theories of Abraham and Bucherer, which could not explain optical experiments accurate to second order in v/c. Einstein's intuition served him well. Undoubtedly having the Kaufmann episode in 1907 in mind, Einstein in 1946 described one of two criteria for assessing a scientific theory thus: "The first point of view is obvious: the theory must not contradict empirical facts. However evident this demand may in the first place appear, its application turns out to be quite delicate."[43]

Abraham, Bucherer, Lorentz, and Poincaré, among others that included Max Planck[44] and Hermann Minkowski,[45] continued to cling to an ether chiefly because: 1) it was the basic ingredient to what appeared to be the realization of the long-sought-after unification of the sciences — an electromagnetic world-picture; 2) something had to mediate disturbances and explain physical effects.

The man whom Einstein held in high esteem, and to whom he referred to in correspondence of 1913 as the "fearsome Abraham"[46] refused to permit Bucherer's data to decide the issue between the competing electron theories; in other words, when his own theory was at stake, Abraham was not a falsificationist. In 1909 Abraham still pursued, as he put it, a version of the "electromagnetic mechanics" that was consistent with the rigid-sphere electron.[47]

In the part of a spirited tirade of 1914 leveled against the special relativity theory Abraham depicted well just how difficult it was to part with a philosophic position and with the great past achievements of physics — that is, the hard-won empirical data and the theories that they engendered: "the theory of relativity excited the young devoted to the study of mathematical physics; under the influence of that theory they filled the halls and corridors of the universities. On the other hand . . . the physicists of the former generation whose philosophy was formed under the influence of Mach and Kirchhoff, remained for the most part skeptical of the audacious innovators who allowed themselves to rely on a small number of experiments still debated by specialists, to overthrow the fundamental tests of every physical measurement."[48]

In conclusion, new archival materials, in conjunction with primary sources, reveal a scenario of science in the making. As was so often the case in turbulent times, the scientists' rules for assessing scientific theories were, as Poincaré wrote, "extremely subtle and delicate, and it is practically impossible to state them in precise language; they must be felt rather than formulated."[49]

Acknowledgements

This essay is based on research supported by the National Science Foundation's History and Philosophy of Science Program, by fellowships from the American Council of Learned Societies and the American Philosophical Society, and a travel grant from the Lorentz Foundation, Leiden University, The Netherlands.

The letters from Lorentz to Poincaré are in the possession of the Poincaré Estate who have granted me sole access to their materials through mid-1981. I thank the Poincaré Estate, Paris, and the Algemeen Rijksarchief, The Hague, for permission to quote from their correspondence.

It is a pleasure to acknowledge Professor Gerald Holton's comments on an early draft of this essay.

NOTES

1. For detailed analyses of the theories of the electron of Abraham, Langevin, Lorentz and Poincaré, as well as of Poincaré's philosophy of science, and for references to other secondary works see my papers listed at the end of this note and the first chapter of my book-length analysis of Einstein's 1905 relativity paper in its historic context. The book is A. I. Miller, *Albert Einstein's Special Theory of Relativity: Emergence (1905) and Early Interpretation (1905-1911)* (Addison-Wesley, Advanced Book Program, Reading, Mass., 1980). The papers are: "A Study of Henri Poincaré's 'Sur la Dynamique de l'Electron'," Arch. History Exact Sci. **10**, (3-5), 207-328 (1973); "On Lorentz's Methodology," Brit. J. Phil. Sci. **25**, 29-45 (1974); "On Einstein, Light Quanta, Radiation and Relativity in 1905," Am. J. Phys. **44**, 912-23 (1976); and "Poincaré and Einstein: A Comparative Study." Boston Studies Phil. Sci. **31** (forthcoming).

2. H. A. Lorentz, "La théorie électromagnétique de Maxwell et son application aux corps mouvants," Arch. Néerl. **25**, 363 (1892); reprinted in *Collected Papers*, Vol. 2, (Nijhoff, The Hague, 1935-39), pp. 164-343.

3. Instead of Lorentz's units I take the liberty to use the Gaussian c.g.s. system of units, whose usefulness in electromagnetic theory was first emphasized by Abraham in papers published in 1902 and 1903. Abraham also coined the term "Maxwell-Lorentz equations," which Einstein used in his 1905 relativity paper. See M. Abraham, "Dynamik des Elektrons," Nachr. Ges. Wiss. Göttingen, pp. 20-41 (1902); "Prinzipien der Dynamik des Elektrons," Phys. Z. **4**, 57-63 (1902); "Prinzipien der Dynamik des Elektrons," Ann. d. Phys. **10**, 105-79 (1903).

4. H. Poincaré, "A propos de la théorie de M. Larmor," L'Éclairage électrique **3**, 5-13, 289-95 (1895); *ibid.*, **5**, 5-14, 385-92 (1895); reprinted in *Oeuvres de Henri Poincaré*, Vol. 9 (Gauthier-Villars, Paris, 1934-1954, pp. 369-426)(this volume is referred to as *Oeuvres*), and "Sur les rapports de la Physique expérimentale et de la Physique mathématique," *Rapports présentés au Congrès international de Physique réuni à Paris en 1900*, Vol. 1 (Gauthier-Villars, Paris, 1900), pp. 1-29; translated as ch. IX and X, pp. 140-182, of H. Poincaré, *Science and Hypothesis* (Dover, New York, 1952).

5. Briefly, Lorentz invented the contraction hypothesis to explain a single experiment, and he derived its mathematical form from Newton's law for the addition of velocities, which the theorem of corresponding states was supposed to have obviated. See my 1974 paper (op. cit. in n. 1) for further discussion of the ad hocness of Lorentz's contraction hypothesis.

6. Thus Poincaré referred to Newton's principle of action and reaction as a "convention."

7. H. Poincaré, "La théorie de Lorentz et le principe de réaction" *Recueil de travaux offerts par les auteurs à H. A. Lorentz* (Nijhoff, The Hague, 1900), pp. 252-78; reprinted in *Oeuvres*, op. cit. in n. 4, pp. 464-88.

8. For brevity I use the term "covariance," which Hermann Minkowski was the first to apply effectively to the transformation properties of equations expressing physical laws. See H. Minkowski, "Die Gründgleichungen für die elektromagnetischen Vorgänge in bewegten Körpern," Nachr. Ges. Wiss. Göttingen, pp. 53-111 (1908).

9. H. Poincaré, "Sur les principes de la méchanique," *Bibliothèque du Congrès international de Philosophie tenu à Paris du 1er au 5 août 1900* (Colin, Paris, 1901), pp. 457-94; an expanded version of this paper is presented on pp. 123-39 of *Science and Hypothesis*, op. cit. in n. 4.

10. Whereas the Michelson-Morley experiment always had been of great concern to Lorentz, in 1892 it did not determine his formulation of the electromagnetic theory. This might seem surprising in the light of Lorentz's strongly empiricist message in 1886; namely, that in a question "as important" as the choice between a pure Fresnel theory, or a hybrid one containing elements of the theories of Fresnel and Stokes, one should not be guided by "considerations of the degree of probability or of simplicity of one or the other hypotheses, but to address onself to experiment" "De l'influence du mouvement de la terre sur les phénomènes lumineux," Versl. Kon Akad. Wetensch. Amsterdam **2**, 297 (1886); reprinted in *Collected Papers*, op. cit. in n. 2, Vol. 4, 153-214.

   (According to G. G. Stokes the ether at the earth's surface was dragged totally, and the dragging decreased with increasing distance from the earth's surface. In the 1886 paper Lorentz proved the inconsistency of Stokes' assuming both a velocity potential and that the relative velocity between the earth and the ether should vanish over the earth's surface. According to Augustin Fresnel a body in motion dragged the excess of ether in its interior, which he assumed to be $1 - n^{-2}$, where n was the body's refractive index.)

   To Lorentz, at that time, Michelson's interferometer experiment of 1881 was of importance since it was the only reliable means to obtain data accurate to second order in v/c. Having shown in 1886 that the Michelson interferometer experiment of 1881 was inconclusive, Lorentz urged its repetition. This was carried out in 1887, and the results apparently excluded a pure Fresnel theory. Yet, as Lorentz explained in 1892, hybrid theories, "being more complicated, are less worthy of consideration," and again in 1897 he wrote that the "Fresnel theory is without doubt simpler." Consequently, the formulation of Lorentz's electromagnetic theory ran counter to his empiricist guideline of 1886, because in the end simplicity was his guide. H. A. Lorentz, "The relative motion of the earth and the ether," Versl. Kon Akad. Wetensch. Amsterdam **1**, 74 (1892); reprinted in *Collected Papers*, op. cit in n. 2, Vol. 4, 219-23;

23

and "Concerning the problem of the dragging along of the ether by the earth," Versl. Kon. Akad. Wetensch. Amsterdam **6**, 266 (1897); reprinted ibid., pp. 237–44.

11. H. Poincaré, "L'état actuel et l'avenir de la Physique mathématique" delivered on 24 September 1904 at the International Congress of Arts and Science at Saint Louis, Mo. and published in Bull. Sci. Mat. **28**, 302–24; in English on pp. 91–111 of *The Value of Science*, translated by George Bruce Halsted (Dover, New York, 1958) from Poincaré's *La Valeur de la Science* (Ernest Flammarion, Paris, 1905). Poincaré, "La dynamique de l'électron," Revue générale des Sciences pures et appliquées **19**, 386–402 (1908); reprinted in *Oeuvres*, op. cit. in n. 4, pp. 551–86. Excerpts from this paper appear in Book III, pp. 199–250 of *Science and Method*, translated by Francis Maitland (Dover, New York, n.d.) from Poincaré's *Science et Méthode* (Ernest Flammarion, Paris, 1908).

12. In my book (op. cit in n. 1) analyzing Einstein's 1905 relativity paper, his thinking toward the relativity of simultaneity is analyzed in the context of the physics of 1905. This analysis is outlined in my paper "The Special Relativity Theory: Einstein's Response to the Physics of 1905," in *Jerusalem, Einstein Centennial Symposium* edited by G. Holton and Y. Elkana (Princeton University Press, Princeton, 1980). Pioneering case studies of Einstein's thinking about the special relativity theory are those of Gerald Holton: *Thematic Origins of Scientific Thought: Kepler to Einstein* (Harvard University Press, Cambridge, Mass. 1973) and *The Scientific Imagination: Case Studies* (Cambridge University Press, Cambridge, England, 1978).

13. W. Wien, "Über die Möglichkeit einer elektromagnetischen Begründung der Mechanik," *Recueil de travaux offerts par les auteurs à H. A. Lorentz* (Nijhoff, The Hague, 1900), pp. 96–107; reprinted in Ann. d. Phys. **5**, 501–513 (1901).

14. W. Kaufmann, "Die magnetische und electrische Ablenkbarkeit der Becquerelstrahlen und die scheinbare Masse der Elektronen," Nachr. Ges. Wiss. Göttingen, pp. 143–55 (1901); "Die elektromagnetische Masse des Elektrons," Phys. Z. **4**, 54–57 (1902); and "Über die 'Elektromagnetische Masse' der Elektronen," Nachr. Ges. Wiss. Göttingen, pp. 90–103 (1903).

15. In the papers cited in n. 3.

16. According to Abraham's restriction of quasi-stationary acceleration, the electron accelerates in such a way that it does not radiate. The criterion for this sort of motion is that the electron accelerates during a time that is much less than the time required for light to traverse its diameter. For details see my 1973 paper and my book cited in n. 1.

17. For example, Abraham's theory of the electron disagreed with the high-precision data of Lord Rayleigh and D. B. Brace [Lord Rayleigh, "Does Motion through the Aether cause Double Refraction?" Phil. Mag. **4**, 678–83, (1902); D. B. Brace, "On Double Refraction in Matter moving through the Aether," Phil. Mag. **7**, 317–29 (1904)]. Rayleigh and Brace sought to detect whether an isotropic substance at rest on the moving earth exhibited double refraction owing to Lorentz's postulate that bodies in motion relative to the ether became deformed — that is, whether a moving isotropic body should respond differently to light propagating through it parallel and transverse to its direction of motion. Rayleigh and Brace detected no anisotropy; Rayleigh's data were accurate to 1 part in 10^{10}, and Brace's to 1 part in 10^{13}. Abraham and Lorentz used their theories of the electron to interpret the results of Rayleigh and Brace as follows [M. Abraham, "Die Grundhypothesen der Elektronentheorie," Phys. Z. **5**, 576–79 (1904) and "Zur Theorie der Strahlung und des Strahlungsdruckes," Ann. d. Phys. **14**, 236–87 (1904); H. A. Lorentz, "Electromagnetic Phenomena in a System Moving with any Velocity Less than that of Light," Proc. Roy. Acad. Amsterdam **6**, 809 (1904), reprinted in *Collected Papers*, op. cit. in n. 2, Vol. 5, 172–97 and in part in *The Principle of Relativity: A Collection of Original Memoirs on the Special and General Theory of Relativity by H. A. Lorentz, A. Einstein, H. Minkowski and H. Weyl*, translated by W. Perrett and G. B. Jeffery (Dover, New York, n.d.), pp. 11–34; and *The Theory of Electrons* (Teubner, Leipzig, 1909); rev. ed., 1916; reprinted from the rev. ed. (Dover, New York, 1952)]. Since the index of refraction depends on the mass of the isotropic substance's constituent electrons, then the index of refraction becomes double-valued owing to the moving electrons developing transverse and longitudinal masses. Abraham's theory predicted a double refraction well within the limits of accuracy of Rayleigh and Brace, while Lorentz's 1904 theory of the electron (see below) predicted no double refraction because of hypotheses concerning how intermolecular forces transform and how moving matter contracts, among others.

18. See, for example, n. 10. Einstein mentioned explicitly only the class of first-order optical experiments ["Zur Elektrodynamik bewegter Körper," Ann. d. Phys. **17**, 891–921 (1905); a new translation is in my book, op. cit. in n. 1. The English version in *The Principle of Relativity*, op. cit. in n. 17, pp. 37–65, contains substantive mistranslations and misprints.]. R. S. Shankland reported the following interview with Einstein on 4 February 1950: "the experimental results which had influenced him [Einstein] most were the observations on stellar aberration and Fizeau's measurements on the speed of light in moving water. 'They were enough,' he said. I reminded him that Michelson and Morley had made a very accurate determination at Case in 1886 of the Fresnel dragging coefficient with greatly improved techniques and showed him their values as given in my paper. To this he nodded agreement, but when I added that it seemed to me that Fizeau's original result was only qualitative, he shook his pipe and smiled. 'Oh it was better than that!'" ["Conversations with Albert Einstein," Am. J. Phys. **31**, 47–57 (1963).]

19. Another physicist who preferred to avoid Lorentz's contraction hypothesis was Emil Cohn, whose work Einstein held in high regard [see Einstein "Über das Relativitätsprinzip und die aus demselben gezogenen Folgerungen," J. d. Radioaktivität **4**, 411–62 (1907)]. The goal of Cohn's research was not an electromagnetic world- picture, but, rather, a consistent version of the electrodynamics of moving bulk matter that was based on the best that the electromagnetic theories of Henrich Hertz and Lorentz could offer: Cohn attempted to suitably modify Hertz's electromagnetic field equations in such a way as to render them in agreement with Lorentz's 1895 theorem of corresponding states [H. A. Lorentz, *Versuch einer Theorie der elektrischen und optischen Erscheinungen in bewegten Körpern* (Brill, Leiden, 1895), reprinted in *Collected Papers*, op. cit. in n. 2, Vol. 5, 1–137; Emil Cohn, "Über die Gleichungen der Elektrodynamik für bewegte Körper," *Recueil de travaux offerts par les auteurs à H. A. Lorentz* (Nijhoff, The Hague, 1900), pp. 516–23, and "Über die Gleichungen des elektromagnetischen Feldes für bewegte Körper," Ann. d. Phys. **7**, 29–56 (1902)].

Cohn emphasized that the term ether is a "metaphorical term [that] should not acquire an importance relative to the theory in question," and consequently in the meanwhile the ether should be taken as a "heuristic concept." The guiding theme in his research, Cohn wrote, was "scientific economy" as this term was used by Ernst Mach [Cohn, "Zur Elektrodynamik bewegter Systeme," Berl. Ber. **40**, 1294–1303 (1904); II, ibid., 1404–16 (1904)].

20. M. Abraham, "Prinzipien der Dynamik des Elektrons," op. cit. in n. 3.

21. Ibid.

22. F. T. Trouton and H. R. Noble, "The Mechanical Forces Acting on a Charged Electric Condenser moving through Space," Phil. Trans. Roy. Soc. London **A202**, 165–81 (1903); Rayleigh, op. cit. in n. 17; Brace, ibid.; Lorentz, ibid. Trouton and Noble attempted to measure the sudden torque on a parallel plate capacitor that was hung by string, owing to the capacitor's being charged or discharged; this torque is related to the earth's velocity relative to the ether. They observed no effect. The hypotheses in Lorentz's 1904 theory of particular importance for explaining this null result were the contraction hypothesis, and that all forces transform like the electromagnetic force.

23. H. Poincaré, "Sur la dynamique de l'électron," Rend. del Circ. Mat. di Palermo **21**, 129–75 (1906), reprinted in *Oeuvres*, op. cit. in n. 4, pp. 494–550; "Sur la dynamique de l'électron," Comptes rendus de l'Académie des Sciences **140**, 1504–8 (1905), reprinted in *Oeuvres*, pp. 489–93.

24. M. Abraham, "Die Grundhypothesen der Elektronentheorie," op. cit. in n. 17.

25. Poincaré, "L'état actuel et l'avenir de la Physique mathématique," op. cit. in n. 11. In geometry there is a principle of the relativity of position that asserts: 1) that the notion of distance has meaning only in terms of relations between bodies, and 2) the equivalence of all points in space. Investigations of the origins of this principle and of whether the nature of physical space could be determined empirically by distance measurements had led Poincaré to extend the principle of the relativity of position into mechanics, where he called it the "principle of relative motion" [op. cit. in n. 7]. In his widely read book of 1902, *Science and Hypothesis* (op. cit. in n. 4), Poincaré discussed this work as well as the possibility that, as he now put it, the "principle of relativity" from geometry could be extended into electromagnetic theory because thus far experiments had not revealed absolute motion. The agreement of Lorentz's theory with this principle, at least to order v/c, pleased Poincaré. Yet owing to his emphasis on empirical data, Poincaré in 1902 had not yet become a full-fledged member of the group to which he referred to as the "partisans of Lorentz." The problem was that V. Crémieu's recent experimental data contradicted Rowland's convection current, which was an essential ingredient of Lorentz's theory. By 1903 these data were found to be inaccurate [see Miller, 1973 op. cit. in n. 1].

After Lorentz's 1904 theory of the electron, Poincaré became a partisan of Lorentz and he considered the principle of relativity to be an extension of the principle of relative motion from mechanics into the physics of the electron. Several of Poincaré's essays on the foundations of geometry are in *Science and Hypothesis*, ch. III–V. These and others of Poincaré's writings on geometry are analyzed in Miller 1973 and 1980 op. cit. in n. 1.

26. The inconsistency in Lorentz's theory concerning m_L was also brought to Lorentz's attention by Abraham in an unpublished letter of Abraham to Lorentz written from Wiesbaden on 26 January 1905 (on deposit at the Algemeen Rijksarchief, The Hague). Abraham developed the calculations in this letter in his 1905 book whose preface was also written at Wiesbaden [*Theorie der Elektrizität: Elektromagnetische Theorie der Strahlung* 1st ed. (Teubner, Leipzig, 1905); 2nd ed., (1908); 3rd ed. (1914)]. For details see Miller 1973 and 1980 op. cit. in n. 1.

27. P. Langevin, "La physique des électrons," delivered on 22 September 1904 at the International Congress of Arts and Science at Saint Louis, Missouri and published in *Revue générale des sciences pures et appliquées*, **16**, 257–76 (1905). Simultaneously and independently A. H. Bucherer in Bonn offered the same model [*Mathematische Einführung in die Elektronentheorie* (Teubner, Leipzig, 1904)].

28. See Fig. 5.3. Abraham's rigid-sphere electron does not transform according to the Lorentz transformations.

29. H. A. Lorentz, "Electromagnetic Phenomena," op. cit. in n. 17.

30. Since the ether system S never moves,, Poincaré could give only a mathematical interpretation to the Lorentz transformation's inversion symmetry: interchanging primes and changing ε to $-\varepsilon$ (where in the Lorentz transformations in Fig. 5.5, ε is the velocity of Σ' relative to S) is the mathematical operation of rotation by 180° about the common y-axes of S and Σ' ["Sur la dynamique de l'électron," 1906 op. cit. in n. 23].

31. In Abraham's opinion, nonelectromagnetic forces were necesssary to bind Lorentz's deformable electron; consequently, Lorentz's theory of the electron was not in the spirit of the electromagnetic world-picture [Abraham, "Die Grundhypothesen der Elektronentheorie," op. cit. in n. 17]. For Poincaré, on the other hand, Lorentz covariance was the guiding theme toward an electromagnetic world-picture, and the internal pressure rendered Lorentz's electron consistent with this requirement — that is, consistent with the principle of relativity. Furthermore, continued Poincaré, since we are "tempted to infer" a relation between the "causes giving rise to gravitation and those which give rise to the [internal pressure]," then Lorentz covariance permits further progress toward the unification of electromagnetism and gravitation [Poincaré, "Sur la dynamique de l'électron," 1906 op. cit. in n. 23.]

32. Historical analysis of archival materials and of primary sources permits me to conjecture that Einstein was aware of most requisite empirical data, as well as of the research toward electromagnetic and mechanical world-pictures. By 1905 his work on the fluctuations of cavity radiation and on the behavior of particles suspended in fluids had revealed to him the failure of Lorentz's electrodynamics and Newtonian mechanics in volumes the size of the electron. As Einstein recalled in his Nobel Prize acceptance address in 1923, the physicists of 1905 were "out of [their] depth." For details see my paper of 1976, op. cit., in n. 1, and especially my book, op. cit. in n. 1. A. Einstein, "Fundamental Ideas and Problems of the Theory of Relativity," delivered 11 July 1923 to the Nordic Assembly of Naturalists in acknowledgement of the Nobel Prize of 1921. Reprinted in *Nobel Lectures: 1901-1921* (Elsevier, New York, 1967), pp. 479-490.

33. Letter on deposit at the Algemeen Rijksarchief, The Hague.

34. W. Kaufmann, "Über die Konstitution des Elektrons," Berl. Ber. **45**, 949-56 (1905), and "Über die Konstitution des Elektrons," Ann. d. Phys. **19**, 487-553 (1906).

35. After pointing out a subtlety involved in defining the quantity force as "mass × acceleration," Einstein (op. cit. in n. 18) made an inappropriate choice for force that resulted in his predicting a transverse mass of $m_o/(1 - v^2/c^2)$, where for Einstein m_o was the electron's inertial mass. Kaufmann (op. cit. in n. 34) asserted that this result should be the same as Lorentz's. Planck ["Das Prinzip der Relativität und die Grundgleichungen der Mechanik," Verh. D. Ges. **6**, 136-41 (1906)] showed that a consistent generalized dynamics could be obtained if force were defined as the rate of change of momentum $\mathbf{p}$, where $\mathbf{p} = m_o\mathbf{v}/\sqrt{1 - v^2/c^2}$, thereby demonstrating quantitatively the mathematical equivalence of Einstein's and Lorentz's predictions for the electron's transverse mass. Kaufmann (1906 op. cit. in n. 34) complimented Einstein's approach to electrodynamics: "It is indeed remarkable," wrote Kaufmann, that from entirely different hypotheses Einstein obtained results which in their "accessible observational consequences" agreed with Lorentz's, but Einstein avoided difficulties of an "epistemological sort," that is, the inclusion of unknown velocities. Kaufmann viewed Einstein's success as based on his raising to a postulate the "principle of relative motion," which he placed at the "apex of all of physics." Then, applying this postulate to the propagation of light, Einstein deduced "a new definition of time and of the concept of 'simultaneity' for two spatially separated points, where Einstein's relationship between the time in two inertial reference systems was identical with Lorentz's local time." Kaufmann went on to emphasize that whereas Lorentz obtained his observationally accessible results through approximations (such as that of quasi-stationary motion), Einstein's kinematics of the rigid body, and his electrodynamics, were exact for inertial systems.

Einstein (op. cit., in n. 18) had deduced the electron's m_T and m_L by relativistic transformations between the electron's instantaneous rest system and the laboratory inertial system of the electron's space and time coordinates, and of the external electromagnetic fields acting on the electron.

In summary, Einstein was considered to have replaced Lorentz's assumptions of how electrons interact with the ether with phenomenological assumptions concerning how clocks were synchronized using light signals; that is, Einstein had improved Lorentz's theory of the electron.

36. In fact, since the first pages of the 1906 version bear the inscription, "Stampato il 14 dicembre 1905," then Poincaré must have read Kaufmann's 1905 paper (op. cit. in n. 34) and then added this final paragraph just before his paper (op. cit. in n.23) went to press in 1906.

37. Poincaré, "La dynamique de l'électron," op. cit. in n. 11.

38. Letter on deposit at the Einstein Archives, Institute for Advanced Study, Princeton, N.J.

39. To the passage in his 1908 paper (op. cit. in n. 11) where Poincaré called for others to repeat Kaufmann's 1905 measurements, he added in a footnote in his book *Science and Method* (op. cit. in n. 11), the following comment: "At the moment of going to press we learn that M. Bucherer had repeated the experiment, surrounding it with new precautions, and that, unlike Kaufmann, he has obtained results confirming Lorentz's views."

Lorentz, too, added a note to a text about to go to press. To a passage in the 1909 edition of *Theory of Electrons*, in which he discussed how well Abraham's theory agreed with the data of Kaufmann, Lorentz appended Note 87: "Recent experiments by Bucherer on the electric and magnetic deflexion of β-rays, made after a method that permits greater accuracy than could be reached by Kaufmann, have confirmed the formula $[m_T]$, so that, in all probability, the only objection that could be raised against the hypothesis of the deformable electron and the principle of relativity has now been removed."

40. C. T. Zahn and A. H. Spees discovered that Bucherer's velocity filters were inadequate ["A Critical Analysis of the Classical Experiments on the Variation of Electron Mass," Phys. Rev. **53**, 511-21 (1938)].

41. Poincaré, however, never elevated the principle of relativity to a convention because of disagreement with the measured value of the result of his Lorentz-covariant gravitational theory for the advance of mercury's perihelion [see Poincaré, *La dynamique de l'électron* (Dumas, Paris, 1913)].

42. Op. cit. in n. 19.

43. A. Einstein, "Autobiographical Notes," in *Albert Einstein: Philosopher-Scientist*, edited by P. A. Schilpp (The Library of Living Philosophers, Evanston, Ill., 1949).

44. In the face of the spirited criticisms and hard data of Abraham and Kaufmann, Planck, at the Meeting of German Scientists and Physicians at Stuttgart, 19 September 1906, could reply only that he remained "sympathetic" toward the "Lorentz-Einstein theory" because he believed in the fecundity of its principle of relativity. ["Die Kaufmannschen Messungen der Ablenkbarkeit der β-Strahlen in ihrer Bedeutung für die Dynamik der Elektronen," Phys. Z. **7**, 753-61 (1906)]. In 1907 Planck used the principle of relativity to formulate a general dynamics that included electromagnetism and thermodynamics ["Zur Dynamik bewegter Systeme," Berl. Ber. **13**, 542-70 (1907); also published in Ann. d. Phys. **26**, 1-34 (1908)]. His goal was to continue extending the principle of relativity in these disciplines in order to determine whether it led to either theoretical inconsistencies or to empirical disagreement. Planck's empiricistic view led him to call for more experiments than Bucherer's of 1908 [A. H. Bucherer, "Messungen an Becquerelstrahlen. Die experimentelle Bestätigung der Lorentz-Einsteinschen Theorie," Phys. Z. **9**, 755-62 (1908)]. As Planck wrote in 1910: "Physical questions, however, cannot be settled from aesthetic considerations, but only by experiment, and this always involves prosaic, difficult and patient work" ["Die Stellung der neueren Physik zur mechanischen Naturanschauung," Phys. Z. **11**, 922-32, reprinted in M. Planck, *A Survey of Physical Theory*, translated by R. Jones and D. H. Williams (Dover, New York, 1960), pp. 27-44].

45. The normally taciturn Minkowski waxed enthusiastic over Bucherer's results. In the discussion session after the lecture by Bucherer (op. cit. in n. 44), Minkowski expressed his "joy" for the vindication of "Lorentz's theory" over the "rigid electron": "The rigid electron is in my view a monster in relation to Maxwell's equations, whose innermost harmony is the principle of relativity . . . the rigid electron is no work hypothesis but a hinderance to work."

46. Quoted in my 1976 paper cited in n. 1.

47. M. Abraham, "Zur elektromagnetischen Mechanik," Phys. Z. **10**, 737-41 (1909). In the preface to the 1914 edition of his *Theorie der Elektrizität*, Abraham stressed that he would not permit current data to decide the issue between his electron theory and Lorentz's. And in this unusually carefully referenced book, Bucherer's data were nowhere mentioned.

48. M. Abraham, "Die neue Mechanik," Scientia **15**, 10-29 (1914).

49. Poincaré, *Science and Method*, op. cit in n. 11.

Looking back on electrodynamics in 1905, we can conjecture that its ambiguities would have been removed without Einstein. The requisite mathematical formalism already existed. Refinements of already performed ether-drift experiments, with additional null experiments, would have given physicists such as Langevin, Lorentz, and Poincaré the license to assert that the Maxwell-Lorentz equations have exactly the same form on the moving earth as they have relative to the ether (perhaps further fine tuning of the Maxwell-Lorentz theory would also have been required). The result of these endeavors could only have been a consistent theory of electromagnetism, and not a special theory of relativity. In his obituary to Langevin, Einstein wrote: "It appears to me as a foregone conclusion that he would have developed the Special Relativity Theory, had that not been done elsewhere" ["Paul Langevin in Memoriam," in A. Einstein, *Ideas and Opinions* (Littlefield, Adams, Totowa, N.J., 1967), pp. 210-11)]. What greater honor could have been bestowed on Langevin's memory? Yet Langevin's 1911 exposition of relativity theory reveals that his view of relativity physics differed fundamentally from Einstein's [P. Langevin, "L'évolution de l'espace et du temps," Scientia **10**, 31-54 (1911)]. Langevin developed Einstein's special relativity theory with an ether whose purpose was to have been the active agent that caused physical effects such as time dilation. But general relativity could not have evolved without Einstein because its formulation required the folding together of physics and geometry. And this was a double-edged masterstroke, for Einstein broke sharply with both the physics and philosophy of his day. The year 1915 was Einstein's second *Annus Mirabilis.*.

Open Discussion

Following Papers by G. HOLTON and A. I. MILLER

Chairman, C. N. YANG

Yang: I find it difficult to refrain from repeating the remark that I once read of Professor Dirac on the development of physics: he said that if he had to choose between beauty and agreement with experiment, he would choose beauty.

P. A. M. Dirac (Florida State U):   You mentioned about choosing between beauty and agreement with experiment. Well that's what people ought to have done when Kaufmann obtained his results.

Holton: We know what Einstein did when he heard about Kaufmann's results — one of the foremost experimentalists of Europe disproving this unknown person's work. Einstein did not respond for nearly two years. Finally, Stark persuaded Einstein to write an article, in December 1907, in which he took this up. This is a revealing response when compared to the responses of Lorentz and Poincaré that Dr. Miller spoke about. Einstein wrote that he had not found any obvious errors in Kaufmann's article, but that the theory that was being proved by Kaufmann's data was a theory of so much smaller generality than his own, and therefore so much less probable, that he would prefer for the time being to stay with it. Actually, it took untl 1916 for a fault in Kaufmann's experimental equipment to be discovered. By that time this whole question was moot.

A. Hermann (U. of Stuttgart):   It is very interesting to see what Planck thought about this contradiction between theory and experiment. Planck was a Platonist and he was always interested in finding out what he called the absolute in nature. When Planck studied Einstein's paper he was fascinated that time and space lost its absolute character, whereas the natural constants c and h were strengthened. There was additional support for the evidence that these natural constants were real absolute characters, and therefore Planck was convinced that this theory must be correct and he took a lot of time to check Kaufmann's experiments. It was Planck who pointed out in 1906 in a speech at Stuttgart that one should reject Kaufmann's experiments. Thus it was Planck who played a very important role in the acceptance of the special theory of relativity. Others who played an important role in the acceptance of the special theory were Felix Klein and Hermann Minkowski, who advanced Klein's ideas about the group character of geometry and physics.

V. Bargmann (Princeton U.):   We have heard about the extremely interesting correspondence between Poincaré and Lorentz. There is one aspect of it that I always found strange. This is the fact that in the famous book by Lorentz on the theory of electrons, Poincaré is hardly mentioned. There is copious mention of Einstein's work, but hardly any of Poincaré's. Now I wonder whether Dr. Miller has any comment on this?

Miller: Lorentz mentions Poincaré in reference to the Poincaré pressure keeping the electron together and then at the end of the 1909 edition he makes mention of Einstein. His interpretation

Essay 2

A Study of Henri Poincaré's
"Sur la Dynamique de l'Électron"

Contents

1. Introduction 208
2. An Electromagnetic World-Picture 211
3. The Electron Theory of MAX ABRAHAM 214
   3.1. The Foundation of ABRAHAM's Theory 214
   3.2. The Calculation of the Electron's Self-Fields 216
   3.3. The Electron's Mass According to ABRAHAM's Theory 218
4. The Electron Theory of H. A. LORENTZ 219
   4.1. The LORENTZ Transformations of 1892 and 1895 219
   4.2. The Contraction Hypothesis of 1895 221
   4.3. The LORENTZ Transformations of 1904 222
   4.4. The Charge Density, Electromagnetic Field and Velocity Transformation
        Equations 224
   4.5. The Contraction Hypothesis of 1904 227
   4.6. The Electron's Mass According to LORENTZ's Theory 228
   4.7. The Unresolved Problems in LORENTZ's Theory 230
5. A Note on the Relation Between POINCARÉ's Epistemology and His Physics . 233
   5.1. Perspective 233
   5.2. POINCARÉ's Epistemology and His Physics 234
6. "Sur la dynamique de l'électron" 246
   6.1. Introduction 246
   6.2. The LORENTZ Transformation [§ 1] 249
   6.3. The Principle of Least Action [§ 2] 255
   6.4. The LORENTZ Transformation and the Principle of Least Action [§ 3] . . 260
   6.5. The LORENTZ Group [§ 4] 261
   6.6. LANGEVIN's Waves [§ 5] 264
   6.7. Contraction of Electrons [§ 6] 266
   6.8. Quasi-Stationary Motion [§ 7] 282
   6.9. Arbitrary Motion [§ 8] 295
7. The Relevance of POINCARÉ's Results to Twentieth-Century Physics 301
   7.1. General Perspective 301
   7.2. LORENTZ's Application of the POINCARÉ Stress 303
   7.3. The Theories of EINSTEIN and LORENTZ 306
   7.4. HERMANN MINKOWSKI's Four-Dimensional Formulation of Relativity . 308
   7.5. MAX VON LAUE's Description of the POINCARÉ Stress within the Context
        of EINSTEINIAN Relativity 309
   7.6. Post-EINSTEINIAN Classical Field Theories of the Electron 314
   7.7. The Elusive 4/3 Factor 317
   7.8. A Quantum Field Theory of the Electron: Quantum Electrodynamics . . 318
8. Concluding Remarks 319
Bibliography 323

from "Archive for History of Exact Sciences", Volume 10, Number 3–5, 1973, P. 207–328 © by Springer-Verlag 1973 Printed in Germany

1. Introduction

HENRI POINCARÉ's major work on a theory of the electron is "Sur la dynamique de l'électron".[1] It is considered, by some, as evidence that POINCARÉ, more than anyone else in the late 19th and early 20th centuries, anticipated EINSTEIN's 1905 theory of relativity.[2]

This study will focus on POINCARÉ's attempt in "Sur la dynamique..." to formulate a purely electromagnetic theory of a deformable electron that is consonant with his conception of the principle of relativity. POINCARÉ believed that if all physical processes could be reduced ultimately to the interaction of charged particles which move about in LORENTZ's all-pervasive ether, then such a theory would be an important step toward a unified description of nature. Thus, the laws of the various branches of physics, and in particular NEWTON's second law, could be derived from those of electromagnetism. This scientific viewpoint (or *Weltbild*) will hereafter be referred to as the "electromagnetic world-picture."[3]

Part 2 of this paper constitutes a survey of efforts toward an electromagnetic world-picture before 1905. The two major theories of the electron at that time were those of MAX ABRAHAM[4,5] and H. A. LORENTZ[6]. "Sur la dynamique..." was

---

[1] H. POINCARÉ, "Sur la dynamique de l'électron," Rend. del Circ. Mat. di Palermo, **21**, 129–175 (1906), submitted 23 July, 1905; reprinted in *Œuvres de Henri Poincaré* (11 vols.; Gauthier-Villars, 1934–1954), Vol. IX, pp. 494–550. This volume of POINCARÉ's collected works will be referred to hereinafter as *Œuvres*. A translation of pp. 129–146, 166–175 corresponding to the Introduction and Sections 1–4, 9 (*Œuvres*, pp. 494–515, 538–550) appears in C. W. KILMISTER, ed., *Special Theory of Relativity* (Pergamon, 1970), pp. 144–185. A brief version of this paper, with the same title, appeared in Comptes rendus de l'Académie des Sciences, **140**, 1504–1508 (1905); reprinted in *Œuvres*, pp. 489–493. All references will be made to Rend. del Circ. Mat. di Palermo (hereafter abbreviated as R.), *Œuvres* (hereafter abbreviated as O.) and, when applicable, to KILMISTER's translation (hereafter abbreviated as K.). Thus, reference to a passage in "Sur la dynamique..." will appear as: R.: O.: K.

Recently H. M. SCHWARTZ published an incomplete translation of Sections 1–8 of "Sur la dynamique..." (H. M. SCHWARTZ, "Poincaré's Rendiconti Paper on Relativity. Part I," Am. J. Phys. **39**, 1287–1294 (1971); "Poincaré's Rendiconti Paper on Relativity. Part II," *Ibid.*, **40**, 862–872 (1972)). For my criticism of this method of historical exposition see A. I. MILLER, "Comment on: 'Poincaré's Rendiconti Paper on Relativity: Part I'," Am. J. Phys. **40**, 923 (1972).

[2] A. EINSTEIN, "Zur Elektrodynamik bewegter Körper," Ann. d. Phys., **17**, 891–921 (1905), submitted 30 June, 1905; reprinted in *The Principle of Relativity: A Collection of Original Memoirs on the Special and General Theory of Relativity by H. A. Lorentz, A. Einstein, H. Minkowski and H. Weyl*, translated by W. PERRETT & G. B. JEFFERY (Dover, n.d.), pp. 37–65. All references will be to the Dover reprint volume which will be designated hereinafter as *P.R.C.*

[3] For a comprehensive survey of the electromagnetic world-picture see R. McCORMMACH, "H. A. Lorentz and the Electromagnetic View of Nature," Isis, **61**, 459–497 (1970).

[4] M. ABRAHAM, "Dynamik des Electrons," Nachr. Ges. Wiss. Göttingen, 20–41 (1902). For an enlightening presentation of ABRAHAM's life and work see S. GOLDBERG, "The Abraham Theory of the Electron: The Symbiosis of Experiment and Theory," Archive for History of Exact Sciences, **7**, 7–25 (1970).

[5] M. ABRAHAM, "Prinzipien der Dynamik des Elektrons," Ann. d. Phys., **10**, 105–179 (1903).

[6] H. A. LORENTZ, "Electromagnetic Phenomena in a System Moving with any Velocity Less than that of Light," Proc. Roy. Acad. Amsterdam, **6**, 809 (1904); Reprinted in *P.R.C.*, pp. 11–34. All references will be to *P.R.C.*

written in part as a response to problems arising from these theories and to inconsistencies in their structure. These theories are discussed in Parts 3 and 4 as part of the background essential to an understanding of "Sur la dynamique..." The other essential part of the background is contained in Part 5's exposition of those portions of POINCARÉ's epistemology relevant to this study. Of particular interest will be the interaction between POINCARÉ's philosophy and his physics — my thesis being that POINCARÉ's physics is a reflection of his philosophy. This theme will be developed further in a future publication, dependent upon this one for technical details. The sequel will necessarily include an analysis of POINCARÉ's use of such concepts as fact, hypothesis, law, postulate, principle and convention. Of particular importance is the difference between a conventional statement in the subdivisions of science which POINCARÉ referred to as geometry, mechanics and the physical sciences.

These two complementary studies are designed to describe the philosophic and scientific dimensions of POINCARÉ. I believe that an analysis that gives equal consideration to historical, philosophical and scientific material can yield a new perspective on such questions as: Why did POINCARÉ write "Sur la dynamique..." in 1905 ? Why was the principle of relativity as he used it not a convention ? If "Sur la dynamique..." contained so many important concepts as a principle of relativity and a discussion of the measurement process using light, then why did POINCARÉ neither produce the theory before EINSTEIN nor later on accept EINSTEIN's formulation?

Indeed, three weeks after POINCARÉ had submitted "Sur la dynamique..." for publication, a paper entitled "Zur Elektrodynamik bewegter Körper"[2] was sent to the *Annalen der Physik* from an employee of the Confederate Patent Office at Bern. The irony is that ALBERT EINSTEIN was not only unaware of LORENTZ's paper of 1904, "Electromagnetic Phenomena in a System Moving with any Velocity Less than that of Light",[6] and a brief version of "Sur la dynamique..." (published with this title on 5 June 1905 in *Comptes rendus*[1]), but also, as he later explained, he may well not even have known of the MICHELSON-MORLEY experiment[7]. EINSTEIN's principle of relativity was a universal one, applicable to *all* branches of physics. Unlike POINCARÉ, EINSTEIN in 1905 was not interested in an electromagnetic world-picture.

In previous studies of these issues, the historical and philosophical aspects of POINCARÉ's work have been emphasized.[8] However, in this study and its sequel

---

[7] For a discussion of this point see the paper of G. HOLTON, "Einstein, Michelson and the 'Crucial' Experiment," Isis, **60**, 133–197 (1969).

[8] (a) E. T. WHITTAKER, *A History of the Theories of Aether and Electricity: The Modern Theories 1900–1926* (Nelson, 1953) esp. Chapter II entitled "The Relativity Theory of Poincaré and Lorentz."

(b) G. H. KESWANI, "Origin and Concept of Relativity," Brit. J. Phil. Sci., **15**, 286–306 (1965); **16**, 19–32 (1965).

(c) S. GOLDBERG, "Henri Poincaré and Einstein's Theory of Relativity," Am. J. Phys., **35**, 934–944 (1967).

(d) S. GOLDBERG, "Poincaré's Silence and Einstein's Relativity: The Role of Theory and Experiment in Poincaré's Physics," Brit. J. Hist. Sci., **5**, 73–84 (1970).

(e) G. HOLTON, "On the Thematic Analysis of Science: The Case of Poincaré and Relativity," Mélanges Alexandre Koyré (Hermann, 1964) 257–268.

14   Arch. Hist. Exact Sci., Vol. 10

I will present a balanced presentation of his philosophy and physics, thereby showing that Poincaré's philosophic and scientific dimensions were not mutually exclusive.

Toward this goal a step-by-step analysis of certain portions of "Sur la dynamique..." will be presented in Part 6. This type of analysis is often the most suitable for studying important scientific papers. In this manner it will be possible to relate Poincaré's methodology to contemporaneous (c. 1905) research, and then, in Part 7, to trace the impact of his results upon elementary particle physics circa 1972. This method also yields a clear picture of the pre-Einsteinian electrodynamics of moving bodies. Furthermore, it allows one to perceive the interplay between physical science and its historic setting.

Of the nine Sections comprising "Sur la dynamique..." the crucial ones are 6, 7 and 8; the step-by-step analysis will be applied only to them. The other Sections, except 9, will be summarized thoroughly and analyzed step-by-step only when necessary; they contain Poincaré's corrections, clarifications and extensions of certain points in Lorentz's paper of 1904. Section 9, "Hypotheses Concerning Gravitation," will be referred to often, but detailed discussion of it is reserved for a future study of gravitational theories before Einstein's.

Sections 6, 7 and 8 deserve a step-by-step analysis because in them Poincaré proved conclusively that of all the electron theories known at the time, only Lorentz's was consonant with the principle of relativity. However, the derivation of Lorentz's theory from a covariant action principle necessitated the addition of a mechanical energy, of unknown origin, to the electromagnetic field Lagrangian. This term, from which the "Poincaré pressure" is derived, simultaneously accounted for the stability of the Lorentz electron, although, as we shall see, Poincaré never explicitly proved this because of certain numerical errors. Moreover, the necessity of the additional energy negated the goal of an electromagnetic world-picture, though Poincaré believed otherwise.

"Sur la dynamique..." is a rare document in the annals of science because it is Poincaré's only detailed work toward a theory of relativity. In it he brought to bear the experience of many years of research on electromagnetic field theory as well as his mathematical insights. Thus it is not only a scientific portrait of Poincaré but also a presentation and analysis of the basic problems confronting what was then the physics of elementary particles.

Poincaré was not a highly innovative physicist; rather, as we shall see in Part 5, his great powers in this discipline lay in criticizing, clarifying and extending previously proposed theories. As a mathematician, however, Poincaré was a legend in his own time[9]. Thus it is reasonable to expect Poincaré's approach to

---

(f) G. Holton, "On the Origins of the Special Theory of Relativity," Am. J. Phys., **28**, 627–636 (1960).

(g) C. Scribner, "Henri Poincaré and the Principle of Relativity," Am. J. Phys., **32**, 672–678 (1964).

(h) T. Kahan, "Sur les Origines de la théorie de la relativité restreinte," Revue d'Histoire des Sciences, *XIII*, 159–165 (1959).

[9] In the report on the Bolyai Prize, which was awarded to Poincaré in 1905, Gustave Rados described him as "incontestably the foremost and most powerful researcher of the time in the domains of mathematics and mathematical physics" (Bull. Sci. Mat., **30**, 105–112 (1906)); reprinted, in part, in Ernest Lebon's *Savant*

a physics problem to be a mathematical one. Indeed, examination of POINCARÉ's studies on electromagnetic field theory reveals this to be the case. However, his essentially mathematical mode of thought was one of the factors that prevented him from drawing pertinent physical conclusions from his research on electromagnetic field theory. Nevertheless, this orientation led to the discovery of several important symmetries which were hidden by the manner in which LORENTZ formulated his electron theory of 1904: the group property of the LORENTZ transformation, the LORENTZ invariance of the principle of least action for the electromagnetic field, the seeds of the four-vector formalism, the notation x, y, z, ict, in order to extend the results of the three-dimensional theory of invariants to four dimensions, a LORENTZ-covariant theory of gravitation. Certain philosophers and historians of science (WHITTAKER[8(a)] and KESWANI[8(b)]) have concluded, on the basis of this list, that POINCARÉ went right to the threshold of relativity theory and that EINSTEIN merely generalized these results. However, the analysis of "Sur la dynamique..." will reveal that this was not the case.[10]

Part 7 discusses the reception of POINCARÉ's results and their elucidation in view of EINSTEIN's theory, particularly by LORENTZ, MINKOWSKI, PLANCK and VON LAUE. To trace the impact of POINCARÉ's research into modern elementary particle physics, I shall present first an exposition of work after 1905 on a classical field description of the electron by DIRAC, EINSTEIN, FERMI, KWAL, LORENTZ, MIE, ROHRLICH and WEYL; and then a brief discussion of the stability problem in quantum electrodynamics.

The concluding remarks, which constitute Part 8, set the stage for the sequel to this study.

2. An Electromagnetic World-Picture

By the end of the nineteenth century it was becoming apparent that an electromagnetic world-picture offered a more promising approach to a unified theory of

---

du Jour: Henri Poincaré Biographie, Bibliographie analytique des écrits (2nd edition, Gauthier-Villars, 1912), pp. 21–26, esp. p. 21. It is of interest to note that POINCARÉ's competitor for this prize was DAVID HILBERT. See CONSTANCE REID's *Hilbert* (Springer, 1970), esp. p. 106, for this episode. TOBIAS DANTZIG, in his biography of POINCARÉ, notes:

> A formal biography of Poincaré would read like a catalogue of academic honors: from degree to degree, from title to title, from grand prize to grand prize. Titular professor at the Sorbonne at 30, member of the Institute at 32; all the homage which his native land could grant was bestowed upon before he turned 40. And not only his native land: universities and learned societies on both sides of the Atlantic showered him with prizes, medals and honorary degrees.
>
> Congresses and conventions, jubilees and festivals vied with each other for his presence, and Poincaré passionately loved to travel. His travelogue would comprise practically every country in Europe and the United States, which, incidentally, he visited at least twice. An estimate of the time consumed in this journeying leaves one aghast. When did this extraordinary man find time to present more than 500 memoirs, publish 30 odd treatises, write scores of tracts on the foundations of science? Yet, he did all this, and more: for, he was active on many scientific commissions, and pronounced scores of eloquent eulogies in memory of illustrious scholars of his time.

(T. DANTZIG, *Henri Poincaré: Critic of Crisis* (Scribner, 1954), p. 3).

[10] My paper on the philosophical side of this question will reinforce this result.

14*

matter than a mechanical world-picture[11]. Indeed, the complex models proposed by the supporters of a mechanical world-picture were overshadowed by the successes of H. A. LORENTZ's electromagnetic theory. In the new research program the laws of mechanics, as well as the laws of other branches of physics, would be reduced to those of electromagnetism. Such a proposal was made by WILHELM WIEN in 1900[12], and was endorsed by LORENTZ[13].

A far-reaching implication of this program is that if matter can ultimately be reduced to electrical charges and an ether, then mass has its origin in the electromagnetic field.[14] That mass has its origin in the electromagnetic field had been suggested already in 1881 by J. J. THOMSON, although he did not express any intent or desire to reduce mechanics to electromagnetism[15]. By means of a hydrodynamic analogy THOMSON demonstrated that the inertia of a charged body increased when the body was set in motion through the ether. However, the additional inertia would be difficult to detect experimentally because it turned out to be a constant. THOMSON's work was elaborated on by HEAVISIDE[16], SEARLE[17] and WIEN[12], who showed the increase in inertia to be velocity-dependent and therefore open to experimental verification.

WALTER KAUFMANN, in 1901[18], utilizing BECQUEREL rays, concluded that the electron's mass was the sum of its mechanical and its apparent masses, the latter

---

[11] A mechanical world-picture is a *Weltbild* in which the laws of physics are reduced to those of mechanics. For discussions of and further references to mechanical models of the ether see Volume I of E. T. WHITTAKER, *History of the Theories of Aether and Electricity: The Classical Theories* (Nelson, 1951) and K. SCHAFFNER *Nineteenth-Century Aether Theories* (Pergamon, 1972).

[12] W. WIEN, "Über die Möglichkeit einer elektromagnetischen Begründung der Mechanik," *Recueil de travaux offerts par les auteurs à H. A. Lorentz* (Martinus Nijhoff, 1900), pp. 96–107; reprinted in Ann. d. Phys., **5**, 501–513 (1901).

[13] H. A. LORENTZ, "Elektromagnetische Theorien physikalischer Erscheinungen," Phys. Zeit., **1**, 498, 514 (1900), in *Collected Papers*, **8**, 333–352: "The Theory of Radiation and the Second Law of Thermodynamics," Versl. Kon. Akad. Wetensch. Amsterdam, **9**, 418 (1900), in *Collected Papers*, **6**, 265–279: "Boltzmann's and Wien's Laws of Radiation," Versl. Kon. Akad. Wetensch. Amsterdam, **9**, 572 (1901), in *Collected Papers*, **6**, 280–292: "Considérations sur la pesanteur," Versl. Kon. Akad. Wetensch. Amsterdam, **8**, 603 (1900), in *Collected Papers*, **5**, 198–215 wherein LORENTZ discussed the possibility of formulating a gravitational theory using concepts based on electromagnetic theory: "Über die scheinbare Masse der Ionen," Phys. Zeit., **2**, 78 (1901), in *Collected Papers*, **3**, 113–116 wherein he noted the possibility that some, but not all, of the electron's mass is of electromagnetic origin and also that the moving electron might undergo a contraction.

[14] For an account of the evolution of the concept of mass from time immemorial to the present see M. JAMMER, *Concepts of Mass* (Harvard U. Press, 1961), esp. Chapter 11 entitled "The Electromagnetic Concept of Mass."

[15] J. J. THOMSON, "On the electric and magnetic effects produced by the motion of electrified bodies," Phil. Mag., **11**, 229–249 (1881).

[16] O. HEAVISIDE, "On the electromagnetic effects due to the motion of electrification through a dielectric," Phil. Mag., **27**, 324–339 (1889).

[17] G. F. C. SEARLE, "On the motion of an electrified ellipsoid," Phil. Mag., **44**, 329–341 (1897).

[18] W. KAUFMANN, "Die magnetische und elektrische Ablenkbarkeit der Becquerelstrahlen und die Scheinbare Masse der Elektronen," Nachr. Ges. Wiss. Göttingen, 143–155 (1901).

being an increasing function of its velocity. Furthermore, the velocity-dependent portion seemed to be consonant with SEARLE's result.

MAX ABRAHAM's theory of the electron[4,5] was the first major effort to account for KAUFMANN's data. In two papers, published in 1902 and 1903, ABRAHAM proposed a theory in which the mechanical properties of a rigid spherical electron could be derived from its own electromagnetic field. An unambiguous identification of the electron's mass as the coefficient of the acceleration in NEWTON's second law (since the intent was to derive the laws of mechanics from those of electromagnetism) necessitated the neglect of the radiation reaction on the electron due to its self-fields. This is known as the quasi-stationary approximation. It is common to all theories of the electron based upon the electromagnetic world-picture and will be discussed further. ABRAHAM's theory appeared to be corroborated when, in 1902, KAUFMANN[19] proclaimed that the electron's entire mass was of electromagnetic origin and that his results were now in agreement with ABRAHAM's.

There was no theory to compete with ABRAHAM's until 1904, when LORENTZ extended the ostensibly all-embracing framework of his electromagnetic theory to the problem of the electron's structure. Whereas ABRAHAM's theory was constructed for a particular problem, LORENTZ's was the logical continuation of his formulations of 1892[20] and 1895[21,22]. The difference between the two theories did not end here; ABRAHAM's electron was a rigid sphere, while LORENTZ's was deformable; ABRAHAM's electron transformed according to the GALILEAN transformations, while LORENTZ's transformed according to a set of equations which accounted for the fact that optical phenomena appeared the same to observers in reference systems which are in uniform relative motion with respect to the ether. LORENTZ referred to this as the theorem of corresponding states. Moreover, LORENTZ claimed that his expression for the transverse mass was also in agreement with KAUFMANN's data.

ABRAHAM, in defense of his theory, questioned the stability of LORENTZ's electron.[23] He recalled that his motive for postulating a rigid electron was that a

---

[19] W. KAUFMANN, "Die elektromagnetische Masse des Elektrons," Phys. Zeit., **4**, 54–57 (1902).

[20] H. A. LORENTZ, "La théorie électromagnétique de Maxwell et son application aux corps mouvants," Arch. néerl., **25**, 363 (1892), in *Collected Papers* 2, 164–343.

[21] H. A. LORENTZ, *Versuch einer Theorie der elektrischen und optischen Erscheinungen in bewegten Körpern* (Brill, 1895), in *Collected Papers*, **5**, 1–137. All references will be to the *Collected Papers*.

[22] For discussions of LORENTZ's work see the article by McCORMMACH mentioned in footnote 3 and

(a) T. HIROSIGE, "Origins of Lorentz's Theory of Electrons and the Concept of the Electromagnetic Field," *Historical Studies in the Physical Sciences*; Volume I. R. McCORMMACH, ed. (U. of Pennsylvania Press, 1969), pp. 151–209.

(b) T. HIROSIGE, "Electrodynamics before the Theory of Relativity, 1890–1905," Japanese Studies in the History of Science, **5**, 1–49 (1966).

(c) S. GOLDBERG, "The Lorentz Theory of Electrons and Einstein's Theory of Relativity," Am. J. Phys., **37**, 982–994 (1969).

(d) K. SCHAFFNER, "The Lorentz Theory of Relativity," Am. J. Phys., **37** 498–513 (1969).

[23] M. ABRAHAM, "Die Grundhypothesen der Elektronentheorie," Phys. Zeit., **5**, 576–579 (1904). For a survey of this paper see S. GOLDBERG, "The Abraham Theory of the Electron ..." *op. cit.* (footnote 4).

flexible one could be held together only by forces that were not electromagnetic. The presence of such forces is inconsistent with a totally electromagnetic world-picture.

In 1904, A. H. BUCHERER[24] and P. LANGEVIN[25] simultaneously and independently proposed another electron theory which also appeared to be in agreement with KAUFMANN's data. These theories will be discussed in Part 6.

More important, however, for an understanding of "Sur la dynamique..." are the theories of ABRAHAM and LORENTZ.

3. The Electron Theory of Max Abraham

3.1. The Foundation of Abraham's Theory. ABRAHAM's theory was built upon LORENTZ's theory of electromagnetism, the basic components of which are a stationary, all-pervasive ether, unaffected by the presence of matter, and five fundamental equations—the four MAXWELL equations, suitably altered to include the presence of charges (which will be referred to as the MAXWELL-LORENTZ equations)

$$\vec{V} \cdot \vec{E} = \varrho, \tag{1a}$$

$$\vec{V} \cdot \vec{B} = 0, \tag{1b}$$

$$\vec{V} \times \vec{E} = -\frac{\partial \vec{B}}{\partial t}, \tag{1c}$$

$$\vec{V} \times \vec{B} = \frac{\partial \vec{E}}{\partial t} + \varrho \vec{v} \tag{1d}$$

(where the units are chosen to conform with those used by POINCARÉ in 1906: HEAVISIDE units with $c = 1$) and the LORENTZ force equation in which the concept of mass is introduced

$$\vec{F} = \vec{E} + \vec{v} \times \vec{B}, \tag{1e}$$

where F is the force per unit charge. The field vectors $\vec{E}$ and $\vec{B}$ are taken here to be only the self-fields; and the charge density ϱ is assumed to be a primitive quantity.

Towards deriving the electron's mechanical properties from its self-fields, ABRAHAM proved first that the equations of electromagnetism could be written in a form resembling LAGRANGE's equation. Then, assuming uniform linear motion for the electron, he was able to identify the electromagnetic field LAGRANGIAN as

$$L = \tfrac{1}{2} \int (B^2 - E^2) \, d\tau, \tag{2}$$

where $d\tau$ is a differential volume element. If the electron's kinetic energy is defined as

$$T = \tfrac{1}{2} \int B^2 \, d\tau$$

---

[24] A. H. BUCHERER, "Mathematische Einführung in die Elektronentheorie" (Teubner, 1904).

[25] P. LANGEVIN, "La physique des électrons," Revue générale des sciences pures et appliquées, **16**, 257–276 (1905). This constitutes the text of the lecture given by LANGEVIN at the International Congress of Arts and Science at Saint Louis, Missouri, in 1904, a conference he attended with POINCARÉ.

(since a magnetic field arises from a charge in motion) and its potential energy as

$$U = \tfrac{1}{2} \int E^2 d\tau$$

(since the COULOMB field accompanies the electron whether or not it is in motion), then (2) is the usual mechanical form for the LAGRANGIAN,

$$L = T - U.$$

ABRAHAM furnished additional proof that (2) is the correct LAGRANGIAN by calculating the momentum $\vec{G}$[26] and energy W of the electromagnetic field according to the procedures of classical mechanics:

$$\vec{G} = \frac{\partial L}{\partial \vec{v}} = \int \vec{E} \times \vec{B} \, d\tau, \tag{3}$$

$$W = -L + \vec{v} \cdot \frac{\partial L}{\partial \vec{v}} = \frac{1}{2} \int (E^2 + B^2) \, d\tau. \tag{4}$$

ABRAHAM's next step was to propose a model for the electron in order to calculate its self-fields and then its momentum from (3). Assuming that the electron's mass was of purely electromagnetic origin, and using (3) and (1e), ABRAHAM

---

[26] Although ABRAHAM was the first to use the concept of electromagnetic field momentum in an electron theory, its importance was first recognized in 1900 by POINCARÉ (H. POINCARÉ, "La théorie de Lorentz et le principe de réaction", in *Recueil de travaux offerts par les auteurs à H. A. Lorentz* (Martinus Nijhoff, 1900), pp. 252–278; reprinted in *Œuvres*, pp. 464–488). Writing (1e) as

$$\vec{F} = \varrho \vec{E} + \varrho \vec{v} \times \vec{B} \tag{26-1}$$

(where only the charge's own fields are taken into account), then substituting for ϱ and $\varrho \vec{v}$ from (1a) and (1d) and then integrating over all space, POINCARÉ demonstrated that the resultant force $\vec{F}_R = \int \vec{F} d\tau$ did not vanish, contradicting NEWTON's law of action and reaction. This violation had already been noted by LORENTZ in 1895 (footnote 21, p. 28), who asserted that NEWTON's third law was not universally valid. POINCARÉ obtained in place of $\vec{F}_R = 0$ the result

$$\vec{F}_R = -\frac{\partial \vec{G}}{\partial t}, \tag{26-2}$$

where $\vec{G}$ is defined in (3). By use of a hydrodynamic analogy (a "fluide fictif") he was able to associate $\vec{G}$ with a fluid momentum. If $\vec{F}_R$ is regarded as the rate of change of mechanical momentum $d\vec{G}_{\text{mech}}/dt$, (26-2) becomes

$$\frac{d}{dt} (\vec{G}_{\text{mech}} + \vec{G}) = \vec{0}, \tag{26-3}$$

thereby maintaining the law of conservation of momentum at the expense of widening its interpretation to include the momentum of non-ponderable matter. This paper will be discussed further in footnote 278. Note that in an electromagnetic world-picture $\vec{G}_{\text{mech}} = 0$ (no mechanical mass), and in the presence of external forces, (26-3) becomes

$$\vec{F}_{\text{ext}} = \frac{d\vec{G}}{dt}$$

where $\vec{G}$ is a function of the electron's self-fields.

obtained a résult for the transverse mass which he then compared with KAUF-MANN's data. Consonant with an electromagnetic world-picture, ABRAHAM chose a rigid spherical electron with either a uniform surface or volume charge distribution. A deformable electron would require the presence of non-electromagnetic forces in order to maintain its stability because of the mutual COULOMB repulsion between its constituent parts.

3.2. The Calculation of the Electron's Self-Fields. It is of interest to discuss ABRAHAM's calculation of the electron's self-fields because of its relation to LORENTZ's theory of the electron in 1904. According to LORENTZ's electromagnetic theory (1892, 1895), the fields and potentials are related by the equations

$$\vec{E} = -\vec{V}\phi - \frac{\partial \vec{A}}{\partial t},$$
$$\vec{B} = \vec{V} \times \vec{A}. \tag{5}$$

By use of (1) and (5) and the LORENTZ gauge condition

$$\vec{V} \cdot \vec{A} + \frac{\partial \phi}{\partial t} = 0, \tag{6}$$

it is possible to derive inhomogeneous wave equations for the potentials

$$\left(V^2 - \frac{\partial^2}{\partial t^2}\right)\vec{A} = -\varrho\vec{v}, \tag{7a}$$

$$\left(V^2 - \frac{\partial^2}{\partial t^2}\right)\phi = -\varrho. \tag{7b}$$

In the case of uniform linear motion in the x-direction with velocity $v_x = v$ (7a) reduces to one equation:

$$\left(V^2 - \frac{\partial^2}{\partial t^2}\right)A_x = -\varrho v. \tag{7a'}$$

Equations (7a') and (7b) can be simplified since the potentials are carried along unchanged by the moving charge. Consequently, the convective derivative of the scalar and vector potentials vanishes:

$$\frac{dq}{dt} = \frac{\partial q}{\partial t} + v\frac{\partial q}{\partial x_r} = 0 \tag{8}$$

where q symbolizes either ϕ or A_x; $x_r = x - vt$, $y_r = y$, $z_r = z$, $t_r = t$ is the relative coordinate of a point on the electron with respect to a system moving with it and x, y, z, t are the coordinates of this point with respect to a reference system fixed in space. Thus, from (8) it follows that

$$\frac{\partial \phi}{\partial t} = -v\frac{\partial \phi}{\partial x_r}, \tag{9a}$$

$$\frac{\partial A_x}{\partial t} = -v\frac{\partial A_x}{\partial x_r}, \tag{9b}$$

and similarly that

$$\frac{\partial^2 \phi}{\partial t^2} = v^2\frac{\partial^2 \phi}{\partial x_r^2}, \tag{10a}$$

$$\frac{\partial^2 A_x}{\partial t^2} = v^2\frac{\partial^2 A_x}{\partial x_r^2}, \tag{10b}$$

By means of (10) the time derivatives can be eliminated in (7b) and (7a'):

$$(1-v^2)\frac{\partial^2 \phi}{\partial x_r^2} + \frac{\partial^2 \phi}{\partial y_r^2} + \frac{\partial^2 \phi}{\partial z_r^2} = -\varrho, \tag{11a}$$

$$(1-v^2)\frac{\partial^2 A_x}{\partial x_r^2} + \frac{\partial^2 A_x}{\partial y_r^2} + \frac{\partial^2 A_x}{\partial z_r^2} = -\varrho v. \tag{11b}$$

But (11b) can be transformed to (11a) by substituting $A_x = v\phi$, thereby reducing the problem to that of finding only the scalar potential. One more change of variables, namely

$$x' = (1-v^2)^{-\frac{1}{2}} x_r,$$
$$y' = y_r, \tag{12}$$
$$z' = z_r,$$

transforms (11a) into the form of POISSON's equation[27],

$$\frac{\partial^2 \phi}{\partial x'^2} + \frac{\partial^2 \phi}{\partial y'^2} + \frac{\partial^2 \phi}{\partial z'^2} = -\varrho. \tag{13}$$

Therefore, the set of transformations (12) can be interpreted as transforming to a system S' wherein the electron is at rest and all lengths in the direction of motion are dilated. Thus in S' the spherical electron becomes an ellipsoid whose major axis is along the direction of motion; S' will be referred to as the instantaneous rest frame. The relation between the volume element in the real coordinate system S (with respect to which a point on the moving electron is defined by x, y and z) and the fictitious system S' can be obtained from (12) as

$$d\tau' = (1-v^2)^{-\frac{1}{2}} d\tau, \tag{14a}$$
$$d\tau = (1-v^2)^{\frac{1}{2}} d\tau'. \tag{14b}$$

Furthermore, since charge is conserved, i.e.

$$\int \varrho \, d\tau = \int \varrho' \, d\tau',$$

we have

$$\varrho' = \varrho(1-v^2)^{\frac{1}{2}}. \tag{15}$$

From (15) it follows that to write (13) entirely in terms of quantities defined in S' necessitates the substitution $\phi' = (1-v^2)^{\frac{1}{2}}\phi$. Then (13) becomes

$$\frac{\partial^2 \phi'}{\partial x'^2} + \frac{\partial^2 \phi'}{\partial y'^2} + \frac{\partial^2 \phi'}{\partial z'^2} = -\varrho' \tag{16}$$

the solution of which is

$$\phi' = \frac{e(1-v^2)^{\frac{1}{2}}}{8\pi r v} \log\left(\frac{1+v}{1-v}\right). \tag{17}$$

This is the potential due to an ellipsoidal distribution of charge having semimajor axis $(1-v^2)^{-\frac{1}{2}} r$ and semiminor axis r. The potential in S can be found from $\phi' = (1-v^2)^{\frac{1}{2}}\phi$ to be

$$\phi = \frac{e}{8\pi r v} \log\left(\frac{1+v}{1-v}\right). \tag{18}$$

---

[27] This transformation was used first by J. J. THOMSON in Phil. Mag. **28**, 1–14 (1889), esp. p. 12. See also J. J. THOMSON, *Notes on Recent Researches in Electricity and Magnetism* (Clarendon, 1893), esp. p. 17.

Since $A_x = v\phi$, the electric and magnetic fields can be calculated from (5) and substituted into (2) to obtain the LAGRANGIAN for the self-fields,

$$L = -\frac{e^2}{16\pi r} \frac{(1-v^2)}{v} \log\left(\frac{1+v}{1-v}\right). \tag{19}$$

The electron's momentum can now be calculated from its self-fields by substituting (19) into (3):

$$G_x = \frac{\partial L}{\partial v} = \frac{e^2}{16\pi r}\left[\frac{1+v^2}{v^2}\log\left(\frac{1+v}{1-v}\right) - \frac{2}{v}\right]. \tag{20}$$

This complicated function of the velocity vanishes as $v \to 0$, as expected, and diverges as the electron's velocity approaches the velocity of light *in vacuo*.

3.3. The Electron's Mass According to Abraham's Theory.

The next step is to calculate the electron's mass using (20). However, an unambiguous identification of the electron's mass as the coefficient of the acceleration requires the quasi-stationary approximation.[28] In this manner it is possible to maintain the "usual" form of NEWTON's second law,

$$\vec{F}_{ext} = \frac{d\vec{G}}{dt}, \tag{21}$$

where $\vec{F}_{ext}$ is the external force acting on the electron and $\vec{G}$ is calculated from the electron's self-fields, as in (20). In particular, for a constant velocity $\frac{d\vec{G}}{dt} = 0$, as in ordinary mechanics. To determine the electromagnetic mass from (21) necess-

---

[28] The force on an electron can be divided into two parts: a part due to external fields and that due to its own field ($\vec{F}_{self}$). Thus,

$$\vec{F}_{ext} + \vec{F}_{self} = m_0\,\vec{a}, \tag{28-1}$$

where m_0 is the mechanical mass (which is zero in ABRAHAM's theory). By expanding the scalar and vector potentials for a moving charge in powers of its velocity v and its higher time derivatives the self-force can be shown to be (for details see H. A. LORENTZ, *The Theory of Electrons* (Brill, 1909; rev. ed., 1915; Dover, 1952); all references will be to the Dover edition hereafter designated as *T.E.*, esp. pp. 48–52, 252–253),

$$\vec{F}_{self} = -m_e\vec{a} + \alpha_2\dot{\vec{a}} + \text{(terms proportional to higher order time derivatives of }\vec{a}). \tag{28-2}$$

The first term is referred to as the inertial reaction, and m_e is a structure dependent constant which is defined as the electromagnetic mass. In pre-EINSTEINIAN electron theories, *i.e.* those of ABRAHAM, BUCHERER-LANGEVIN, and LORENTZ, m_e is computed from the time rate of change of the electromagnetic field momentum and is, therefore, a two component function of the velocity. The second term in (28-2) is the force exerted on the electron by its own radiation field and α_2 is a structure independent constant. The higher-order terms have no physical interpretation. Equation (28-1) then becomes

$$\vec{F}_{ext} = (m_0 + m_e)\vec{a} - \alpha_2\dot{\vec{a}} - \text{(higher-order time derivatives of }\vec{a}). \tag{28-3}$$

Therefore, in order to maintain NEWTON's second law in a description of the electron by means of an electromagnetic field theory, the radiation reaction force and the higher-order time derivatives of $\vec{a}$ must be neglected. This is known as the quasi-stationary approximation. More will be said about the quasi-stationary approximation when we discuss "Sur la dynamique ..." in Part 6.

itates writing it in terms of components parallel and perpendicular to the direction of the electron's motion. If

$$\vec{F}_{ext} = \frac{d\vec{G}_\parallel}{dt} + \frac{d\vec{G}_\perp}{dt},$$ (22)

then

$$\frac{d\vec{G}_\parallel}{dt} = m_\parallel \vec{a}_\parallel$$

$$\frac{d\vec{G}_\perp}{dt} = m_\perp \vec{a}_\perp$$ (23)

where $m_\parallel$ and $m_\perp$ are the longitudinal and transverse masses, respectively. The equations for $m_\parallel$ and $m_\perp$ from ABRAHAM's theory are

$$m_\parallel = \frac{dG_\parallel}{dv} = \frac{e^2}{8\pi r v^3}\left[\frac{2v}{1-v^2} - \log\left(\frac{1+v}{1-v}\right)\right],$$ (24a)

$$m_\perp = \frac{dG_\perp}{dv} = \frac{e^2}{16\pi r v^3}\left[(1+v^2)\log\left(\frac{1+v}{1-v}\right) - 2v\right].$$ (24b)

KAUFMANN considered (24b) to be in good agreement with his data[29]. The expansion of (24) in powers of v,

$$m_\parallel = \frac{e^2}{6\pi r}\left[1 + \frac{6}{5}v^2 + \frac{9}{7}v^4 + \cdots\right],$$

$$m_\perp = \frac{e^2}{6\pi r}\left[1 + \frac{2}{5}v^2 + \frac{9}{35}v^4 + \cdots\right],$$ (25)

reveals that for $v \ll 1$, $m_\parallel = m_\perp = \frac{e^2}{6\pi r}$ (this value is dependent on the charge distribution).

Thus, utilizing the LAGRANGIAN of the electromagnetic field (2), ABRAHAM had apparently succeeded in accounting for the electron's mechanical properties. Mechanics had been absorbed into electromagnetism and mass had been deprived of its substantiality.

4. The Electron Theory of H. A. Lorentz

4.1. The Lorentz Transformations of 1892 and 1895.

In 1904 LORENTZ published the extension of his electromagnetic framework to a theory of the electron[6]. The basic difference between LORENTZ's and ABRAHAM's theories was that LORENTZ's did not admit classical rigid bodies. LORENTZ's development of his equations of relativistic transformation has been sketched in footnote 22. It still demands a detailed treatment; however, an outline will have to suffice here.

LORENTZ, in the *Versuch* of 1895[21], sought to account systematically for optical experiments accurate to order v. First, however, he demonstrated that the solutions to certain problems in electromagnetic theory were facilitated by use of the S' system. He transformed the fundamental equations to a coordinate system S_r which moves with a uniform velocity v with respect to frame S fixed in the

---

[29] Whereas in 1902 ABRAHAM had reservations as to the validity of his results, in his paper of 1903 he was confident. The reason was that KAUFMANN claimed that his more recent experimental results were in perfect agreement with ABRAHAM's theory.

ether. The equations of transformation are the usual GALILEAN ones:

$$x_r = x - vt,$$

$$y_r = y,$$

$$z_r = z,$$ (26)

$$t_r = t.$$

LORENTZ used the symbols (x), (y), (z), (t_2) for x_r, y_r, z_r, t_r respectively. He then proved that the components of force exerted by the ether on an electron are[30]

$$F_x = (1 - v^2) \frac{\partial \phi}{\partial x_r},$$

$$F_y = \frac{\partial \phi}{\partial y_r},$$ (27)

$$F_z = \frac{\partial \phi}{\partial z_r}.$$

The scalar potential of the moving electron is the one derived in Part 3.2 above (and by LORENTZ in 1892):[20]

$$\phi = (1 - v^2)^{-\frac{1}{2}} \phi'.$$ (28)

Therefore, the spatial coordinates in S' are related to those in S_r (and, therefore, to S by means of (26)) by (12). At this point in the *Versuch* the coordinate of time remained unchanged. Consequently, in 1895 LORENTZ had already "derived" the spatial part of his equations of relativistic transformation.

The forces in S and S' can be related by substituting (28) and (12) into (27):

$$F_x = F_x',$$

$$F_y = (1 - v^2)^{\frac{1}{2}} F_y',$$ (29)

$$F_z = (1 - v^2)^{\frac{1}{2}} F_z'.$$

LORENTZ then gave what appeared to be physical significance to S'. However, he was merely discussing a possible physical interpretation for a mathematical device which simplifies the calculation of the force acting on a moving electron. We shall see that LORENTZ maintained this point of view in his paper of 1904, *i.e.* that x, y, z, t are the true space-time coordinates.

LORENTZ went on to demonstrate that the fundamental equations are unchanged in S_r to order v if the relative time coordinate is changed to

$$t' = t - v x$$ (30)

(where he used x instead of x_r because terms of order v^2 were neglected) instead of $t_r = t$. He called this new time coordinate the local time. Hence, the introduction of the coordinate of local time accounts for the null result of the aberration

---

[30] *Cf.* LORENTZ, *Versuch, op. cit.* (footnote 21), p. 36. These equations can be obtained by writing (1e) in terms of the scalar and vector potentials by means of (5). The forces in S and S_r are the same because the two systems are related by a GALILEAN transformation.

experiments. Therefore, if the state of a system characterized by $\vec{E}$ and $\vec{B}$, as functions of x, y, z, t, exists in S there will exist in S_r a corresponding state characterized by $\vec{E_r}(=\vec{E}+\vec{v}\times\vec{B})$ and $\vec{B_r}(=\vec{B}-\vec{v}\times\vec{E})$ as functions of x_r, y_r, z_r, t'.[31] The meaning of this statement, known as the theorem of corresponding states, is that the equations of electromagnetism are unchanged to order v when transformed to a system moving with uniform linear relative motion with respect to the ether. Thus, to order v, optical phenomena occur on the moving earth as if it were at rest.

Hence, LORENTZ in the *Versuch* used two sets of coordinate systems for two types of problems. The fictitious S' system was used to facilitate the calculation of the force on a moving electron. S' is related to S_r by means of the equations

$$x' = (1-v^2)^{-\frac{1}{2}} x_r, \tag{31a}$$

$$y' = y_r, \tag{31b}$$

$$z' = z_r. \tag{31c}$$

$$t' = t_r. \tag{31d}$$

Then, to account systematically for the null result of optical experiments accurate to first order in v, LORENTZ utilized a modified GALILEAN transformation to relate S_r and S:

$$x_r = x - vt, \tag{32a}$$

$$y_r = y, \tag{32d}$$

$$z_r = z, \tag{32c}$$

$$t' = t - vx. \tag{32d}$$

4.2. The Contraction Hypothesis of 1895. LORENTZ, however, had to return to the relation between S' and S_r to give theoretical suppord to his *ad hoc* explanation for the null result of the MICHELSON & MORLEY experiment, accurate to order v^2. This relation is given by the set of transformation equations (31). LORENTZ assumed that if molecular forces transform as do electrical forces (see (29)), then molecular dimensions transform according to (31). By inverting (31) to

---

[31] LORENTZ's original statement of this theorem is found on p. 84 of the *Versuch*. His use of the notation x, y, z instead of $(x), (y), (z)$ (in my notation $(x)=x_r, (y)=y_r, (z)=z_r$) can be traced to a rather abrupt change of variables on p. 32: "... the new ones [the coordinates in S_r] for simplification will not be designated as $(x), (y), (z)$ but as x, y, z." LORENTZ, in 1892, had already introduced a time variable similar to the local time:

$$t' = t - \frac{v}{(1-v^2)} x_r$$

(see p. 297 of the work mentioned in footnote 20). The purpose of introducing this mathematical substitution was to change a certain linear second-order partial differential equation to the form of an inhomogeneous wave equation. Since this study focuses on theories of the electron, I have taken the liberty of discussing LORENTZ's theorem in terms of the microscopic electric and magnetic fields. LORENTZ, in 1895, proved this theorem for the macroscopic fields either inside of a neutral conductor or in charge-free space.

obtain the length in S_r (and thereby in S), LORENTZ was led to assert that "the displacement... would naturally bring about a shortening of the interferometer arm in the direction of motion"[32] by the amount $(1-v^2)^{\frac{1}{2}}$, thus accounting for the null result of MICHELSON & MORLEY. It is important to note that LORENTZ's inversion of (31 a) to obtain $x_r = (1-v^2)^{\frac{1}{2}} x'$ indicates that to him S' was simply a mathematical artifact and therefore was not equivalent to the real system S. We shall see that LORENTZ's simultaneous use of the three coordinate systems S, S_r and S', with S_r acting as an intermediary, led him into severe technical difficulties.

POINCARÉ's comment on the *ad hoc* nature of the hypothesis of contraction was that "An explanation was necessary [for the null result of the MICHELSON & MORLEY experiment] and was forthcoming; they always are; hypotheses are what we lack the least."[33]

4.3. The Lorentz Transformations of 1904. That LORENTZ was especially sensitive to POINCARÉ's criticism is evident from the introduction to the paper of 1904:

> Poincaré has objected to the existing theory of electric and optical pheno-
> mena in moving bodies that, in order to explain Michelson's negative result, the
> introduction of a new hypothesis has been required, and that the same neces-
> sity may occur each time new facts will be brought to light. Surely this course
> of inventing special hypotheses for each new experimental result is somewhat
> artificial. It would be more satisfactory if it were possible to show by means of
> certain fundamental assumptions and without neglecting terms of one order
> of magnitude or another, that many electromagnetic actions are entirely inde-
> pendent of the motion of the system.[34]

However, by the time that LORENTZ was finished he had made at least eleven "fundamental assumptions";[8(f)] nevertheless, POINCARÉ was satisfied, for reasons to be discussed in Part 5.

First LORENTZ used (26) to transform the "fundamental equations of the theory of electrons"[35] (1) to a reference system in uniform linear motion with respect to

---

[32] H. A. LORENTZ, *Versuch ... op. cit.* (see footnote 21), p. 124. Indeed, since only the second term of the expansion of $\sqrt{1-v^2}$ is required to "account" for MICHELSON and MORLEY's null result, this represented a significant step towards a second-order theory. For a discussion of G. F. FITZGERALD's simultaneous and independent postulation of the contraction effect see A. F. BORK, "The 'Fitzgerald' Contraction", Isis, **57**, 199–207 (1966) and S. G. BRUSH, "Note on the History of the Fitzgerald-Lorentz Contraction", Isis, **58**, 230–232 (1967).

[33] H. POINCARÉ, *Science and Hypothesis* (Dover, 1952), p. 172. Hereinafter *Science and Hypothesis* will be referred to as *S.H.*

[34] LORENTZ, *P.R.C.*, p. 13. LORENTZ had already indicated in 1899 that he was working towards a higher-order theory in "Théorie simplifiée des phénomènes électriques et optiques dans des corps en mouvement," Versl. Kon. Akad. Wetensch. Amsterdam, **7**, 507 (1899), in *Collected Papers*, **5**, 139–155. For discussions of this paper see the papers cited in footnote 3, pp. 473–474, footnote 22(b), pp. 24–27, and footnote 22(d), pp. 505–506. It is important to note that SCHAFFNER throughout his paper also stresses LORENTZ's non-reciprocal interpretation of his transformation. For a translation of LORENTZ's paper see K. SCHAFFNER, *Nineteenth-Century ... op. cit.* in footnote 11, pp. 255–273.

[35] *Ibid.*, p. 13.

the ether. As before, v is the velocity of the moving coordinate system S_r (LORENTZ refers to it as Σ) in the x-direction. The subscript r, although it was not used in the paper of 1904, will be used here to clarify LORENTZ's reasoning.[36] The space-time coordinates of a point on the electron with respect to S are x, y, z, t. In addition, the electron is assumed to be in motion with respect to S_r such that a point on it has a velocity $\vec{u}$. Then the velocity of this point with respect to S is $v_x = u_x + v$, $v_y = u_y$ and $v_z = u_z$. The postulation of the equations of relativistic transformation is prefaced by the statement: "We shall further transform these formulae [the GALILEAN transformations] by a change of variables"[37]

$$x' = klx_r, \tag{33a}$$

$$y' = ly_r, \tag{33b}$$

$$z' = lz_r, \tag{33c}$$

$$t' = \frac{l}{k} t - klvx_r \tag{33d}$$

where $k = (1 - v^2)^{-\frac{1}{2}}$ and l is a coefficient to be determined later. The factor l is assumed to be a function of v that reduces to one for $v \ll 1$ in order that in this limit (33) reduce to the GALILEAN transformation. The frame of reference to which x', y', z', t' refer is called Σ'. This frame is obtained from S_r by changing all lengths in the direction of motion by lk and all lengths perpendicular to it by l. This two-step transformation is similar to the one used in the paper of 1895. In fact, because of the presence of l, (33 a–c) are generalizations of (31 a–c). However, (33 d) is the version of the local time valid to all orders in v, whereas (32 d) is correct only to order v. The Σ' system is a generalization of the S' system. Thus, (33) represent a synthesis of the *two* sets of transformation equations (31) and (32) used by LORENTZ in the *Versuch*. Thus, it is possible to trace the evolution of LORENTZ's equations of relativistic transformation from his papers of 1892[20] and 1895[21] through 1904. Moreover, it will become increasingly evident that the Σ' system, like S' in 1895, possesses no physical reality. It was obtained from the independent variables x, y, z, t by a scale transformation in the case of (33 a–c) and a new form for the transformation of time (33 d) which mixes t and x as well as scaling them.

---

[36] A clear understanding of LORENTZ's notation is necessary in order to follow his two steps of transformation. It was his style to change notation with very little warning (see also footnote 31). In this paper he transforms the fundamental equations to Σ. Consequently, the equations for $\vec{V} \cdot \vec{E} = \varrho$ and $(\vec{V} \times B)_x = \frac{\partial E_x}{\partial t} + \varrho v$ on p. 14 of the work of 1904 should be read as

$$\frac{\partial E_x}{\partial x_r} + \frac{\partial E_y}{\partial y_r} + \frac{\partial E_z}{\partial z_r} = \varrho,$$

$$\frac{\partial B_z}{\partial y_r} - \frac{\partial B_y}{\partial z_r} = \left(\frac{\partial}{\partial t_r} - v \frac{\partial}{\partial x_r} \right) E_x + \varrho (v + u_x).$$

LORENTZ acknowledged in *T.E.* that equations analogous to (33) had been used by W. VOIGT in 1887. See *T.E.*, p. 198. See also J. LARMOR, *Aether and Matter* (Cambridge U. Press, 1900), esp. Chapters X and XI.

[37] LORENTZ, *P.R.C.*, p. 14.

4.4. The Charge Density, Electromagnetic Field and Velocity Transformation Equations.

LORENTZ went on to demonstrate that if the MAXWELL-LORENTZ equations (1 a–d) were to retain their form in Σ' then the relation between the fields, velocities and charge densities in S and Σ' would have to be [38]

$$E'_x = \frac{1}{l^2} E_x,$$

$$E'_y = \frac{k}{l^2} (E_y - v B_z), \tag{34a}$$

$$E'_z = \frac{k}{l^2} (E_z + v B_y),$$

---

[38] These equations are derived in the following manner. From (33) we have

$$\frac{\partial}{\partial x_r} = k l \frac{\partial}{\partial x'}, \quad \frac{\partial}{\partial y_r} = l \frac{\partial}{\partial y'}, \quad \frac{\partial}{\partial z_r} = l \frac{\partial}{\partial z'}, \quad \frac{\partial}{\partial t_r} = \frac{l}{k} \frac{\partial}{\partial t'}.$$

Substituting this change of variables into the equations in footnote 36 gives

$$k l \frac{\partial E_x}{\partial x'} + l \frac{\partial E_y}{\partial y'} + l \frac{\partial E_z}{\partial z'} = \varrho, \tag{38-1}$$

$$l \frac{\partial B_z}{\partial y'} - l \frac{\partial B_y}{\partial z'} = \frac{l}{k} \frac{\partial E_x}{\partial t'} - v k l \frac{\partial E_x}{\partial x'} + \varrho (v + u_x). \tag{38-2}$$

Elimination of $\partial E_x / \partial x'$ in (38-2) by means of (38-1) gives

$$l \frac{\partial}{\partial y'} (B_z - v E_y) - l \frac{\partial}{\partial z'} (B_y + v E_z) = \frac{l}{k} \frac{\partial E_x}{\partial t'} + \varrho u_x. \tag{38-3}$$

The other equations in (34) are derived in a similar fashion. Now (38-3) will be the same as its counterpart in S if

$$B'_z = \frac{k}{l^2} (B_z - v E_y),$$

$$B'_y = \frac{k}{l^2} (B_y + v E_z),$$

$$E'_x = \frac{E_x}{l^2}.$$

Then

$$\frac{\partial B'_z}{\partial y'} - \frac{\partial B'_y}{\partial z'} = \frac{\partial E'_x}{\partial t'} + \varrho' u'_x$$

where

$$\varrho' u'_x = \frac{k}{l^3} \varrho u_x. \tag{38-4}$$

Since charge is conserved,

$$\int \varrho \, dx_r dy_r dz_r = \int \varrho' dx' dy' dz'$$

or

$$\int \varrho \, \Delta \, dx' dy' dz' = \int \varrho' dx' dy' dz' \tag{38-5}$$

where it is assumed that $\varrho = \varrho_r$, and where Δ is the Jacobian for the transformation from S_r to S'. From (38-5) we have

$$\varrho' = \varrho \Delta, \tag{38-6}$$

and from (33)

$$\Delta = \frac{\partial (x', y', z')}{\partial (x_r, y_r, z_r)} = k l^3. \tag{38-7}$$

Therefore

$$\varrho' = \frac{\varrho}{k l^3}.$$

Substituting (38-6) and (38-7) into (38-4) yields an equation for u'_x:

$$u'_x = k^2 u_x.$$

$$B'_x = \frac{1}{l^2} B_x,$$

$$B'_y = \frac{k}{l^2} (B_y + v E_z),$$ (34b)

$$B'_z = \frac{k}{l^2} (B_z - v E_y),$$

$$u'_x = k^2 u_x,$$
$$u'_y = k u_y,$$ (34c)
$$u'_z = k u_z,$$

$$\varrho' = \frac{\varrho}{k l^3}.$$ (34d)

The field transformation equations are correct, but the ones for velocity and charge density are not. This will be seen to be a consequence of LORENTZ's two-step method of transformation which uses S, S_r and Σ'. Furthermore, the asymmetrical occurrence of the space-time coordinates in (33) is a direct result of relating Σ' to S_r, instead of to S.

Removing this asymmetry enabled POINCARÉ to derive the correct set of transformation equations for the velocity and charge density. Indeed, the only way that (34c) can be "derived" from (33) is to assume that $u_x \ll 1$.[39] *Thus*

---

[39] This statement can be proved as follows. From (33) we have

$$d x' = k l d x_r,$$
$$d t' = \frac{l}{k} d t [1 - k^2 v u_x]$$ (39-1)

where

$$u_x = \frac{d x_r}{d t}.$$ (39-2)

Then

$$u'_x = \frac{d x'}{d t'} = \frac{k^2 u_x}{[1 - k^2 v u_x]}.$$ (39-3)

Consequently, if $u_x \ll 1$, (39-3) becomes

$$u'_x = k^2 u_x.$$

The equations for u'_y and u'_z follow in a similar manner. LORENTZ then used (33) to derive the transformation equations for the acceleration. Using (33), one obtains

$$\frac{d^2 x'}{d t'^2} = k^2 \frac{d^2 x_r}{d t^2} \frac{1}{(d t'/d t)}$$

$$= \frac{k^3}{l} \frac{d^2 x_r}{d t^2} \quad (\text{if } u_x \ll 1).$$

Then by inversion

$$a_x = \frac{l}{k^3} a'_x.$$

By a similar method the x and y components are found to be

$$a_y = \frac{l}{k^2} a'_y,$$

$$a_z = \frac{l}{k^2} a'_z.$$

It will be shown that these transformations are valid in S', where u_x is identically zero.

It is important to note that the accelerations and forces are the same in S and S_r because these systems are related by a GALILEAN transformation.

Lorentz, *in his paper of 1904, was really concerned with electromagnetic phenomena in systems moving with a speed much less than that of light, not with electromagnetic phenomena in systems moving with any speed less than that of light.* It was Poincaré who extended Lorentz's results to the latter systems. The equation for transforming the charge density, as we shall see, is valid only in the particular instance that Σ' is the electron's instantaneous rest frame (S' as defined in Part 3.2). The field equations (34a–b), valid to all orders in v, are the generalization of the ones from the paper of 1895 which were valid only to order v: $\vec{E}_r = \vec{E} + \vec{v} \times \vec{B}, \vec{B}_r = \vec{B} - \vec{v} \times \vec{E}$.

The fundamental equations in Σ' are

$$\vec{V}' \cdot \vec{E}' = (1 - v u'_x) \varrho', \tag{35a}$$

$$\vec{V}' \cdot \vec{B}' = 0, \tag{35b}$$

$$\vec{V}' \times \vec{E}' = -\frac{\partial \vec{B}'}{\partial t'}, \tag{35c}$$

$$\vec{V}' \times \vec{B}' = \frac{\partial \vec{E}'}{\partial t'} + \varrho' \vec{u}', \tag{35d}$$

$$F_x = l^2 [E'_x + (\vec{u}' \times \vec{B}')'_x + v(u'_y E'_y + u'_z E'_z)], \tag{35e}$$

$$F_y = \frac{l^2}{k} [E'_y + (\vec{u}' \times \vec{B}')_{y'} - v u'_x E'_y], \tag{35f}$$

$$F_z = \frac{l^2}{k} [E'_z + (\vec{u}' \times \vec{B}')_{z'} - v u'_x E'_z]. \tag{35g}$$

At this point Lorentz chose to ignore the incorrectness of (35a) and (35e–g). Indeed, according to (35a), a coordinate system can be found wherein charge is not conserved,[40] thus making it possible in principle to determine the absolute velocity of the earth. This is, however, contrary to the results of Michelson & Morley, Trouton & Noble, Rayleigh & Brace and the aberration experiments. It is also at variance with Lorentz's reason for writing this paper, *i.e.*, "... to show by means of certain fundamental assumptions and without neglecting terms of one order of magnitude or another that many electromagnetic actions are entirely independent of the motion of the system [which they are not according to (35a, e–g) wherein v appears]."[34] Thus, Lorentz proved the covariance of the Maxwell-Lorentz equations only for charge-free space.

Nevertheless, Lorentz's conclusions were unaffected by these errors because all relevant calculations were done in an "electrostatic system, *i.e.*, a system

---

[40] Charge conservation is "built into" (1a–d). Taking the divergence of (1d) yields

$$\vec{V} \cdot (\vec{V} \times \vec{B}) = 0 = \frac{\partial}{\partial t} (\vec{V} \cdot \vec{E}) + \vec{V} \cdot (\varrho \vec{v}),$$

and then substituting for $\vec{V} \cdot \vec{E}$ from (1a) gives

$$\frac{\partial \varrho}{\partial t} + \vec{V} \cdot (\varrho \vec{v}) = 0.$$

On the other hand, applying this procedure to (35a) and (35d) gives the incorrect result

$$\frac{\partial}{\partial t} ([1 - v u'_x] \varrho') + \vec{V} \cdot (\varrho \vec{u}') = 0.$$

having no other motion but the translation with the velocity v."[41] Since $\vec{u}'=\vec{u}=0$, Σ' becomes the frame S' which was defined in Part 3.2. Equation (35a) now becomes

$$\vec{V}'\cdot\vec{E}'=\varrho'.\tag{35a'}$$

Hence, ϱ' now satisfies conservation of charge. Equations (35e–g) become

$$F_x=l^2F_x',$$
$$F_y=\frac{l^2}{k}F_y',\tag{36}$$
$$F_z=\frac{l^2}{k}F_z',$$

which are in agreement with the results of the paper of 1895[21] $\left(\text{see }(29)\right)$. In fact, it is highly probable that LORENTZ was guided by the results of that paper.

4.5. The Contraction Hypothesis of 1904. LORENTZ then posited the contraction hypothesis so that an electron, which is a sphere when at rest in S, maintains its shape in S'. This is required by the theorem of corresponding states. His argument for the contraction was similar to the one from 1895. In the S' frame the electron is stretched in the direction of motion by an amount kl $\left(\text{see }(33a)\right)$; its dimensions perpendicular to the direction of motion are changed by l (see (33b–c)). In LORENTZ's notation the dimensions change by (kl, l, l). The contraction hypothesis asserts that the electron's dimensions change by $\left(\frac{1}{kl}, 1/l, 1/l\right)$. Therefore, the electron retains its spherical shape in the "imaginary system $[S']$"[42] because the contraction cancels the dilation.

LORENTZ then assumed that all forces transform in the same way as electromagnetic forces and that "there is but one configuration of equilibrium."[43] Thus, if an electron is in equilibrium in S, then it should also be in equilibrium in S'. We shall see that it was POINCARÉ who first perceived the full meaning of this statement. The reason for the contraction is similar to that given in the paper of 1895: since S' is obtained from S by the deformation (kl, l, l), then, conversely, to determine lengths in S one inverts (33a–c) to obtain $\left(\frac{1}{kl}, 1/l, 1/l\right)$.

Hence, "the translation will produce the deformation $\left(\frac{1}{kl}, 1/l, 1/l\right)$."[44]

---

[41] LORENTZ, *P.R.C.*, p. 17.

[42] *P.R.C.*, p. 22.

[43] *Loc. cit.*

[44] *Loc. cit.* This is not exactly the way in which LORENTZ drew this conclusion on p. 22 of *P.R.C.* I have presented it in this manner in order to avoid the confusion which exists in the relevant portions of the paper of 1904. LORENTZ asserted that the system S' can be assumed to be at absolute rest since the velocity of the earth with respect to the ether cannot be detected. However, the real electrostatic system Σ is in motion with the absolute velocity v. He concludes that "if the velocity v is imparted to it $[S']$ will of itself change into the system Σ. In other terms, the translation will produce the deformation $\left(\frac{1}{kl}, 1/l, 1/l\right)$." (p. 22.) This confused state of affairs, wherein three coordinate systems are in use with Σ playing the role of an intermediary, is obviously caused by the fictitious role ascribed to $\Sigma'(S')$ by interpretation of (33) as merely a coordinate transformation.

15*

4.6. The Electron's Mass According to Lorentz's Theory. LORENTZ could then calculate, with relative ease (compared to ABRAHAM), the electromagnetic momentum. In the S system,

$$G_x = \int (E_y B_z - E_z B_y) d\tau. \tag{37}$$

Transforming the fields and volume element to S', wherein $\vec{B}' = 0$, gives

$$G_x = klv \int (E_y'^2 + E_z'^2) d\tau'. \tag{38}$$

Since the electron in S' is spherical, its electric field can easily be calculated. For a uniform density of surface charge G_x becomes

$$G_x = \frac{2}{3} klv \int E'^2 d\tau' = \frac{e^2}{6\pi r} klv \tag{39}$$

(from spherical symmetry $E_x'^2 = E_y'^2 = E_z'^2 = \frac{1}{3} E'^2$) or in general,[45]

$$\vec{G} = \frac{e^2}{6\pi r} kl \vec{v}. \tag{40}$$

The longitudinal and transverse masses can be obtained by assuming that the electron's entire mass is of electromagnetic origin and the motion is quasi-stationary. The result is

$$m_\parallel = \frac{e^2}{6\pi r} \frac{d}{dv} (klv), \tag{41a}$$

$$m_\perp = \frac{e^2}{6\pi r} kl. \tag{41b}$$

The derivative in (41a) cannot be evaluated because the functional dependence of l on v is as yet undetermined. The evaluation can be accomplished with the aid of the following statement: "... if we start from any given state of motion in a system without translation, we may deduce from it a corresponding state that can exist in the same system after a translation has been imparted to it ...".[46] Therefore, to a state in S characterized by $\vec{E}$ and $\vec{B}$ as functions of x, y, z, t there corresponds a state in S' characterized by $\vec{E}'$ and $\vec{B}'$ as functions of x', y', z', t'. This is the generalization of the theorem of corresponding states from the *Versuch* (1895).

The quantity l can be determined by noting that the equivalent states can exist only if the forces in S and S' are related by[47]

$$F(S) = \left(l^2, \frac{l^2}{k}, \frac{l^2}{k}\right) F(S') \tag{42}$$

(recall (36)); then

$$m\vec{a}(S) = \left(l^2, \frac{l^2}{k}, \frac{l^2}{k}\right) m\vec{a}(S'). \tag{43}$$

---

[45] LORENTZ, on p. 18 of *P.R.C.*, considers (40) to be a relation between G_x in Σ and $\Sigma'(S')$. However, (40) can also be interpreted as relating G_x in S and S' since Σ serves as an intermediary.

[46] *P.R.C.*, p. 25.

[47] In this paper LORENTZ related forces, accelerations and masses in $\Sigma(S_r)$ and S. However, the forces are indentical in these two systems since these systems are related by the GALILEAN transformation. Consequently, I have chosen to eliminate the (S_r) system.

LORENTZ used (33) to obtain the transformation equations for the acceleration [39]

$$\vec{a}(S) = \left(\frac{l}{k^3}, \frac{l}{k^2}, \frac{l}{k^2} \right) \vec{a}(S').$$ (44)

Comparing (43) and (44) gives

$$m(S) = (k^3 l, kl, kl) m(S').$$ (45)

The result of comparing (45) with (41 a) is

$$\frac{d}{dv}(klv) = k^3 l,$$ (45')

which can only be satisfied if $l = 1$ (models for which $l \neq 1$ will be discussed in Part 6 in conjunction with POINCARÉ's work). Consequently, the longitudinal and transverse masses in LORENTZ's theory are

$$m_\parallel = \frac{e^2}{6\pi r} k^3,$$ (46a)

$$m_\perp = \frac{e^2}{6\pi r} k.$$ (46b)

These results should be compared with ABRAHAM's complicated expressions (24). In doing so we should keep in mind that the comparative simplicity of LORENTZ's results was obtained by positing the contraction hypothesis.

Since $l = 1$ the field transformation equations (34a–b) no longer have to be inverted to calculate $\vec{E}$ and $\vec{B}$ in terms of $\vec{E}'$ and $\vec{B}'$. One need only change v to $-v$. However, this operation, which implies that S and S' are equivalent systems, is tantamount to rejecting the ether; LORENTZ never accepted this interpretation. For POINCARÉ, as will become evident in Part 6, this symmetry was purely a mathematical nicety.

Expanding $m_\perp$ in terms of v as

$$m_\perp = \frac{e^2}{6\pi r} \left[1 + \frac{1}{2} v^2 + \frac{3}{8} v^4 + \cdots \right]$$

shows that it differs very little from ABRAHAM's in the limit of low velocities:

$$m_\perp \text{(ABRAHAM)} = \frac{e^2}{6\pi r} \left[1 + \frac{2}{5} v^2 + \frac{9}{35} v^4 + \cdots \right].$$

LORENTZ went on to assert the necessity that masses transform like the electromagnetic mass of the electron in order that the theorem of corresponding states remain valid.

Aware that he was in competition with a theory the results of which "have been confirmed in a most remarkable way by KAUFMANN's measurements of the deflexion of radium-rays in electric and magnetic fields," [48]—LORENTZ went on to demonstrate that his equation for $m_\perp$ was in equally good agreement with KAUFMANN's data.

---

[48] *P.R.C.*, p. 31. The data in question are from W. KAUFMANN, "Die elektromagnetische Masse ..." *op. cit.* (footnote 19) and W. KAUFMANN, "Über die 'Elektromagnetische Masse' der Elektronen," Nachr. Ges. Wiss. Göttingen, 90–103 (1903).

However, LORENTZ realized that he had constructed his theory at the price of piling up hypotheses:

> It need hardly be said that the present theory is put forward with all due reserve.[49]
> ·
> Our assumption about the contraction of the electrons cannot in itself be pronounced to be either plausible or inadmissible. What we know about the nature of electrons is very little, and the only means of pushing our way farther will be to test each hypothesis as I have here made.[50]

Yet, LORENTZ leaves an impression of confidence in his hypotheses, especially in the contraction hypothesis which is necessary to account for the null result of MICHELSON & MORLEY.

4.7. The Unresolved Problems in Lorentz's Theory.

Thus, circa 1904, ABRAHAM and LORENTZ claimed to be able to account for the dynamics of the electron on a purely electromagnetic basis. Fundamental to both theories was the assumption that the electron's energy and momentum was identical with the (self) electromagnetic energy and momentum. However, to "derive" NEWTON's second law, and thereby to calculate unambiguously the electron's mass, appeared to necessitate taking account of only quasi-stationary motion. In ABRAHAM's theory, G is calculated from the LAGRANGIAN of the electromagnetic field on the assumption that the electron be a rigid sphere which undergoes an imaginary contraction in an imaginary coordinate system, while in LORENTZ's theory G is calculated from its electromagnetic definition under the assumption that the deformable electron undergoes a real contraction in the direction of motion but maintains its spherical shape in an imaginary coordinate system. The portions of both theories amenable to experiment $(m_\perp)$ were in agreement with the available data, although KAUFMANN asserted that ABRAHAM's agreed better.

There were, however, two basic faults with LORENTZ's theory, which was clearly the more ambitious of the two: 1) The MAXWELL-LORENTZ equations are in general not left unchanged under the transformation (33). This was, as POINCARÉ discovered, a technical problem. 2) More serious is the instability of the deformable electron caused by the COULOMB repulsion of its constituent parts. Therefore, it appears necessary to add cohesive mechanical forces to LORENTZ' model. Since this is the major problem that confronted POINCARÉ, we must develop it.

ABRAHAM had already emphasized, in 1903, that a model with a deformable electron would necessitate the inclusion of cohesive mechanical forces.[51] In 1904, ABRAHAM felt obliged to publish a short paper in which he compared the structure of his theory with LORENTZ's. Once again he questioned the stability of the deformable electron and noted the need for the addition of mechanical forces in

---

[49] *P.R.C.*, p. 29.

[50] *P.R.C.*, p. 30.

[51] *Cf.* the work cited in footnote 5, pp. 108–109. In this paper ABRAHAM proved also that the stable orientation for an ellipsoidal charge distribution moving with a constant linear velocity is that in which its major axis is along the direction of motion (pp. 174–179). However, this proof could not be extended to deformable electrons such as those included in LORENTZ's theory of 1904.

LORENTZ's theory.[52] Then, in 1905, in the second volume of his book *Theorie der Elektrizität*,[53] ABRAHAM elaborated on the problems which he had raised in 1903 and 1904. He made the important observation that LORENTZ had derived the basic results of his theory ($m_\parallel$ and $m_\perp$) solely from the electron's momentum and had not discussed its energy. Furthermore, ABRAHAM found that the $m_\parallel$ calculated from the energy W of the LORENTZ electron $\left(m_\parallel = \dfrac{1}{v}\dfrac{dW}{dv}\right)$ disagreed with the one calculated from its momentum $\left(m_\parallel = \dfrac{dG}{dv}\right)$.[54, 55] Therefore, the entire energy of the LORENTZ electron cannot be accounted for by electromagnetic forces alone.

Let us now derive the equation for W in LORENTZ's theory and then calculate $m_\parallel$ from it. The work done by a force acting along the direction of the electron's motion is[56]

$$m_\parallel\, v\dot{v}\,dt = \frac{e^2}{6\,\pi r}\,(1-v^2)^{-\frac{3}{2}} v\dot{v}\,dt. \tag{47}$$

However, the electromagnetic energy W of a moving LORENTZ electron is[57]

$$W = \frac{e^2}{6\,\pi r}\,(1-v^2)^{-\frac{1}{2}} - \frac{e^2}{24\,\pi r}\,(1-v^2)^{\frac{1}{2}}. \tag{48}$$

The change in W in a time dt is

$$dW = \frac{e^2}{6\,\pi r}\,(1-v^2)^{-\frac{3}{2}} + \frac{e^2}{24\,\pi r}\,(1-v^2)^{-\frac{1}{2}} v\dot{v}\,dt, \tag{49}$$

and consequently[58]

$$\frac{1}{v}\frac{dW}{dv} = m_\parallel + \frac{1}{4}\,m_\perp \neq \frac{dG}{dv}. \tag{50}$$

The incorrect term $\frac{1}{4}m_\perp$ in (50) comes from the second term on the right-hand side of (48). Thus, the two terms on the right-hand side of (48), which are a consequence of associating the electron's energy with the energy of its self-fields, are not the total energy of the LORENTZ electron—hence, the need for mechanical forces whose work on the electron will provide the necessary counterterm to cancel $\dfrac{-e^2}{24\,\pi r}\,(1-v^2)^{\frac{1}{2}}$ in (48).[59]

---

[52] M. ABRAHAM, "Die Grundhypothesen ..." *op. cit.* (footnote 23), p. 578.

[53] M. ABRAHAM, *Theorie der Elektrizität*, (2 Vol.; Teubner, 1905), Volume 2.

[54] *Ibid.*, p. 205.

[55] The time rate of change of the total electromagnetic field energy can be written as

$$\frac{dW}{dt} = \frac{dW}{dv}\frac{dv}{dt}.$$

This represents the work done per unit time by the force. Therefore, $(1/v)\,dW/dt$ represents the force which implies that $(1/v)\,dW/dv = m_\parallel$ (with the assumption of quasi-stationary motion).

[56] This method of deriving W is due to LORENTZ (see LORENTZ, *T.E.*, p. 213).

[57] This result can be obtained from the potential formalism outlined in Part 3.2. In LORENTZ's theory ϕ' represents the potential due to a spherical electron with a uniform surface charge distribution.

[58] This equation corresponds to equation (126b) on p. 205 of the work cited in footnote 53.

[59] LORENTZ considered the resolution of this inconsistency in his theory to be of fundamental importance. See H. A. LORENTZ, "Ergebnisse und Probleme der Elektrontheorie," in Elektrotechn. Verein zu Berlin, 1904 (Springer, 1905), in *Collected Papers*, 8, 76–124.

The root of this inconsistency can be seen clearly from Abraham's derivation of (50). He went directly to the heart of the matter by showing that $\vec{G}$ as calculated from its defining equation in terms of electromagnetic fields $\left(\vec{G} = \int \vec{E} \times \vec{B}\, d\tau\right)$ is not equal to the $\vec{G}$ obtained from Lagrangian mechanics $\left(\vec{G} = \dfrac{\partial L}{\partial v}\right)$. The electro-magnetic field Lagrangian (19) can be rewritten as

$$L = -(1-v^2)^{\frac{1}{2}} \int \frac{\varrho'\phi'}{2}\, d\tau', \tag{51}$$

but $\frac{1}{2}\int \varrho'\phi'\,d\tau'$ is the electrostatic energy W'. Then (51) becomes

$$L = -(1-v^2)^{\frac{1}{2}}\, W' \tag{52}$$

where in Lorentz's theory W' is

$$W' = \tfrac{1}{2}\int \varrho'\phi'\,d\tau' = \tfrac{1}{2}\int E'^{2}\,d\tau'.$$

The integral in this equation has already been evaluated in (39), giving

$$W' = \frac{e^2}{8\pi r}. \tag{53}$$

Since the electron's inertia in its rest frame m'_e must be due solely to its electro-static field, then $W' = m'_e$ (remember $c = 1$). Therefore, $\vec{G}$ obtained from (52) as

$$G = \frac{\partial L}{\partial v} = \frac{m'_e v}{(1-v^2)^{\frac{1}{2}}}, \tag{54}$$

is not equal to (40), which can be written as

$$G = \frac{4}{3}\,\frac{m'_e v}{(1-v^2)^{\frac{1}{2}}}, \tag{55}$$

which implies that

$$\frac{dG}{dv} \neq \frac{1}{v}\frac{dW}{dv}.$$

This can be proven also directly from the basic relationship between W, L and $\vec{G}$:

$$W = vG - L. \tag{56}$$

Differentiating (56) with respect to v gives

$$\frac{1}{v}\frac{dW}{dv} = \frac{dG}{dv} + \frac{1}{v}\left[G - \frac{dL}{dv}\right], \tag{57}$$

but G in this case is not equal to dL/dv. Substituting for G in the second term on the right-hand side of (57) its value from (54) gives (50). Therefore, if L from (52) were multiplied by $\frac{4}{3}$,[60] the G's in (54) and (55) would be equal. The second term

---

[60] That this factor is ingrained in any theory which makes use of the electromagnetic field momentum as a kinematical quantity, *i.e.*, a theory whose goal is an electromagnetic world picture, can be seen as follows. The electromagnetic field momentum is defined as

$$\vec{G} = \int \vec{E} \times \vec{B}\, d\tau.$$

From the Biot-Savart law $\vec{B} = \vec{v} \times \vec{E}$ (for $v \ll 1$),

$$\vec{G} = \int \vec{E} \times (\vec{v} \times \vec{E})\, d\tau.$$

For a symmetrical charge distribution $\vec{G}$ becomes

$$\vec{G} = \tfrac{4}{3} m'_e \vec{v}.$$

on the right-hand side of (50), and consequently the one in (48), would then vanish. However, the vanishing of this term implies the existence of a counterterm which can only be obtained at the expense of the introduction of non-electromagnetic forces, thereby negating a purely electromagnetic world-picture. Nevertheless, the construction of a LAGRANGIAN formulation of LORENTZ's theory, which is in agreement with the kinematical quantities obtained from electromagnetic theory, is crucial in order to be able to subsume mechanics into the framework of a theory of electrons which maintains the principle of relativity. This is the problem that POINCARÉ took up in 1905.

5. A Note on the Relation Between Poincaré's Epistemology and His Physics

5.1. Perspective. Parts 3 and 4 have indicated the state of affairs at the time HENRI POINCARÉ began examining critically the foundations and consequences of all possible electron theories. The universal scope of LORENTZ's theory and its theorem of corresponding states particularly interested POINCARÉ. In 1904 he referred to this theorem, in a broader context, as the principle of relativity.[61]

Even before 1904, however, POINCARÉ had concluded as a result of analyzing the foundations of geometry and classical mechanics that only relative, not absolute, quantities should appear in these theories. POINCARÉ referred to this result in different terms: in geometry, in 1899, as the principle of relativity;[62] in classical mechanics first as the law of relativity (1899)[63] and then as the principle of relative motion (1900).[64] These statements, rooted firmly in empiricism and then generalized by means of mathematics so as to be above experimental verifi-

---

[61] H. POINCARÉ, *The Value of Science*, translated by G. B. HALSTED (Dover, 1958), p. 98. (Chapters VII–IX, pp. 91–111 of *The Value of Science* are a translation of POINCARÉ's lecture entitled "L'état actuel et l'avenir de la physique mathématique" which was delivered on 24 September, 1904, at the International Congress of Arts and Science at Saint Louis, Missouri and published in Bull. Sci. Mat., **28**, 302–324). Hereafter *The Value of Science* will be referred to as *V.S.*

[62] H. POINCARÉ, "Des fondements de la Géométrie: À propos d'un Livre de M. Russell," Rev. Mét. Mor. **7**, 251–279 (1899). This is a book review of BERTRAND RUSSELL's *An Essay on the Foundations of Geometry* (Cambridge, 1897; Dover, 1952). Pages 265–269 of POINCARÉ's review constitute pages 75–79 of Chapter V, "Experiment and Geometry", of POINCARÉ's *S. H.* POINCARÉ used the term "principle of relativity" while discussing the concept of spatial measurement in projective and metrical geometries (pp. 255–266). RUSSELL had used the terms "homogeneity of space" and "relativity of space" in his book (*cf.* p. 148). Thus, the term "relativity" was already in use in 1899.

[63] H. POINCARÉ, *S.H.*, *op. cit.* (footnote 62), p. 76 of Chapter V, "Experiment and Geometry": "... the state of the bodies and their mutual distances at any moment will solely depend on the state of the same bodies and on their mutual distances at the initial moment, but will in no way depend on the absolute initial position of the system and of its absolute initial orientation. This is what we shall call, for the sake of abbreviation, *the law of relativity*." (Italics in original.)

[64] H. POINCARÉ, *S.H.*, *op. cit.* (footnote 62), p. 111 (Chapters VI, VII and pp. 89–122 and pp. 135–138 of Chapter VIII are a translation of POINCARÉ's "Sur les principes de la Mécanique," *Bibliothèque du Congrès international de Philosophie tenu à Paris du 1ᵉʳ au 5 août* 1900 (Colin, 1901), pp. 457–488, 491–494; pp. 123–135 of *S.H.* are an expanded version of pp. 488–491 of the "Sur les principes ...", while pp. 138–139 were written expressly for *S.H.*).

cation, POINCARÉ defined as conventions. He considered, for example, the null result of the first-order aberration experiments and the second-order MICHELSON & MORLEY experiment, to determine the motion of the earth through the ether, as non-fortuitous.[65] Hence, he reasoned that higher-order experiments should fail as well. The implication is that in the realm of microphysics, which POINCARÉ called the physical sciences,[66] there exists a general law of nature to which he referred, in 1904, as the principle of relativity. This principle in the physical sciences was not a convention simply because of the lack of empirical evidence: the higher-order experiments to detect the earth's motion through the ether were non-existent; and furthermore, if KAUFMANN's data were correct, then LORENTZ's theory and the principle of relativity which was inextricably linked to it would have been invalidated.

Therefore, it seemed logical to POINCARÉ to examine all possible theories of the electron in order to find those which conformed to the principle of relativity. This is the theme of "Sur la dynamique de l'électron." POINCARÉ found that of those proposed (ABRAHAM, BUCHERER and LANGEVIN) only LORENTZ's was consonant with the principle of relativity. However, agreement could be accomplished only at the expense of introducing forces of a non-electromagnetic origin to hold together LORENTZ's deformable electron. The unexpected result, therefore, was the impossibility—using LORENTZ's 1904 formalism—of constructing a purely electromagnetic field theoretical description of a deformable electron.

Thus, ABRAHAM was correct in asserting that a theory of a deformable electron had no place in an electromagnetic world-picture. But ABRAHAM, LORENTZ and POINCARÉ could not perceive the larger issue of a universal theory of relativity. This theory could be obtained only by deductive methods. To the extent that ABRAHAM, LORENTZ and POINCARÉ were committed proponents of the electromagnetic world-picture, they could neither discover nor accept the universal theory of relativity.

Unlike ABRAHAM, LORENTZ and most of the other proponents of an electromagnetic world-picture, POINCARÉ was a philosopher-scientist. Thus, an understanding of his concept of what constitutes a physical theory and of why he was unsuccessful in discovering a universal theory of relativity calls for an elaboration of his epistemology.

An important task in this discussion of POINCARÉ's philosophy of science will be to relate what POINCARÉ defined as the principle of relativity in the physical sciences to what he defined as the principle of relative motion in classical mechanics and the principle of relativity in geometry.

5.2. Poincaré's Epistemology and His Physics. POINCARÉ's conventionalistic epistemology is rooted in inductivism; thus "[e]xperiment is the sole source of truth"[67] because its results reveal objective knowledge of the physical world.

---

[65] Cf. H. POINCARÉ, S.H., op. cit. (footnote 62), esp. p. 172. Chapters IX and X, pp. 140–182 of S.H. are a translation of POINCARÉ's "Sur les rapports de la Physique expérimentale et de la Physique mathématique," *Rapports présentés au Congrès international de Physique réuni à Paris en 1900* (Gauthier-Villars, 1900) Vol. 1, pp. 1–29.

[66] Cf. H. POINCARÉ, S.H., op. cit. (footnote 62), esp. p. xxvi.

[67] H. POINCARÉ, S.H., op. cit. (footnote 65), p. 140.

Objective reality in POINCARÉ's philosophy of science is the relation between sensations:

> External objects, for instance, for which the word *object* was invented, are really *objects* and not fleeting and fugitive appearances, because they are not only groups of sensations, but groups cemented by a constant bond. It is this bond, and this bond alone, which is the object in itself, and this bond is a relation.[68]

Thus POINCARÉ concluded:

> In sum, the sole objective reality consists in the relations of things [*i.e.*, sensations] whence results the universal harmony.[69]

Indeed, to POINCARÉ, "the aim of science is not things themselves ... but the relations between things; outside those relations there is no reality knowable."[70] These relations are obtained from empirical data because "[e]xperiment teaches us relations between bodies; this is the fact in the rough; these relations are extremely complicated."[71]

To illustrate this statement let us consider POINCARÉ's discussion of the axioms of geometry. POINCARÉ asserted that instead of studying the relative displacements of two real solid bodies A and B, which are the facts in the rough, one could imagine replacing these bodies with two ideal solids A' and B'. The result of studying the relations between A' and B' is embodied in the axioms of geometry:

> *The geometrical axioms are therefore neither synthetic à priori intuitions nor experimental facts.* They are conventions. Our choice among all possible conventions is *guided* by experimental facts; but it remains *free*, and is only limited by the necessity of avoiding every contradiction, and thus it is that postulates may remain rigorously true even when the experimental laws which have determined their adoption are only approximate. In other words, *the axioms of geometry ... are only definitions in disguise.*[72]

Thus, the axioms of geometry, although obtained inductively from empirical data, are not experimentally verifiable: they do not refer to the real world. This is a property common to all conventions.

[68] H. POINCARÉ, *V.S.*, *op. cit.* (footnote 61), pp. 137–138. (Part III, pp. 112–142 of *V.S.* is an expanded verion of POINCARÉ's "Sur la valeur objective de la Science," Rev. Mét. Mor. **10**, 263–293 (1902); italics in original). This is in contrast to P. DUHEM's philosophy of science, according to which one cannot know physical reality but can only approach it asymptotically. See P. DUHEM, *The Aim and Structure of Physical Theory* (Atheneum, 1962), translated by P. P. WIENER.

[69] *Ibid.*, p. 140.

[70] H. POINCARÉ, *S.H.*, *op. cit.* (footnote 62), p. xxiv.

[71] H. POINCARÉ, *V.S.*, *op. cit.* (footnote 68), p. 125. For POINCARÉ, in contrast to DUHEM, "[t]here is no precise frontier between the fact in the rough and the scientific fact; it can only be said that such an enunciation of fact is *more crude* or, on the contrary, *more scientific* than such another" (p. 122, italics in original). For DUHEM's rejoinder to this article see P. DUHEM, *The Aim and Structure* ..., *op. cit.*, (footnote 68), esp. pp. 149–153.

[72] H. POINCARÉ, *S.H.*, *op. cit.* (footnote 62), p. 50. (Chapter III of *S.H.*, entitled "Non-Euclidean Geometries," pp. 35–50, is a translation of POINCARÉ's "Les Géométries non-euclidiennes," Revue générale des Sciences pures et appliquées, **2**, 769–774 (1891); italics in original.)

The conventions of classical mechanics such as Newton's three principles, the conservation of energy and the principle of relative motion, according to which " [t]he movement of any system whatever ought to obey the same laws, whether it is referred to fixed axes or to the movable axes which are implied in uniform motion in a straight line"[64] were, like the geometrical axioms, obtained inductively. Unlike the geometrical axioms, however, the conventions of classical mechanics were obtained from the empirical data by means of mathematical physics.

By the end of the nineteenth century new developments in the various branches of physics, especially in electromagnetism, necessitated the extension of the conventions of classical mechanics into the domain of the physical sciences.[73] Here, due to lack of sufficient empirical data, Newton's principles, the principle of conservation of energy and the principle of relative motion could no longer be considered as conventions. Nevertheless, their usefulness in the domain of classical mechanics, establishing relationships among phenomena, suggested strongly the possibility of their validity in the physical sciences: "They were obtained in the search for what was common in the enunciation of numerous physical laws ..."[74]

Consonant with his conventionalistic epistemology, Poincaré believed that in the microscopic domain Lorentz's theory was the most satisfactory one to test the validity of the principles of physics:

> The most satisfactory theory is that of Lorentz; it is unquestionably the theory that best explains the known facts, the one that throws into relief the greatest number of known relations, the one in which we find most traces of definitive construction.[75]

Poincaré elaborated on this statement in the concluding Chapter of *Science and Hypothesis*:

> Lorentz's theory is very attractive. It gives a very simple explanation of certain phenomena, which the earlier theories—even Maxwell's in its primitive form— could only deal with in an unsatisfactory manner; for example, the aberration of light, the partial impulse of luminous waves, magnetic polarisation, and Zeeman's experiment.[76]

Despite his provisional acceptance of Lorentz's theory, Poincaré was nevertheless in full agreement with Lorentz on the indispensability of the ether, though motion with respect to it is undetectable due to the compensating effects of the local time and the Lorentz contraction. On the existence of the ether, Poincaré stated, in 1900:

> We know the origin of our belief in the ether. If light takes several years to reach us from a distant star, it is no longer on the star, nor is it on the earth. It must be somewhere, and supported, so to speak, by some material agency.[77]

---

[73] H. Poincaré, *S.H.*, *op. cit.* (footnote 62), p. xxvi.
[74] H. Poincaré, *S.H.*, *op. cit.* (footnote 65), p. 166.
[75] H. Poincaré, *S.H.*, *op. cit.* (footnote 65), p. 175.
[76] H. Poincaré, *S.H.*, *op. cit.* (footnote 62), p. 243. This Chapter entitled "Electrodynamics" will be discussed in more detail below.
[77] H. Poincaré, *S.H.*, *op. cit.* (footnote 65), p. 169.

Two years later POINCARÉ elaborated on this statement:

> It may be said, for instance, that the ether is no less real than any external
> body; to say this body exists is to say there is between the color of this body, its
> taste, its smell, an intimate bond, solid and persistent; to say the ether exists
> is to say there is a natural kinship between all the optical phenomena, and
> neither of the two propositions has less value than the other.[78]

Thus, the ether provides a medium for the propagation of electromagnetic waves,
thereby establishing relations between things. Consequently, the ether is real and
not a convention.

That LORENTZ's theory could account for the null result of the aberration
experiments and the MICHELSON & MORLEY experiment was of particular importance to POINCARÉ, in whose philosophy and physics there was no place for absolute
motion:

> Experiments have been made that should have disclosed the terms of the first
> order; the results were nugatory. Could that have been by chance ? No one
> has admitted this; a general explanation was sought, and Lorentz found it.
> He showed that the terms of the first order should cancel each other, but not
> the terms of the second order. Then more exact experiments were made, which
> were also negative; neither could this be the result of chance. An explanation
> was necessary, and was forthcoming; they always are; hypotheses are what we
> lack the least.[79]

Thus POINCARÉ was satisfied with LORENTZ's explanation of the aberration experiments, but critical of the *ad hoc* manner in which the second-order result was
accounted for. Nevertheless, the failure of the first-order and second-order experiments to detect the earth's motion through the ether implied the existence of a
general law of nature:

> I do not believe that more exact observations will ever make evident anything
> else but the relative displacements of material bodies... The mutual destruction
> of [higher-order terms in v] will be rigorous and absolute.[80]

To illustrate the extent to which POINCARÉ's philosophy and physics were
intertwined, let us now sketch the evolution of these ideas. In particular, let
us go back to the year 1895, when POINCARÉ discussed these ideas in more nascent
form in "À propos de la théorie de M. Larmor".[81] In this paper he proposed certain
conditions which an electrodynamical theory of moving bodies should satisfy[82]:

1. It should be consonant with the result of FIZEAU's experiment, "that is to
say of the *partial* entrainment of light waves, or what is the same of transverse
electromagnetic waves." (Italics in original.)

---

[78] H. POINCARÉ, *V.S.*, *op. cit.* (footnote 68), pp. 139–140.

[79] H. POINCARÉ, *S.H.*, *op. cit.* (footnote 65), p. 172.

[80] *Ibid.*, p. 172.

[81] H. POINCARÉ, "À propos de la théorie de M. Larmor," L'Éclairage électrique,
3, 5–13, 285–295 (1895); *ibid.*, 5, 5–14, 385–392 (1895); reprinted in *Œuvres*, pp. 369–
426. All references will be to the *Œuvres*.

[82] H. POINCARÉ, "À propos ..." *op. cit.* (footnote 81), p. 395.

2. It should be consonant with the principle of conservation of electricity and magnetism.[83]

3. It should be consonant with the principle of action and reaction.

Of the theories that POINCARÉ discussed (those of HELMHOLTZ, HERTZ, LARMOR, LORENTZ and THOMSON), only LORENTZ's satisfied the first two criteria. POINCARÉ went on to note the necessity of accepting provisionally the least defective *(la moins défectueuse)* theory—LORENTZ's.[84] However, for POINCARÉ it was "very difficult to admit that the principle of reaction is violated, even in appearance ..."[85] We shall discuss further the third condition in a moment.

POINCARÉ went on to infer, from the failure of the first-order and second-order experiments to detect the influence of the earth's motion on optical phenomena, the impossibility of observing "the absolute motion of matter, or better yet the relative motion of ponderable matter with respect to the ether."[86] Thus, POINCARÉ concluded that only "the motion of ponderable matter with respect to ponderable matter" is observable.[87]

In 1899 POINCARÉ commented again on LORENTZ's theory. By this time he had evidently thought a great deal about LORENTZ's *Versuch* and in particular about the nature of LORENTZ's hypothesis of contraction. He asserted that the hypothesis of contraction is a *"coup de pouce"* provided by nature to prevent the detection, to order v^2, of absolute motion.[88] This being the case, POINCARÉ asked whether "a new *coup de pouce*, a new hypothesis" must be posited to explain the expected null results to each higher order in v—to which he answered emphatically: "Of course not ..."[89] For POINCARÉ the true theory of the electrodynamics of moving bodies will contain rigorously the principle *(principe)* that only the relative motion of ponderable bodies is observable. He went on to assert that at that time no such theory existed; however, LORENTZ's was the most satisfactory. Indeed, POINCARÉ concluded, that perhaps "without very profound modifications" this theory may ultimately be completely satisfactory.[90]

POINCARÉ, however, did not consider LORENTZ's local time coordinate as a *"coup de pouce,"* or *ad hoc* hypothesis. The probable reason is that it enabled him to prove mathematically to first order in v the covariance of the MAXWELL-

---

[83] Conservation of charge, in the context that POINCARÉ used it, meant that the charge on a moving body moved along with the body, *i.e.*

$$\frac{d}{dt} \left(\int \varrho \, d\tau \right) = 0.$$

That this condition was "built into" LORENTZ's theory is explained in footnote 40. HELMHOLTZ's theory, for example, violates this conservation law (see p. 391 of "À propos ...") Conservation of magnetism means that there exist no real magnetic charges, *i.e.* $\vec{V} \cdot \vec{B} = 0$.

[84] H. POINCARÉ, "À propos ..." *op. cit.* (footnote 81), p. 409.

[85] *Loc. cit.*

[86] H. POINCARÉ, "À propos ..." *op. cit.* (footnote 81), p. 412.

[87] *Loc. cit.*

[88] H. POINCARÉ, *Électricité et Optique* (Gauthier-Villars, 1901), esp. p. 536. This volume prints lectures POINCARÉ gave at the Sorbonne in 1888, 1890 and 1899. They served to introduce the majority of French physicists to MAXWELL's electromagnetic field theory.

[89] *Ibid.*, p. 536; italics in original.

[90] *Loc. cit.*

LORENTZ equations[91]. On the other hand, LORENTZ was forced to posit the hypothesis of contraction to account for the absence of effects due to the second order terms (in v) which had simply been discarded. POINCARÉ interpreted the local time coordinate as merely a convenient shift in the origin of the time variable.[92]

Using an equation identical to (32d), he estimated the difference between the local-time and true time for two clocks situated 1 km apart on the moving earth as $\frac{1}{3} \times 10^{-9}$ sec. Thus, he asserted that the difference between the local time t' and the *true* time t was too small to be observed.[93]

We have seen that by 1900 POINCARÉ had arrived at the concept of a physics of principles wherein NEWTON's three principles, the principle of conservation of energy and the principle of relative motion were accorded the status of conventions. In that year POINCARÉ published a predominantly scientific article in a LORENTZ "Festschrift" which is of interest because it illustrates clearly the reflection of POINCARÉ's conventionalism in his physics.[94] In this paper, POINCARÉ returned to the objection that he had raised in 1895 concerning LORENTZ's theory, namely, that it is in violation of NEWTON's principle (the principle of action and reaction). But NEWTON's principle is a convention and must, therefore, be preserved in this most satisfactory of all existing electromagnetic theories. POINCARÉ argued for the validity of NEWTON's principle in LORENTZ's theory on the grounds that the ether provided the necessary compensating mechanisms for instantaneous action and reaction.[95] Furthermore he proved, using a method analogous to the one in footnote 26, the necessity of ascribing a momentum to the electromagnetic field in order to preserve the law of conservation of momentum in LORENTZ's theory. Consequently, the NEWTONIAN version of this law had to be widened to include non-ponderable media. Thus, in 1900, POINCARÉ believed that in electromagnetic theory, as in classical mechanics, conservation of momentum is a direct result of the principle of action and reaction.

POINCARÉ went on to emphasize the importance of NEWTON's principle by asking the question: "Why does the principle of action and reaction impose itself on our thinking?"[96] In reply he proved that the principle of action and reaction can be obtained, for an arbitrary conservative system, from the principles of the conservation of energy and relative motion; both of these statements are conventions.[97]

As a note of caution POINCARÉ emphasized that in LORENTZ's theory the principle of relative motion is not imposed *a priori*; rather, it is verified *a posteriori*[98]. POINCARÉ ascribed the theory's "imperfect" verification of this principle to the following compensating effects: neglect of terms higher than first order in v; and use of the local time coordinate.

---

[91] *Cf. ibid.*, esp. pp. 528–533 for POINCARÉ's proof of LORENTZ's theorem of corresponding states.

[92] *Ibid.*, p. 530.

[93] *Ibid.*, p. 530. For v use the earth's orbital velocity about the sun (30 km/sec).

[94] H. POINCARÉ, "La théorie de Lorentz et le principe de réaction," *op. cit.* (footnote 26).

[95] See also *S.H.*, *op. cit.* (footnote 65), esp. p. 170.

[96] H. POINCARÉ, "La théorie ..." *op. cit.* (footnote 26), p. 481.

[97] See *ibid.*, pp. 481–482 for the details.

[98] *Ibid.*, p. 483.

In conclusion POINCARÉ noted that the principle of action and reaction and the principle of relative motion are inextricably linked; thus, if the former principle were found to be invalid, then we should have to "modify profoundly all our ideas on electrodynamics."[99]

BERTRAND RUSSELL in his incisive review of POINCARÉ's *Science and Hypothesis* asserted that: "... this ... book consists in the main of previous articles somewhat rewritten ..."[100] This is by and large true; nevertheless, there is something to be said about rethinking one's philosophy and then setting it down in a more systematic and crystallized form. As evidenced by the *Author's Preface*, POINCARÉ realized by 1902 that he must distinguish clearly between conventional statements in geometry, classical mechanics and the physical sciences.[66] This was made apparent in "Electrodynamics" which is the concluding chapter of *Science and Hypothesis*.[101] Here POINCARÉ, in response to an experiment suggested by his colleague GABRIEL LIPPMANN, provided a glimpse as to how the principle of the relativity of space from geometry would eventually find its way via classical mechanics into the physical sciences. Consider two positively charged conductors at rest on the earth's surface. As a consequence of the earth's motion through the ether, these conductors (assuming the validity of ROWLAND's experimental results) constitute convection currents; thus they should attract one another. A measurement of the attractive force would enable the earth's absolute velocity to be inferred.[102] POINCARÉ chose to resolve this problem within the context of LORENTZ's theory: "'No!' replied the partisans of Lorentz. 'What we could measure in that way is not their absolute velocity, but their relative velocity *with respect to the ether*, so that the principle of relativity is safe.'"[103] POINCARÉ, however, was fully aware that at this time LORENTZ's theory was consonant with the relativity of space only to first order in v. Perhaps this was why he referred to the "partisans of Lorentz" who already in 1902 had full confidence that LORENTZ's theory would someday conform rigorously to a version in the physical sciences of the principle of relativity from geometry. Consonant with his conventionalistic epistemology, POINCARÉ was not completely committed to LORENTZ's theory. Moreover, experimental evidence had accumulated recently which affected directly the very basis of LORENTZ's theory—the fundamental equations. VICTOR CRÉMIEU, at the Sorbonne, had concluded on the basis of a series of experiments begun in 1899 that contrary to ROWLAND's result, a charged body in motion does not produce any magnetic effect.[104] This result, if true, would have been disastrous to LORENTZ's theory, wherein the convection current term (ϱv) is indispensable. POINCARÉ first mentioned

---

[99] *Ibid.*, p. 488.

[100] B. RUSSELL, Mind, **14**, 412–418 (1905), p. 412.

[101] H. POINCARÉ, *S.H.*, *op. cit.* (footnote 62), pp. 225–244. Apparently POINCARÉ wrote this chapter expressly for *S.H.*

[102] Despite my search of the literature I have not been able to find the publication of LIPPMANN's experiment. For POINCARÉ's description see *S.H.*, *op. cit.* (footnote 62), esp. pp. 243–244.

[103] *Ibid.*, pp. 243–244; italics in original.

[104] For the state of CRÉMIEU's results *circa* 1901 and for references to his earlier work, see V. CRÉMIEU, "Sur les expériences de M. Rowland relatives à l'effet magnétique de la 'convection électrique'," *Comptes rendus de l'Académie des Sciences*, **132**, 797–800 (1901). This paper was presented by POINCARÉ.

CRÉMIEU's results in the Foreword to *Électricité et Optique*. POINCARÉ noted that he had learned of CRÉMIEU's results after the book had gone to press; moreover, if these results were true, POINCARÉ continued, they "would completely modify our ideas on the electrodynamics of moving bodies."[105] POINCARÉ concluded *Science and Hypothesis* with a similar statement:

> This quiescence [in the subject of electromagnetism] has been recently disturbed by the experiments of M. Crémieu, which have contradicted, or at least have seemed to contradict, the results formerly obtained by Rowland. Numerous investigators have endeavoured to solve the question, and fresh experiments have been undertaken. What result will they give? I shall take care not to risk a prophecy which might be falsified between the day this book is ready for the press and the day on which it is placed before the public.[106]

This interesting debate has apparently not been studied by most historians of science.[107] For this reason, and in particular for the purpose of illuminating the central role played by POINCARÉ in the scientific community, let us review this episode.

In accord with his epistemology, POINCARÉ took a keen interest in CRÉMIEU's experiments. He presented many of CRÉMIEU's subsequent papers at French scientific meetings and also was in close correspondence with physicists both in England and the United States on this problem. CRÉMIEU, by 1901, was in the midst of a controversy with HAROLD PENDER, a British physicist, who had obtained a positive result in a similar set of experiments.[108]

By the fall of 1902 POINCARÉ had decided that it was of the utmost importance to settle the issue conclusively. He proposed that CRÉMIEU and PENDER repeat their experiments in the same laboratory in order to observe each others' techniques. Lord KELVIN proposed that this collaborative effort be carried out at the Sorbonne. The Johns Hopkins University put at PENDER's disposal all necessary equipment and the Carnegie Institution paid all his travel expenses. The Institute of France defrayed all remaining costs of experimentation. This international effort, begun in January 1903, proved conclusively that CRÉMIEU's results were invalid. CRÉMIEU had coated his revolving metal disks with a dielectric material,

---

[105] H. POINCARÉ, *Électricité et Optique, op. cit.* (footnote 88), p. II.

[106] H. POINCARÉ *S.H., op. cit.* (footnote 62), p. 244.

[107] For example E. T. WHITTAKER in his two volume treatise *A History of Theories of Aether and Electricity* (Nelson, Vol. 1, 1951, Vol. 2, 1953) made no explicit reference to this episode. T. HIROSIGE discussed the CRÉMIEU-PENDER episode briefly; however he made no mention of POINCARÉ's role (T. HIROSIGE, "Electrodynamics Before the Theory of Relativity, 1890–1905," *op. cit.* (footnote 21(b)), esp. p. 7). For more details see J. D. MILLER, "Rowland and the Nature of Electric Currents," Isis, **63**, 5–27 (1972), esp. pp. 24–27.

[108] For an excellent review of both PENDER's and CRÉMIEU's work *circa* 1901 see H. PENDER, "On the Magnetic Effect of Electrical Conduction," Phil. Mag., **2**, 179–208 (1901). PENDER noted that he was assisted by ROWLAND himself in the early stages of this work. It is of interest to note that WHITTAKER did refer to this paper (see Volume I, p. 306 of *A History of ..., op. cit.* (footnote 107)). Thus, he was probably aware of the CRÉMIEU-PENDER controversy, but for some reason considered it as an unimportant episode in the history of electromagnetism. One must bear in mind, however, that WHITTAKER's treatise has a somewhat biased view of the history of physics: *cf.* Chapter II of Volume II, entitled "The Relativity Theory of Poincaré and Lorentz."

causing a reduction in the expected magnetic effect.[109] Thus, at least with respect to a reference system fixed in the ether, the MAXWELL-LORENTZ equations were saved. One alternative would have been a theory such as HELMHOLTZ's which violated the law of conservation of charge.[110] By September of the following year POINCARÉ had become an ardent follower of LORENTZ.

On 24 September 1904 POINCARÉ delivered an important address entitled "L'état actuel et l'avenir de la Physique mathématique" at the International Congress of Arts and Science at St. Louis, Missouri. In addition to praising LORENTZ's recently published theory of the electron ("Electromagnetic Phenomena in a System Moving with any Velocity Less than that of Light"[6]), POINCARÉ elaborated on the principle of relativity in the physical sciences "according to which the laws of physical phenomena must be the same for a stationary observer as for an observer carried along in a uniform motion of translation; so that we have not and can not have any means of discerning whether or not we are carried along in such a motion."[111]

POINCARÉ began his discussion of the principle of relativity by referring first to what "Rowland has taught us"[112] and then to LIPPMANN's experiment—although he did not use LIPPMANN's name. However, POINCARÉ asserted immediately that as of 1904 all attempts to measure the absolute velocity of the earth had failed "and it is precisely to explain this obstinacy that the mathematicians are forced today to employ all their ingenuity".[113] POINCARÉ went on to explain that LORENTZ solved this problem "only by accumulating hypotheses"; furthermore, POINCARÉ continued, "the most ingenious idea was that of local time".[114] The discussion that followed of the synchronization of two clocks by observers who travel with them and exchange light signals is indicative of POINCARÉ's strictly mathematical interpretation of LORENTZ's local time coordi-

---

[109] For details as well as a fairly complete set of references see H. PENDER & V. CRÉMIEU, "On the Magnetic Effect of Electrical Convection," Phil. Mag., 6, 442–464 (1903). My information on POINCARÉ's organization of the CRÉMIEU-PENDER collaboration is from p. 443 of this paper. A brief version of this paper was published under the title "Nouvelles recherches sur la convection électrique," in Comptes rendus de l'Académie des Sciences, 136, 548–550 (1903), and was presented by POINCARÉ. Evidence for the widespread publicity that this collaboration received can be found, for example, in a paper by WILLIAM SUTHERLAND, "The Crémieu-Pender Discovery," Phil. Mag. 7, 405–407 (1904). SUTHERLAND began his paper by discussing the "happy collaboration of Messrs. Crémieu and Pender" (p. 405). See SUTHERLAND's paper for details of the CRÉMIEU-PENDER effect, which is not germane to this discussion. It is of interest to note that WHITTAKER (Volume I, p. 306, of A History of ...) referred to a slightly different version of CRÉMIEU-PENDER paper, which was published in the Philosophical Magazine, namely, "Recherches Contradictoires sur l'effet Magnétique de la Convection Électrique," Journal de Physique théorique et appliquée, 2, 641–666 (1903). WHITTAKER noted simply that CRÉMIEU and PENDER had repeated ROWLAND's experiment under improved conditions.

[110] H. POINCARÉ, "À propos ..." op. cit. (footnote 81), esp. p. 391.

[111] H. POINCARÉ, V.S., op. cit. (footnote 61), p. 94. POINCARÉ listed the other principles (not conventions) of the physical sciences as: CARNOT's principle, the principle of conservation of energy, the principle of the equality of action and reaction, the principle of conservation of mass and the principle of least action.

[112] H. POINCARÉ, V.S., op. cit. (footnote 61), p. 98; italics added.

[113] Ibid., p. 99.

[114] Loc. cit.

nate. POINCARÉ emphasized that only clocks at rest with respect to the ether can be synchronized in this manner to register the true time. However, clocks in motion with respect to the ether register the local time, not the true time. As to the discrepancy between the true time and the local time, "it matters little, since we have no means of perceiving it".[115] Thus, whereas in 1899 POINCARÉ's reason for neglecting the difference between true and local time was a mathematical one (valid only to first order in v), in 1904 it appeared to have a physical basis (valid mathematically to all orders in v); namely, that the principle of relativity prevents us from knowing whether we are in uniform motion with respect to the ether or at rest in the ether.

However, even if we ignore the word ether, this argument is not as EINSTEINIAN as it appears. The reason is that in the MAXWELL-LORENTZ theory of electromagnetism the speed of light (in vacuum) is *really c* $(3 \times 10^8 \text{ m/sec})$ only with respect to reference systems at rest in the ether. When measured by observers in reference systems in motion with respect to the ether, its measured value is also c because the local time and the contraction of length act as compensatory mechanisms: the latter effect compensates for the retardation of clocks in motion with respect to the ether compared to those at rest in the ether.[116] Most important, it can be inferred from POINCARÉ's discussion of 1904 regarding synchronization of clocks, as well as from his future discussions,[117] that he considered the retardation of moving clocks as experimentally unobservable. Thus, it is not a real effect; rather, the hypothesis of a local time coordinate is purely mathematical. The hypothesis of the contraction of length, however, is physical because it was inferred from MICHELSON & MORLEY's experiment. Furthermore, LORENTZ and POINCARÉ considered the hypothesis of the contraction of length as independent of the hypothesis of the equations for the LORENTZ transformation. The latter hypothesis was considered as a mathematical device. (This statement will be reinforced in Part 6.) For POINCARÉ the postulation of these two compensatory effects, one purely mathematical and the other real, was a necessary prerequisite for a principle of relativity.

POINCARÉ was also deeply interested in problems concerning gravitation—in particular, the speed at which the gravitational field is propagated. He noted in "L'état actuel et l'avenir de la Physique mathématique" that if gravity is transmitted with a velocity much greater than that of light, as in LAPLACE's theory,

---

[115] *Loc. cit.*

[116] For a detailed proof see H. LORENTZ, *T.E.*, pp. 224–226.

[117] (a) H. POINCARÉ, *Science and Method*, translated by F. MAITLAND (Dover, n.d.), esp. pp. 219–222 (Book III, pp. 199–250 of *Science and Method* is composed of excerpts from POINCARÉ's "La dynamique de l'électron," Revue générale des Sciences pures et appliquées, **19**, 386–402 (1908), reprinted in *Œuvres*, pp. 551–586). Hereinafter *Science and Method* will be referred to as *S.M.*

(b) H. POINCARÉ, *La dynamique de l'électron* (Dumas, 1913), esp. pp. 39–47.

POINCARÉ bases his discussion on the fact that the speed of light is not the same in all coordinate systems. It is c with respect to systems fixed in the ether. However, because of the relativistic compensating effects of the local time and the LORENTZ contraction, a moving observer will measure the speed of light as c.

For an *almost* EINSTEINIAN discussion of the notion of simultaneity see:

(c) H. POINCARÉ, "La mesure du temps," Rev. Mét. Mor. **6**, 371–384 (1898), translated in *V.S., op. cit.* (footnote 19), pp. 26–36.

16*

then one could use it as a signal to check whether two clocks in motion with respect to the ether were properly synchronized by the method of exchanging light signals.[118] Thus, it would be possible to "observe discrepancies which would render evident the common translation of the two [clocks]."[119] To rescue the principle of relativity, "which is irresistibly imposed upon our good sense,"[120] POINCARÉ formulated in "Sur la dynamique ..." a LORENTZ-covariant gravitational theory in which the gravitational field, just as the electromagnetic field, propagates with the speed of light. For POINCARÉ, unlike EINSTEIN circa 1905, the problems of gravitation and of the electrodynamics of moving bodies were inextricably linked. This too is consonant with POINCARÉ's epistemology, according to which a principle (in this case the principle of relativity) must agree with *all* empirical data before it can become a convention; then it applies to all of physical theory, *i.e.*, to non-inertial as well as to inertial reference systems.

In view of POINCARÉ's criticism of the manner in which LORENTZ accounted for the MICHELSON & MORLEY experiment, it seems surprising that he was now satisfied with a theory which succeeded at the expense of accumulating hypotheses. However, POINCARÉ emphasized that LORENTZ was forced to accumulate hypotheses because the principle of relativity is "imposed upon our good sense,"[121] and besides, "it is not merely a principle which it is a question of saving, it is the indubitable results of the *experiments* of Michelson."[122]

POINCARÉ was also impressed that LORENTZ's hypotheses could be generalized to provide relations among a great many phenomena, *e.g.*, all forces transform like those of electromagnetism. Furthermore, the large number of hypotheses in LORENTZ's theory made it a "supple theory," meaning that tampering with one of its hypotheses would not destroy the entire theoretical structure.[123]

POINCARÉ then returned to LIPPMANN's proposed experimental method for measuring the earth's velocity with respect to the ether. He explained that on the basis of LORENTZ's theory the experimental effect, *i.e.*, the attractive force between the conductors at rest on the earth's surface—cannot be observed. POINCARÉ's reasoning went as follows: Two charges of like sign at rest with respect to the ether repel one another. The attractive force arising from their motion through the ether acts to reduce the COULOMB repulsion. To measure this reduced repulsion requires the use of other forces. But according to (42) all forces transverse to the direction of motion are "reduced in the same proportion, we perceive nothing."[124] POINCARÉ had become one of the "partisans of LORENTZ."[125]

In "L'état actuel ..." POINCARÉ also rejected NEWTON's third law because, due to the finite speed of light, the recoil of the emitter will not be instantaneously compensated for by a receiver (if one exists). Furthermore, if electromagnetic

---

[118] See H. POINCARÉ, *V.S.*, *op. cit.* (footnote 61), esp. p. 100.

[119] *Loc. cit.*

[120] *Loc. cit.*

[121] *Ibid.*, p. 98.

[122] *Ibid.*, p. 103; italics added.

[123] POINCARÉ first discussed the criterion of suppleness in "La théorie de Lorentz et le principe de réaction," *op. cit.* (footnote 26), esp. p. 464.

[124] H. POINCARÉ, *V.S.*, *op. cit.* (footnote 61), esp. p. 100.

[125] H. POINCARÉ, *S.H.*, *op. cit.* (footnote 101), pp. 243–244.

energy is dissipated en route, then the reaction will not be equal to the recoil.[126] Thus, contrary to NEWTONIAN mechanics, conservation of momentum is valid but the law of action and reaction is invalid—whereas in NEWTONIAN physics the third law was postulated to yield conservation of momentum. *In this manner the law of conservation of momentum transcends Newton's third law.* POINCARÉ concluded that to hypothesize that an invisible ether compensated somehow for the visible motion of ponderable matter (*e.g.*, the recoil of an emitter) "is [to be] able to explain everything ... [however], it does not enable us to foresee anything, it does not enable us to decide between the various hypotheses, since it explains everything beforehand. It therefore becomes useless."[127] It was, in fact, an *ad hoc* hypothesis just as was LORENTZ's statement in 1895 of the hypothesis of contraction. Obviously, LORENTZ's electron theory of 1904 had convinced POINCARÉ of the importance of the field momentum and in turn of the invalidity of NEWTON's third law in the physical sciences. The reason is that in 1900, in the same paper in which he castigated LORENTZ for putting forth *ad hoc* hypotheses, POINCARÉ had asserted: "According to Lorentz, we do not know what the movements of the ether are; and because we do not know this, we may suppose them to be movements compensating those of matter, and reaffirming that action and reaction are equal and opposite."[128] Now that LORENTZ had eliminated his *ad hoc* hypothesis it was time for POINCARÉ to follow suit, which he did by stripping the principle of action and reaction of its conventional status in the physical sciences. Nevertheless, with respect to the status of the "new mechanics" POINCARÉ asserted in 1904: "I hasten to say ... that we are not yet there, and as yet nothing proves that the principles will not come forth from out the fray, victorious and intact."[129]

By 1908 POINCARÉ believed that the "new mechanics" would be firmly established: "We must [accept] ... Lorentz's theory, and consequently *give up the principle of reaction.*"[130]

With respect to NEWTON's first law POINCARÉ asserted in 1908: "In the new Dynamics the Principle of Inertia is still true ..."[131]

In the analysis of "Sur la dynamique ..." we shall see what became of the second law in the physical sciences.

In 1904, POINCARÉ was guardedly optimistic about the future of LORENTZ's theory; *e.g.*, "[f]rom these results, *if they are confirmed* would arise a new mechanics [in which] no velocity could surpass that of light."[132] It was, therefore, necessary to take into account KAUFMANN's results and to resolve the aforementioned theoretical difficulties with LORENTZ's theory.

POINCARÉ immediately set to work on these problems upon returning home to France from the United States.[133] The result was "Sur la dynamique de l'électron".

---

[126] H. POINCARÉ, *V.S.*, *op. cit.* (footnote 61), esp. p. 101.

[127] *Ibid.*, p. 102.

[128] H. POINCARÉ, *S.H.*, *op. cit.* (footnote 65), p. 176.

[129] H. POINCARÉ, *V.S.*, *op. cit.* (footnote 61), p. 111.

[130] H. POINCARÉ, *S.M.*, *op. cit.* (footnote 117 (a)), p. 225; italics in original.

[131] *Ibid.*, p. 229.

[132] H. POINCARÉ, *V.S.*, *op. cit.* (footnote 61), p. 104; italics in original.

[133] See V. VOLTERRA, J. HADAMARD, P. LANGEVIN and P. BOUTROUX, *Henri Poincaré* (Librairie Félix Alcan, 1914), esp. pp. 169–170 for LANGEVIN's account of the trip that he and POINCARÉ took to the International Congress of Arts and Science.

6. "Sur la dynamique de l'électron"

6.1. Introduction. (Section headings in the remainder of Part 6 will conform to POINCARÉ's; the Section number in "Sur la dynamique ..." will appear in square brackets.)[134]

Generalizing from the null result of the ether drift experiments, POINCARÉ asserted:

> This impossibility of experimentally demonstrating the absolute motion of the Earth appears to be a general law of Nature; it is reasonable to assume the existence of this law, which we shall call the *postulate of relativity* and to assume that it is universally valid. Whether this postulate, which so far is in agreement with experiment, be later confirmed or disproved by more accurate tests, it is, in any case, of interest to see what consequences follow from it.[135]

(POINCARÉ, for some unstated reason, chose to refer in this paper to the principle of relativity as the postulate of relativity; clearly they are identical. I shall hereafter refer only to the principle of relativity.) Thus the tone of the introduction is one of guarded optimism. The reason for this is found at the end of the introduction:

> I have therefore not hesitated to publish these incomplete results, even though at the present time the entire theory may seem to be threatened by the discovery of cathode rays.[136]

Indeed, the degree of incompleteness of these results will manifest itself in Sections 6–8 of "Sur la dynamique ...". POINCARÉ's cautiousness illustrates the weight placed on KAUFMANN's experimental results because if they are correct, then LORENTZ's theory and, therefore, the principle of relativity would have to be revised. This constitutes further proof for my assertion that the principle of relativity is not a convention.

POINCARÉ explained that his criticism of the *ad hoc* manner in which the LORENTZ contraction was put forward in 1895 had been taken into account in LORENTZ's paper of 1904:

> Lorentz has sought to extend and modify the hypothesis so as to make it fully compatible with the postulate of relativity. This he has succeeded in doing, in his paper "Electromagnetic phenomena in a system moving with any velocity smaller than that of light" (*Proceedings of the Section of Sciences, Koninklijke Akademie van Wetenschappen te Amsterdam* 6, 809–831, 1904).[137]

He went on to note that

> In view of the importance of this problem, I resolved to examine it further. The results which I have obtained agree with those of Lorentz in all the

---

[134] R. 129–132: O. 494–498: K. 145–149. For code see footnote 1.

[135] R. 129: O. 494: K. 145. KILMISTER translated "*le Postulat de Relativité*" as the relativity postulate; however, it should be translated as I have done here as the postulate of relativity. Italics in original.

[136] R. 132: O. 498: K. 149.

[137] R. 130: O. 495: K. 146.

principal points, and I have needed only to modify and augment them in certain details.[138]

POINCARÉ's modification and augmentation of LORENTZ's results was the derivation of the correct transformation equations for the velocity and charge density. Using the corrected equations, he was able to complete the proof of the covariance of the MAXWELL-LORENTZ equations under the LORENTZ transformation.

POINCARÉ then referred to the nature of the equations of transformation and the theorem of corresponding states:

> Lorentz's concept may be summarised thus: if a common translatory motion may be imparted to the entire system without any alteration of the observable phenomena, then the equations of an electromagnetic medium are unaltered by certain transformations, which we shall call *Lorentz transformations*. In this way two systems, of which one is fixed and the other is in translatory motion, become exact images of each other.[139]

In addition to LORENTZ's work of 1904, continued POINCARÉ, another theory of the electron was simultaneously and independently proposed by BUCHERER[24] and LANGEVIN[25]. In this theory the electron undergoes both a contraction and a dilation in such a way that its volume remains constant. Furthermore, the prediction of this theory is "in agreement with the experiments of Kaufmann, as is Abraham's original hypothesis of the rigid electron."[140]

> The advantage of Langevin's theory is that it involves only the electromagnetic forces and the constraints; but it is not compatible with the postulate of relativity. This was shown by Lorentz, and I have likewise proved it by a different method, based upon the use of group theory.[141]

POINCARÉ went on to prove that although LANGEVIN's theory circumvents certain of the difficulties in LORENTZ's, it requires that $l = k^{-\frac{1}{2}}$. However, LORENTZ had proven that only $l = 1$ is compatible with the theorem of corresponding states. POINCARÉ obtained this result by making use of the group-theoretical properties of the LORENTZ transformation.

The main goal of POINCARÉ's work was to demonstrate conclusively that of all possible models for the electron only LORENTZ's was compatible with the principle of relativity. However,

> in order to maintain [it] free from unacceptable contradictions, a special force must be invoked to account both for the contraction and for the constancy of two of the axes. I have attempted to determine this force, and have found that *it can be regarded as a constant external pressure acting upon an electron capable of deformation and compression, the work done being proportional to the change in the volume of the electron.*[142]

---

[138] R. 130: O. 495: K. 146.
[139] *Loc. cit.*; italics in original.
[140] R. 130: O. 496: K. 147.
[141] *Loc. cit.*
[142] *Loc. cit*; italics in original.

The "unacceptable contradictions" are due to the unstable nature of the deformable electron in LORENTZ's theory. It will be seen that, contrary to what is sometimes attributed to this paper, POINCARÉ never computed the counter term necessary to cancel the second term on the right-hand side of (48), nor did he reduce the factor of $\frac{4}{3}$ in G to unity. Rather, he proved the necessity of introducing mechanical stresses into LORENTZ's theory to account for the inertia of a deformable electron in a manner consonant with the principle of relativity:

> Then, if the inertia of matter is exclusively of electromagnetic origin, as has been customarily supposed since Kaufmann's experiment, and if all forces (other than the constant pressure to which I have just alluded) are of electromagnetic origin, the postulate of relativity can be accepted as strictly valid. I show this by means of a very simple calculation based upon the principle of least action.[143]

However, the presence of these stresses negates a purely electromagnetic theory of the electron's inertia. Nevertheless, they are needed to construct a LANGRANGIAN formulation of LORENTZ's theory.

> In Lorentz's view, all forces, no matter how originating, are affected by the Lorentz transformation (and therefore by a translatory motion) in the same manner as the electromagnetic forces.[144]

The final section of "Sur la dynamique ..." (Section 9—"Hypotheses Concerning Gravitation") is representative of POINCARÉ's mathematical insight into the LORENTZ transformation. As an aid in the search for quantities that are invariant under the LORENTZ group, POINCARÉ introduced the notion of a four-dimensional space spanned by space-time coordinates x, y, z, ict. As a result of inspecting all possible invariants of the LORENTZ group that can be constructed from the space-time coordinates, velocities and forces, POINCARÉ concluded that it is reasonable to assume that gravity is propagated through the ether with the velocity of light. He then went on to construct the first LORENTZ-covariant theory of gravitation wherein the force equation is analogous to the LORENTZ force.[145]

However, from POINCARÉ's point of view, the important result of his investigations into electromagnetism and gravitation is that both fields propagate through the ether with the speed of light:

> If we assume the postulate of relativity, we find a quantity common to the law of gravitation and the laws of electromagnetism, and this quantity is the velocity of light; and this same quantity appears in every other force, of whatever origin. There can be only two explanations.
>
> Either, everything in the universe is of electromagnetic origin; or, this constituent which appears common to all the phenomena of physics has no

---

[143] *Loc. cit.*

[144] *Loc. cit.*

[145] For a qualitative account of this theory and of the state of POINCARÉ's research in gravitational theory *circa* 1908 see H. POINCARÉ, "La dynamique de ..." *op. cit.* (footnote 117 (a)), esp. pp. 571–586 (*Œuvres*) and pp. 235–250 (*S.M.*), *op. cit.* (footnote 117 (a)).

real existence, but arises from our methods of measurement. What are these methods ? One might first reply, the bringing into juxtaposition of objects regarded as invariable solid things; but this is no longer so in our present theory, if the Lorentz contraction is assumed. In this theory, two lengths are by definition equal if they are traversed by light in the same time.[146]

The speed of light plays the role of a connector of things—of sensations (see Part 5)—and is therefore of great importance. This quantity, common to both theories, accounts for the non-fortuitous formal analogy between POINCARÉ's relativistic gravitational force and the LORENTZ force. However, this does not diminish the importance of the ether which acts to support the fields in transit: "If the gravitational attraction is propagated with the velocity of light, this cannot occur by mere chance, but must be dependent on the ether."[147]

POINCARÉ's almost EINSTEINIAN discussion of the role played by the velocity of light in the measurement of length and time was first presented in 1898.[117(c)] However, as discussed previously (see Part 5.2), in this and subsequent expositions[117(a,b)] it soon becomes apparent that, for POINCARÉ, the velocity of light is not c in all reference frames but only appears to be, due to the relativistic compensating effects of the proper time and the LORENTZ contraction. This is not surprising because, as we shall see, POINCARÉ, like LORENTZ, did not ascribe any physical reality to the Σ' frame.

6.2. The Lorentz Transformation [§ 1].[148] A word must be said about POINCARÉ's notation. The only way in which it differed from the modern relativistic and electromagnetic terminology is that he used d for both regular and partial derivatives. He reserved the usual notation for the partial derivative ∂ for his variational principle. In order to avoid confusion between d and ∂ I shall take the liberty in this one instance to modernize his notation. The symbols d and ∂ will be used for regular and partial derivatives, respectively, and δ will be used in discussing the differential operation used in his variational principle.

If the electric field components are denoted by (f, g, h), the magnetic field by (α, β, γ), the vector potential by (F, G, H), the scalar potential by ψ, the electric charge density by ϱ, the electron's velocity by (ξ, η, ζ), the total electric current (displacement plus convection) by (u, v, w) the fundamental equations are

$$u = \frac{\partial f}{\partial t} + \varrho \xi = \frac{\partial \gamma}{\partial y} - \frac{\partial \beta}{\partial z}, \tag{58a}$$

$$\alpha = \frac{\partial H}{\partial y} - \frac{\partial G}{\partial z}, \tag{58b}$$

$$f = -\frac{\partial F}{\partial t} - \frac{\partial \psi}{\partial x}, \tag{58c}$$

$$\frac{\partial \alpha}{\partial t} = \frac{\partial g}{\partial z} - \frac{\partial h}{\partial y}, \tag{58d}$$

$$\frac{\partial \varrho}{\partial t} + \Sigma \frac{\partial (\varrho \xi)}{\partial x} = 0. \tag{58e}$$

---

[146] R. 131: O. 498: K. 149.
[147] R. 131: O. 497: K. 148.
[148] R. 132–136: O. 498–503: K. 150–156.

$$\Sigma \frac{\partial f}{\partial x} = \varrho, \tag{58f}$$

$$\frac{\partial \psi}{\partial t} + \Sigma \frac{\partial F}{\partial x} = 0, \tag{58g}$$

$$\square = V^2 - \frac{\partial^2}{\partial t^2} = \Sigma \frac{\partial^2}{\partial x^2} - \frac{\partial^2}{\partial t^2}, \tag{58h}$$

$$\square \psi = -\varrho, \quad \square F = -\varrho \xi, \tag{58i}$$

$$X = \varrho f + \varrho (\eta \gamma - \zeta \beta). \tag{58j}$$

The scalar product of two vectors is denoted by Σ. Equations (58a, b, d, f, j) are equivalent to (1a–e). Equation (58i) is equivalent to (7) and (58g) is the LORENTZ gauge condition (6) while (58e) is the equation of conservation of charge.

Before completing LORENTZ's proof of the covariance of the fundamental equations under the LORENTZ transformation, POINCARÉ stated:

> These equations can be subjected to a remarkable transformation discovered by Lorentz, the significance of which is that it explains why no experimental demonstration of the absolute motion of the universe is possible. If we put

$$x' = kl(x + \varepsilon t), \tag{59a}$$

$$y' = ly, \tag{59b}$$

$$[\S\ 1\text{-}3]$$

$$z' = lz, \tag{59c}$$

$$t' = kl(t + \varepsilon x) \tag{59d}$$

where l and ε are any constants, and

$$k = \frac{1}{\sqrt{(1 - \varepsilon^2)}},$$

and if we also put

$$\square' = \Sigma \frac{d^2}{dx'^2} - \frac{d^2}{dt'^2},$$

then

$$\square' = \square \cdot l^{-2}. \text{ [149]}$$

(From now on references to equations that appear in POINCARÉ's paper will be placed in square brackets [ ] containing the section number as well as the equation number.) Equation (59) is the modern version of the LORENTZ transformation. It is symmetric with respect to the appearance of the space-time coordinates because of POINCARÉ's omission of the Σ system. It should be compared to LORENTZ's version (33) which evolved from his papers of 1892 and 1895. However, it will soon become apparent that this symmetry does not imply the equivalence of S and Σ'. POINCARÉ, like LORENTZ, took the ether-fixed system S as the true one; Σ' was considered to be a fictitious system used to simplify the calculation of fields and forces. There is another basic distinction between (59) and (33) which will lead to certain disconcerting differences in sign between LORENTZ's and POINCARÉ's re-

---

[149] R. 132: O. 498–499: K. 150–151.

sults. Equation (33) becomes, upon substituting for x_r, y_r, z_r in terms of x, y, z,

$$x' = kl(x - vt),$$
$$y' = ly,$$
$$z' = lz, \tag{33'}$$
$$t' = kl(t - vx).$$

It is evident upon comparing (33') and (59) that Poincaré denoted the relative velocity between S and Σ' as $-\varepsilon$ whereas Lorentz denoted it by $+v$. Consequently, reversals of sign in the field transformation and momentum equations are to be expected.

Poincaré then discussed how a moving sphere of radius r becomes an ellipsoid:

Let a sphere be carried along with the electron in a uniform translatory motion, and let the equation of this moving sphere be

$$(x - \xi t)^2 + (y - \eta t)^2 + (z - \zeta t)^2 = r^2; \tag{60}$$

the volume of the sphere is then $\frac{4}{3}\pi r^3$.

The foregoing transformation will change the sphere into an ellipsoid, whose equation is easily found. From the equations (3), it immediately follows that

$$x = \frac{k}{l}(x' - \varepsilon t'), \quad t = \frac{k}{l}(t' - \varepsilon x'), \quad y = \frac{y'}{l}, \quad z = \frac{z'}{l}. \quad [\S 1 - 3']^{150} \tag{61}$$

Substituting (61) into (60) gives the form of the ellipsoid at $t' = 0$ as

$$k^2 x'^2 (1 + \xi \varepsilon)^2 + (y' + \eta k \varepsilon x')^2 + (z' + \zeta k \varepsilon x')^2 = l^2 r^2;$$

therefore its volume is

$$\frac{4}{3} \pi r^3 \frac{l^3}{k(1 + \varepsilon \xi)}. \tag{62}$$

Now, the total charge e on the spherical electron is $\frac{4}{3}\pi r^3 \varrho$, and from (62) it is $\frac{4}{3}\pi \frac{r^3 l^3 \varrho'}{k(1 + \varepsilon \xi)}$ on the deformed electron. However, since charge is conserved,

$$\varrho' = \frac{k}{l^3} [\varrho + \varepsilon (\varrho \xi)] \tag{63}$$

which is the equation of transformation for the charge density.

Poincaré then derived the equations for transformation of velocity in (what is now considered to be) the usual way:

$$\xi' = \frac{dx'}{dt'} = \frac{d(x + \varepsilon t)}{d(t + \varepsilon x)} = \frac{\xi + \varepsilon}{1 + \varepsilon \xi}, \tag{64a}$$

$$\eta' = \frac{dy'}{dt'} = \frac{dy}{kd(t + \varepsilon x)} = \frac{\eta}{k(1 + \varepsilon \xi)}, \tag{64b}$$

$$\zeta' = \frac{dz'}{dt'} = \frac{dz}{kd(t + \varepsilon x)} = \frac{\zeta}{k(1 + \varepsilon \xi)}; \tag{64c}$$

---

150 R. 133: O. 499: K. 151.

therefore

$$\varrho'\xi' = \frac{k}{l^3}\,(\varrho\xi + \varepsilon\varrho)\,,\tag{65a}$$

$$\varrho'\eta' = \frac{1}{l^3}\,\varrho\eta\,,\tag{65b}$$

$$\varrho'\zeta' = \frac{1}{l^3}\,\varrho\zeta\,.\tag{65c}$$

Poincaré's equation (64) should be compared to the Lorentz equation (34c), the latter being incorrect because of Lorentz's attempt to relate S_r and Σ' instead of S and Σ'. Equation (63) reduces to Lorentz's equation (34d) in S', where $\varepsilon = -\xi$.

It is important to take note of the way in which Poincaré wrote (63) and (65). He stated these equations in this way because he noticed that the charge density and convection current transform as do t and x, y, z, respectively. In the course of this paper he systematically searched for other quantities that transform in this manner, because from them one can construct quantities that are invariant under the Lorentz transformation, e.g. the quantity $x^2 + y^2 + z^2 - t^2$. This procedure, which is in essence the four-vector formalism, was formally discovered two years later by another mathematician, Hermann Minkowski[151]. Minkowski's work will be discussed in Part 7.4.

Poincaré now asserted: "Here I must for the first time indicate a disagreement with Lorentz's analysis ..."[152] He then indicated the difference between his transformation equations for ϱ and ξ, η, ζ and those derived by Lorentz. At this point one expects a demonstration of the covariance of the Maxwell-Lorentz equations (using Poincaré's velocity and current transformation equations), and then a comment that ϱ' satisfies the equation of continuity. However, Poincaré, the mathematician, chose another method to prove that ϱ' in (63) is the correct one. This procedure uses the identical transformation propoerties of t, x, y, z and ϱ, $\varrho\xi$, $\varrho\eta$, $\varrho\zeta$:

For, let λ be an undetermined coefficient, and D the Jacobian of

$$t + \lambda\varrho\,, \qquad x + \lambda\varrho\xi\,, \qquad y + \lambda\varrho\eta\,, \qquad z + \lambda\varrho\zeta \qquad\qquad [\S\,2\text{-}5]$$

with respect to t, x, y, z. Then

$$D = D_0 + D_1\lambda + D_2\lambda^2 + D_3\lambda^3 + D_4\lambda^4,$$

with

$$D_0 = 1\,, \qquad D_1 = \frac{d\varrho}{dt} + \Sigma\,\frac{d(\varrho\xi)}{dx} = 0\,.$$

Let $\lambda' = l^2\lambda$; then the four functions

$$t' + \lambda'\varrho'\,, \qquad x' + \lambda'\varrho'\xi'\,, \qquad y' + \lambda'\varrho'\eta'\,, \qquad z' + \lambda'\varrho'\zeta' \qquad\qquad [\S\,2\text{-}5']$$

---

[151] H. Minkowski: (a) "Das Relativitätsprinzip," lecture delivered before the Math. Ges. Göttingen, on 5 Nov. 1907 published in Ann. d. Phys., **47**, 927–938 (1915). (b) "Die Grundgleichungen für die elektromagnetischen Vorgänge in bewegten Körpern," Nachr. Ges. Wiss. Göttingen, 53–111 (1908). (c) "Raum und Zeit," lecture delivered at the Congress of Scientists, Cologne, 21 Sept. 1908 translated in P.R.C., op. cit. (footnote 2), pp. 75–91.

[152] R. 133: O. 500: K. 152.

are related to the functions (5) by the same linear relationships as those which exist between the old and new variables. If, therefore, D' denotes the Jacobian of the functions (5') with respect to the new variables, then

$$D' = D, \qquad D' = D_0' + D_1' \lambda' + \cdots + D_4' \lambda'^4,$$

whence

$$D_0' = D_0 = 1, \qquad D_1' = l^{-2} D_1 = 0$$

$$= \frac{d\varrho'}{dt'} + \Sigma \frac{d(\varrho' \xi')}{dx'}, \qquad \qquad \text{q.e.d.}^{153}$$

(Note: the line above (5') should read "Let $\lambda' = l^4 \lambda \ldots$".) Therefore, if the equation of continuity holds in a frame fixed in the ether, it must also be valid in Σ' because Σ' is related to S by the Lorentz transformation.

Transforming (58a–j) from S to Σ' by means of (59), and assuming the validity of the principle of relativity, Poincaré derived the transformation equations for the potentials, fields and forces in one step instead of two:

$$\Psi' = \frac{k}{l} (\Psi + \varepsilon F),$$

$$F' = \frac{k}{l} (F + \varepsilon \Psi),$$

$$G' = \frac{1}{l} G, \qquad \qquad [\S\, 1\text{-}7] \qquad \qquad (66)$$

$$H' = \frac{1}{l} H,$$

$$f' = \frac{1}{l^2} f, \quad g' = \frac{k}{l^2} (g + \varepsilon \gamma), \quad h' = \frac{k}{l^2} (h - \varepsilon \beta),$$

$$\qquad \qquad \qquad \qquad \qquad \qquad [\S\, 1\text{-}9] \quad (67)$$

$$\alpha' = \frac{1}{l^2} \alpha, \quad \beta' = \frac{k}{l^2} (\beta - \varepsilon h), \quad \gamma' = \frac{k}{l^2} (\gamma + \varepsilon g),$$

$$X' = \frac{k}{l^5} (X + \varepsilon \Sigma X \xi),$$

$$Y' = \frac{1}{l^5} Y, \qquad \qquad [\S\, 1\text{-}11] \qquad \qquad (68)$$

$$Z' = \frac{1}{l^5} Z.$$

(Note that (35) differs from (67) since v is replaced by $-\varepsilon$.) The manner in which (66) and (68) are stated is, once again, indicative of Poincaré's mathematical insight. He noticed that Ψ, F, G, H transform as do t, x, y, z, and the force components X, Y, Z transform as do x, y, z. In Section 9 Poincaré found that if the fourth component of the force is taken as $T = \Sigma X \xi$, then T, X, Y, Z transform as do t, x, y, z.[154] Recognizing that (68) does not agree with Lorentz's results (see (34e–g)), he asserted that: "It occurs evidently, because the formulae for

---

153 R. 134: O. 500–501: K. 152–153.
154 R. 169: O. 543: K. 177. Thus, Poincaré was led to assert that $\Sigma X^2 - T^2$ as well as $\Sigma X x - T t$ are Lorentz invariants.

ξ', η', ζ' are not the same, whereas those for the electric and magnetic fields are the same.''[155]

Having corrected LORENTZ's technical errors, POINCARÉ turned to the main problem:

> *If the inertia of the electrons is of purely electromagnetic origin, and if more-over they are subject only to forces of electromagnetic origin, the condition of equilibrium requires that, within the electrons,*

$$X = Y = Z = 0.$$

From the relations (11), these are clearly equivalent to

$$X' = Y' = Z' = 0.$$

Thus the equilibrium conditions are unaffected by the transformation.

Unfortunately, such a simple hypothesis is inadmissible. For, if we assume that $\xi = \eta = \zeta = 0$, the conditions $X = Y = Z = 0$ will imply that $f = g = h = 0$, and therefore

$$\Sigma \frac{df}{dx} = 0, \quad \text{i.e. } \varrho = 0.$$

Similar results would be obtained in the general case. Hence we must assume that there are not only electromagnetic forces but also either forces or con-straints. We then have to determine the conditions governing these forces or constraints such that the equilibrium of the electrons is unaffected by the transformation. This will be done in a subsequent section.[156]

This is a restatement of the problem described in Part 4.7, *i.e.*, the deformable electron is unstable under purely electromagnetic forces. POINCARÉ's statement

---

[155] R. 136: O. 503: K. 156. In order to compare his result with LORENTZ's, POIN-CARÉ wrote the transformation equations for the force per unit charge $\vec{F} = \vec{E} + \vec{v} \times \vec{B}$ as

$$X'_1 = \frac{k}{l^5} \frac{\varrho}{\varrho'} (X_1 + \varepsilon \Sigma X_1 \xi),$$

$$Y'_1 = \frac{1}{l^5} \frac{\varrho}{\varrho'} Y_1, \qquad [\S 1\text{-}11']$$

$$Z'_1 = \frac{1}{l^5} \frac{\varrho}{\varrho'} Z_1$$

and then rewrote LORENTZ's results (35 e–g) in his notation as

$$X_1 = l^2 X_1 - l^2 \varepsilon (\eta' g' + \zeta' h'),$$

$$Y_1 = \frac{l^2}{k} Y'_1 + \frac{l^2 \varepsilon}{k} \xi' g', \qquad [\S 1\text{-}11'']$$

$$Z_1 = \frac{l^2}{k} Z'_1 + \frac{l^2 \varepsilon}{k} \xi' h'.$$

The difference between [11''] and [11'] is obvious.

[156] R. 136: O. 503: K. 156; italics in original.

of the problem is that if an electron is in equilibrium, then $X = Y = Z = 0$. But if the electron is at rest, then $\vec{F} = \varrho \vec{E} = 0$; therefore, $\vec{E} = 0$ because the existence of a charge density was assumed. This leads to a contradiction because the equation $\vec{V} \cdot \vec{E} = \varrho$ with $\vec{E} = 0$ implies that $\varrho = 0$. This contradictory result can be viewed in another way; namely that the LORENTZ force for an electron at rest is not zero because of the electron's own COULOMB field. Therefore, the deformable electron is not in equilibrium in its rest frame (S') and will tend to explode under the action of the repulsive COULOMB forces between its constituent parts. The problem, therefore, is to find cohesive mechanical forces, which have the proper properties for a LORENTZ transformation. The mathematical method by which POINCARÉ chose to find these forces is the principle of least action.

6.3. The Principle of Least Action [§ 2].[157]

Lorentz's derivation of his equations from the principle of least action is well known. I shall, however, discuss this point further (although I have nothing essential to add to Lorentz's analysis), since I prefer to present it in a slightly different form, which will be of use later.[158]

LORENTZ's "well known" derivation of his equations was really not at all a derivation and, furthermore, was not meant to be one.[159] LORENTZ stated in "Contributions to the Theory of Electrons" (1903) that he merely wanted to illustrate "that the fundamental equations may be transformed in such a way that we arrive at theorems of the same mathematical form as the general principles of mechanics."[160] This was the limit of LORENTZ's efforts towards a mechanical explanation of the fundamental equations. However, this type of formulation must be possible in order to complete the reductionist program of an electromagnetic world-picture. Had LORENTZ tried to formulate his theory in this manner in 1904 he would have noted, and possibly corrected, the defect that was quantitatively pointed out by ABRAHAM in 1905. In the same paper LORENTZ went on to express his displeasure with a mechanical world-picture:

> The physicists who have endeavoured, by means of certain hypotheses on the mechanism of electromagnetic phenomena, to deduce the fundamental equations from the principles of dynamics, have encountered considerable difficulties, and it is best, perhaps, to leave this course, and to adopt the equations [1]—or others, equivalent to them—as the simplest expression we may find for the laws of electromagnetism.[161]

[157] R. 136–142: O. 503–510: K. 156–165. The technical details in this Section will not be presented since they are not directly related to the question of the electron's stability. Furthermore, as we shall see, a detailed exposition would require a digression into fluid mechanics.

[158] R. 136: O. 503–504: K. 156–157.

[159] For LORENTZ's description of his variational principle see (a) the work cited in footnote 20; (b) H. A. LORENTZ, "Weiterbildung der Maxwellschen Theorie. Elektronentheorie," Encykl. math. Wiss. Vol. V2, Art. 14 (Teubner, 1904); (c) H. A. LORENTZ, "Contributions to the Theory of Electrons," Proc. Roy. Acad. Amsterdam, **5**, 608 (1903), in *Collected Papers*, **3**, 132–154.

[160] See 159 (c), p. 136.

[161] *Ibid.*, p. 136.

This is what LORENTZ did in his major papers on the electrodynamics of moving bodies (1895 and 1904).

Briefly, LORENTZ's derivation of the fundamental equations from a variational principle is as follows: starting from the "kinetic energy" of the field

$$T = \tfrac{1}{2} \int B^2 d\tau$$

and using as input three of the MAXWELL-LORENTZ equations

$$\vec{V} \cdot \vec{B} = 0, \qquad\qquad\qquad (69\,\text{a})$$

$$\vec{V} \times \vec{B} = \frac{\partial \vec{E}}{\partial t} + \varrho \vec{v}, \qquad\qquad (69\,\text{b})$$

$$\vec{V} \cdot \vec{E} = \varrho \qquad\qquad\qquad (69\,\text{c})$$

and charge conservation[162], LORENTZ could derive, by suitably varying the kinetic energy, the fourth MAXWELL equation

$$\vec{V} \times \vec{E} = -\frac{\partial \vec{B}}{\partial t}$$

and the LORENTZ force equation

$$\vec{F} = \varrho \vec{E} + \varrho \vec{v} \times \vec{B}.$$

In addition, he was able to demonstrate that the electromagnetic field formalism could be put into a form resembling D'ALEMBERT's principle.[163]

The only way that all of the MAXWELL-LORENTZ equations can be derived from a variational principle is to take for the LAGRANGIAN

$$L = \int \varrho \vec{A} \cdot \vec{v} d\tau - \int \varrho \phi d\tau + \tfrac{1}{2} \int (E^2 - B^2) d\tau. \qquad (70)$$

This had already been done by K. SCHWARZSCHILD in 1903.[164] However, this type of LAGRANGIAN is not consonant with a research program whose goal is to construct an electron solely from the LAGRANGIAN for the electromagnetic field.

POINCARÉ, like LORENTZ felt that "To demonstrate the possibility of a mechanical explanation of electricity we need not trouble to find the explanation itself; we need only know the expression of two functions T and U, which are the two parts of energy, and to form with these two functions Lagrange's equations, and then to compare these equations with the experimental laws."[165]

---

[162] LORENTZ continually emphasized that the electron's charge is conserved, *i.e.* that the divergence of the total current vanishes (see footnote 40): "The variations to be considered here are not wholly arbitrary. We shall limit our choice by supposing in the first place that each element of volume of an electron preserves its charge during the displacements ..." (*op. cit.* footnote 159 (c), p. 137). However, as we have seen, initially he was willing to gloss over this important point in his paper of 1904.

[163] *Op. cit.* footnote 159 (c), pp. 136–147.

[164] K. SCHWARZSCHILD, "Zur Elektrodynamik; I," Nachr. Ges. Wiss. Göttingen, 125–131 (1903); "II," *Ibid.*, 132–141 (1903); "III," *Ibid.*, 245–278 (1903). In particular, see "I" for a detailed exposition of his variational principle. SCHWARZSCHILD's LAGRANGIAN was first written in covariant form by M. BORN in "Die träge Masse und das Relativitätsprinzip," Ann. d. Phys., **28**, 571–584 (1909).

[165] H. POINCARÉ, *S.H.*, *op. cit.* (footnote 62), p. 223.

Poincaré knew what T and U were but preferred initially to write the action in a slightly different form:[166]

$$J = \int dt\, d\tau \left[\tfrac{1}{2}\Sigma f^2 + \tfrac{1}{2}\Sigma\alpha^2 - \Sigma F u\right] \qquad [\S\ 2\text{-}1]. \qquad (71)$$

The input equations to his variational principle are

$$\Sigma \frac{\partial f}{\partial x} = \varrho,$$

$$\alpha = \frac{\partial H}{\partial y} - \frac{\partial G}{\partial z}, \qquad (72)$$

$$u = \frac{\partial f}{\partial t} + \varrho\xi.$$

Then, by varying J, he showed that

$$u = \frac{\partial\gamma}{\partial y} - \frac{\partial\beta}{\partial z}. \qquad (73)$$

Substitution of this into (71) yields

$$J = \int dt\, d\tau \left[\tfrac{1}{2}\Sigma f^2 - \tfrac{1}{2}\Sigma\alpha^2\right] \qquad (74)$$

where Poincaré preferred to write the Lagrangian of the electromagnetic field as $U - T$ instead of $T - U$. Actually nothing new appears so far because the total current is precisely the curl of B. However, by means of (71), which is equivalent to (74), Poincaré had proven that J is independent "of δF and therefore of $\delta\alpha$."[167] Returning to (71), but now constraining the variation by adding to it the first of equations (72) multiplied by a Lagrange multiplier Ψ, he obtained

$$\delta J = \int dt\, d\tau\, \Sigma\delta f \left(f + \frac{\partial F}{\partial t} + \frac{\partial\Psi}{\partial x}\right) + \int dt\, d\tau \left(\Psi\delta\varrho - \Sigma F\delta(\varrho\xi)\right). \qquad (75)$$

(F and α are not varied since δJ is independent of δF and $\delta\alpha$.) "If now it be assumed that the electrons undergo no variation, then $\delta\varrho = \delta(\varrho\xi) = 0$ and the

---

[166] The action J is defined as

$$J = \int L\, dt,$$

where definite limits are assumed for t. For a discussion of the fixed endpoint variational principle used by Poincaré see H. Goldstein, *Classical Mechanics* (Addison-Wesley, 1965), esp. pp. 30–55.

[167] R. 138: O. 505: K. 159. By means of varying F, G, H and using the second of equations (72) Poincaré shows that

$$\delta J = -\int dt\, d\tau\, \Sigma\delta F \left(u - \frac{\partial\gamma}{\partial y} + \frac{\partial\beta}{\partial z}\right) = 0$$

(see R. 138: O. 505: K. 158). Since F, G, H are independent field coordinates, each of the coefficients of $\delta F, \delta G, \delta H$ must vanish. Consequently,

$$u = \frac{\partial\gamma}{\partial y} - \frac{\partial\beta}{\partial z}, \text{etc.}$$

and δJ is independent of δF and $\delta\alpha$ (from the second of equations (72)).

second integral vanishes. For δJ to be zero, we must have ..."[168]

$$f = -\frac{\partial F}{\partial t} - \frac{\partial \Psi}{\partial x} \tag{76}$$

which is (58c)—note that the LAGRANGE multiplier is the scalar potential. This completes the "derivation" of the MAXWELL-LORENTZ equations from a variational principle.[169]

The last term on the right hand side of (75) contains the LORENTZ force equation. This term is, therefore, equal to the work done on the electron by the force F. If U, V, W denote the displacements of a volume element $d\tau$ of the electron, the virtual work done by F is

$$-\int \Sigma X \delta U d\tau \tag{77}$$

and consequently

$$\delta J = -\int \Sigma X \delta U d\tau dt \tag{78}$$

where from (75)

$$\delta J = \int dt d\tau \left(\Psi \delta \varrho - \Sigma F \delta (\varrho \xi) \right). \tag{79}$$

The remainder of Section 2 is concerned with transforming $\Psi \delta \varrho$ and $\Sigma F \delta(\varrho \xi)$ in (79) into terms proportional to δU, δV, δW. The LORENTZ force can be identified by equating this result to (78).

Although the details of this derivation are not germane to the problem of stability, the concept employed is of interest. POINCARÉ utilized a method that he had used successfully in previous work on electromagnetic theory — the hydrodynamic analogy.[170] To transform (79) into (78), POINCARÉ derived two different forms of the equation of continuity. He considered the electron to behave like a compressible fluid. If a point on the electron moves from x_0, y_0, z_0 to x, y, z, the relation between these two points is

$$\begin{aligned} x &= x_0 + U, \\ y &= y_0 + V, \\ z &= z_0 + W. \end{aligned} \tag{80}$$

---

[168] R. 138–139: O. 506: K. 160.

[169] Equation (76) written in vector form is

$$\vec{E} = -\frac{\partial \vec{A}}{\partial t} - \vec{V}\psi.$$

Taking the curl of both sides gives

$$\vec{V} \times \vec{E} = -\frac{\partial}{\partial t}(\vec{V} \times \vec{A}),$$

but

$$\vec{V} \times \vec{A} = \vec{B};$$

therefore

$$\vec{V} \times \vec{E} = -\frac{\partial \vec{B}}{\partial t}.$$

[170] Cf. H. POINCARÉ, "À propos ..." op. cit. (footnote 81), pp. 414–416. See also H. POINCARÉ, "La théorie de Lorentz ..." op. cit. (footnote 26).

The total time rate of change of the JACOBIAN Δ $(\Delta = \partial(x, y, z)/\partial(x_0, y_0, z_0))$ relating the volume elements of the electron at these two points is

$$\frac{1}{\Delta} \frac{\delta \Delta}{\delta t} = \vec{V} \cdot \vec{v} \tag{81}$$

(where δ is used instead of POINCARÉ's notation for the convective derivative which is the usual partial derivative sign ∂ and $\vec{v}$ is the velocity of a small piece of fluid (electron)). In fluid dynamics Δ is interpreted as the ratio of the volume occupied by a small piece of fluid at time t to its volume at time $t = 0$. Further indication that POINCARÉ had a fluid dynamical interpretation in mind is this very interesting slip: "Since the mass of the electron is constant"[171]

$$\frac{\delta(\varrho \Delta)}{\delta t} = 0. \tag{82}$$

POINCARÉ meant charge and not mass even though this equation also represents conservation of mass for a fluid.[172] The equation for conservation of electric charge (58e) is then derived from (81) and (82). The second version of (58e) is obtained from the usual variational process wherein the path is varied but the endpoints remain fixed[173]. The result is

$$\delta \varrho + \Sigma \frac{\partial}{\partial x} (\varrho \delta U) = 0. \tag{83}$$

By means of (58) and (83) POINCARÉ derived the LORENTZ force equation (58j).[174]

Up to this point POINCARÉ gave merely an elegant exposition of LORENTZ's variational procedure. However, in the following sections, he demonstrated the importance, from the point of view of the principle of relativity, of expressing electromagnetic field theory in a LAGRANGIAN formulation; an importance that transcends the statement that if a T and a U can be found for the electromagnetic field, then mechanical explanations are possible. The usefulness of the LAGRANGIAN formulation had been demonstrated by ABRAHAM, but its full meaning and the symmetry it contained was first uncovered by POINCARÉ using the mathematical statement of the principle of relativity, the LORENTZ transformation.

---

[171] R. 140: O. 508: K. 162.

[172] Cf. J. SERRIN, "Mathematical Principles of Classical Fluid Mechanics," Handbuch der Physik: Volume VIII, Part 1 (Springer, 1959).

[173] Cf. H. GOLDSTEIN, Classical Mechanics, op. cit. (footnote 166), esp. pp. 31–33.

[174] Equations (53e) and (83) are used to derive

$$\int dt\, d\tau\, \Psi\, \delta \varrho = \int dt\, d\tau\, \Sigma \varrho\, \frac{\partial \Psi}{\partial x}\, \delta U$$

$$- \int dt\, d\tau\, \Sigma F \delta(\varrho \xi) = \int dt\, d\tau\, \Sigma \varrho\, \delta U \left(\frac{\partial F}{\partial t} + \beta \zeta - \gamma \eta \right).$$

The LORENTZ force is obtained by equating the sum of these two equations to (78):

$$X = \varrho f - \varrho (\beta \zeta - \gamma \eta).$$

There is a notational error on R. 142: O. 510: K. 165—wherein the charge density ϱ was omitted from this equation.

17*

6.4. The Lorentz Transformation and the Principle of Least Action [§ 3].[175]

Let us consider whether the principle of least action can explain the success of the Lorentz transformation.[176]

Beginning with this statement Poincaré demonstrated that the form of the action for the electromagnetic field (74) is not fortuitous. Using the Lorentz transformation (59) and the field transformation equations (67), he proved that

$$\int dt\, d\tau\,(\tfrac{1}{2}\Sigma f^2 - \tfrac{1}{2}\Sigma \alpha^2) = \int dt'\, d\tau'\,(\tfrac{1}{2}\Sigma f'^2 - \tfrac{1}{2}\Sigma \alpha'^2) \tag{84}$$

or

$$J = J', \tag{85}$$

which means that the principle of least action is invariant under the Lorentz transformation.[177]

Since δJ can also be written as

$$\delta J = -\int \Sigma X\, \delta U\, dt\, d\tau \tag{86}$$

(similarly $\delta J' = -\int \Sigma X'\, \delta U'\, dt'\, d\tau'$) and (85) implies that $\delta J = \delta J'$, the problem remains to determine the transformation properties of X, Y, Z such that the equality (84) is maintained by (86). Using (59) and the equations for the transformation of velocity (64), Poincaré determined how $\delta U, \delta V, \delta W$[178] are related to $\delta U', \delta V', \delta W'$. From this relation he found the necessary condition on X, Y, Z

---

[175] R. 142–144: O. 510–513: K. 166–169.
[176] R. 142: O. 510: K. 166.
[177] This proof depends on the important equation

$$l^4 dt\, d\tau = dt'\, d\tau'$$

which implies that in Lorentz's theory, where $l = 1$, $dt\, d\tau$ is an invariant volume element. This equation can be derived by straightforward application of the Lorentz transformation (59):

$$d\tau'\, dt' = \frac{\partial(x', y', z', t')}{\partial(x, y, z, t)}\, d\tau\, dt.$$

Since the Jacobian is

$$\frac{\partial(x', y', z', t')}{\partial(x, y, z, t)} = \begin{vmatrix} kl & 0 & 0 & kl\varepsilon \\ 0 & l & 0 & 0 \\ 0 & 0 & l & 0 \\ kl\varepsilon & 0 & 0 & kl \end{vmatrix} = l^4,$$

then

$$dt'\, d\tau' = l^4 dt\, d\tau.$$

[178] Using the Lorentz transformation to relate the displacement equations for the electron in S and Σ', Poincaré showed that

$$\left.\begin{aligned} \delta U &= \frac{k}{l}\,\delta U' + \frac{k}{l}\,\varepsilon\xi\,\delta U' \\[2mm] \delta V &= \frac{k}{l}\,\delta V' + \frac{k}{l}\,\varepsilon\eta\,\delta U' \\[2mm] \delta W &= \frac{k}{l}\,\delta W' + \frac{k}{l}\,\varepsilon\zeta\,\delta U' \end{aligned}\right\} \quad \text{(R. 144: O. 512: K. 168)}$$

to maintain $\delta J = \delta J'$ to be (68). The transformation equation for the force can therefore be derived from the principle of least action: "Thus the principle of least action leads to the same results as does the analysis given in Section 1."[179] Therefore, it is the LORENTZ invariance of the principle of least action which accounts for the reason why the fundamental equations have the same form in S and Σ'.[180] Even if POINCARÉ had fully appreciated the power of the principle of least action he would, nevertheless, have failed to exploit it fully because of his incorrect use of LORENTZ'S transformations.

POINCARÉ concludes by noting that although $\Sigma f'^2 - \Sigma \alpha'^2$ is invariant under the LORENTZ transformation, the sum of these two terms (which is proportional to the energy density) is not.[181]

It is important to note that POINCARÉ's proof of the LORENTZ invariance of the principle of least action is valid for any value of l in (59). From there POINCARÉ went on to deduce the value of l for LORENTZ's theory.

6.5. The Lorentz Group [§ 4].[182]

It is noteworthy that the Lorentz transformations form a group. For, if we put

$$x' = kl(x + \varepsilon t), \qquad y' = ly, \qquad z' = lz, \qquad t' = kl(t + \varepsilon x),$$

and

$$x'' = k'l'(x' + \varepsilon' t'), \qquad y'' = l'y', \qquad z'' = l'z', \qquad t'' = k'l'(t' + \varepsilon' x'),$$

with

$$k^{-2} = 1 - \varepsilon^2, \qquad k'^{-2} = 1 - \varepsilon'^2,$$

we find that

$$x'' = k''l''(x + \varepsilon'' t), \qquad y'' = l''y,$$
$$z'' = l''z, \qquad t'' = k''l''(t + \varepsilon'' x),$$

(where ξ, η, ζ, are the components of the velocity of the electron's center of mass). By use of these equations it is easy to prove that

$$\Sigma X \delta U = \frac{1}{l} [k X \delta U' + Y \delta V' + Z \delta W'] + \frac{k \varepsilon}{l} \delta U' \Sigma X \xi,$$

but in order to maintain

$$\int \Sigma X' \delta U' dt' d\tau' = \int \Sigma X \delta U dt d\tau$$

it is necessary to assume that the force components X, Y, Z transform according to (68).

[179] R. 144: O. 513: K. 168.

[180] This point was emphasized by H. A. LORENTZ in the notes and commentaries which he wrote on pp. 683–701 of the Œuvres (published as H. A. LORENTZ. "Deux mémoires d'Henri Poincaré sur la physique mathématique," Acta mathématica, **38**, 293 (1914), in Collected Papers, **8**, 258–273): "At first, he is not satisfied to see that the relativistic transformations leave unchanged the form of the equations of electromagnetism. He explains the success of the substitutions by remarking that these equations can be included in a form of the principle of least action and that the fundamental equation which expresses the principle as well as the operations by which one deduces the field equations are the same in the systems x, y, z, t and x', y', z', t'."

[181] The proof of this statement is a straightforward exercise in the use of (67).

[182] R. 144–146: O. 513–515: K. 169–172.

with

$$\varepsilon'' = \frac{\varepsilon + \varepsilon'}{1 + \varepsilon \varepsilon'}, \qquad l'' = l l', \qquad k'' = k k' (1 + \varepsilon \varepsilon') = \frac{1}{\sqrt{(1 - \varepsilon''^{2})}}.^{183}$$

Thus, the product of two LORENTZ transformations is also a LORENTZ transformation. POINCARÉ then demonstrated that the set of LORENTZ transformations $A_1, A_2, \ldots A_n$ possesses the following properties:

(a) $A_1(A_2 A_3) = (A_1 A_2) A_3$;
(b) there exists an identity transformation I such that $I A_1 = A_1$;
(c) there exists an inverse transformation.

Any set possessing these three properties and the one demonstrated at the beginning of this section is a group. In particular, POINCARÉ showed that the LORENTZ group is a one-parameter continuous group.[184]

---

[183] R. 144–145: O. 513: K. 169. (The inhomogeneous LORENTZ group, *i.e.* the group of rotations and displacements in four-dimensional space, is referred to as the POINCARÉ group.) This property of the LORENTZ transformation is now part of the repertoire of most undergraduate physics majors. The crucial point is that if one considers three coordinate systems S, S', S'', where S' and S'' move along a common axis with uniform speeds ε with respect to S and ε' with respect to S', respectively, then the two LORENTZ transformations from S to S' and from S' to S'' can be replaced by a single LORENTZ transformation from S to S'' with the relative speed of S'' with respect to S of

$$\varepsilon'' = \frac{\varepsilon + \varepsilon'}{1 + \varepsilon \varepsilon'}.$$

For EINSTEIN's proof that (59) form a group see the reference in footnote 2, p. 51.

[184] Briefly, he does this in the following manner (for a detailed, relativistically oriented exposition of continuous groups, see H. M. SCHWARTZ, *Introduction to Special Relativity* (McGraw-Hill, 1968) esp. pp. 67–79). The LORENTZ transformation (59) can be written as

$$\begin{pmatrix} x' \\ y' \\ z' \\ t' \end{pmatrix} = \begin{pmatrix} kl & 0 & 0 & kl\varepsilon \\ 0 & l & 0 & 0 \\ 0 & 0 & l & 0 \\ kl\varepsilon & 0 & 0 & kl \end{pmatrix} \begin{pmatrix} x \\ y \\ z \\ t \end{pmatrix} \qquad (184\text{-}1)$$

or $\boldsymbol{R'} = \boldsymbol{T R}$ (where $\boldsymbol{R'}, \boldsymbol{T}, \boldsymbol{R}$ are matrices). Consider the case in which ε is infinitesimal. Then $l = k = 1$ and

$$x' = x + \delta x, \qquad y' = y + \delta y, \qquad z' = z + \delta z, \qquad t' = t + \delta t \qquad (184\text{-}2)$$

where

$$\delta x = \varepsilon t, \qquad \delta y = 0, \qquad \delta z = 0, \qquad \delta t = \varepsilon x. \qquad (184\text{-}3)$$

Consequently, from (184-1), the relation between $\boldsymbol{R'}$ and $\boldsymbol{R}$ for an infinitesimal continuous parameter ε is

$$\boldsymbol{R'} = (1 + \boldsymbol{T_1}) \boldsymbol{R} \qquad (184\text{-}4)$$

where

$$\boldsymbol{T_1} = \begin{pmatrix} 0 & 0 & 0 & \varepsilon \\ 0 & 0 & 0 & 0 \\ 0 & 0 & 0 & 0 \\ \varepsilon & 0 & 0 & 0 \end{pmatrix}. \qquad (184\text{-}5)$$

$\boldsymbol{T_1}$ is interpreted as the generator for an infinitesimal LORENTZ transformation along the x-axis. Similarly, the generators for infinitesimal LORENTZ transformations along

Consider the S and Σ' systems to be rotated by $180°$ about their y-axes. This transformation must also belong to the LORENTZ group and can be written as:

$$x' = kl(x - \varepsilon t), \quad y' = ly, \quad z' = lz, \quad t' = kl(t - \varepsilon x) \quad [\S\,4\text{-}2].$$

The functional dependence of l on ε has been assumed to be one in which l is unchanged when ε is replaced by $-\varepsilon$ in order that $[\S\,4\text{-}2]$ belong to the LORENTZ group. The transformation inverse to (63) is[185]

$$x' = \frac{k}{l}(x - \varepsilon t), \quad y' = \frac{y}{l}, \quad z' = \frac{z}{l}, \quad t' = \frac{k}{l}(t - \varepsilon x) \quad [\S\,4\text{-}3].$$

the y and z axes are

$$T_2 = \begin{pmatrix} 0 & 0 & 0 & 0 \\ 0 & 0 & 0 & \eta \\ 0 & 0 & 0 & 0 \\ 0 & \eta & 0 & 0 \end{pmatrix}, \quad T_3 = \begin{pmatrix} 0 & 0 & 0 & 0 \\ 0 & 0 & 0 & 0 \\ 0 & 0 & 0 & \zeta \\ 0 & 0 & \zeta & 0 \end{pmatrix},$$

respectively (where η and ζ are the velocities in these directions). Then by forming what is now known as the commutator of T_1 and T_2 it can be shown easily that

$$[T_1, T_2] = T_1 T_2 - T_2 T_1 = \begin{pmatrix} 0 & \varepsilon\eta & 0 & 0 \\ -\varepsilon\eta & 0 & 0 & 0 \\ 0 & 0 & 0 & 0 \\ 0 & 0 & 0 & 0 \end{pmatrix}$$

which is the operator that generates an infinitesimal rotation through an angle $\varepsilon\eta$ about the z axis. Analogous results can be obtained from $[T_1, T_3]$ and $[T_2, T_3]$. The operator corresponding to a null velocity of transformation, *i.e.* $\varepsilon = 0$ and consequently $l = 1 + \delta l$, is easily shown from (184-1) to be a multiple of the unit matrix (designated by POINCARÉ as T_0; see R. 145: O. 514: K. 170). This (roughly) completes the proof that the LORENTZ group is a one-parameter continuous group which contains the following infinitesimal transformations that leave the MAXWELL-LORENTZ equations unchanged:

"(1) the transformation T_0 which commutes with every other (it is a multiple of the identity transformation)

(2) the three transformations T_1, T_2, T_3

(3) the three rotations $[T_1, T_2], [T_2, T_3], [T_1, T_3]$"

(R. 146: O. 514: K. 171). In particular, POINCARÉ noted that any transformation belonging to this group can be resolved into the succession of a scale transformation

$$x' = lx, \quad y' = ly, \quad z' = lz, \quad t' = lt$$

and a linear transformation that transforms the quadratic form

$$x^2 + y^2 + z^2 - t^2 \tag{184-6}$$

into itself, *i.e.* (59) with $l = 1$. It is necessary to separate out the factor of l since (59) with $l \neq 1$ would not leave the above quadratic form unchanged. This basic importance of (184-6), which to POINCARÉ was merely a quadratic form, was noted one year later in 1907 by MINKOWSKI (see (a) of footnote 151).

[185] The transformation

$$x' = \frac{k}{l}(x - \varepsilon t), \quad y' = \frac{y}{l}, \quad z' = \frac{z}{l}, \quad t' = \frac{k}{l}(t - \varepsilon x)$$

can be written as

$$R' = T'R$$

Since [§ 4-3] as well as [§ 4-2] belong to the LORENTZ group, it is necessary that $l=1$.[186] Thus, the transformations (59) can be inverted by simply changing v to $-v$. For POINCARÉ, however, this important symmetry was a mathematical one: it had no physical interpretation.

6.6. Langevin's Waves [§ 5].[187]

M. Langevin has put forth a particularly elegant formulation of the formulas which define the electromagnetic field produced by the motion of a single electron.[188]

In this section POINCARÉ showed how the results of LANGEVIN's electromagnetic field formalism,[189] particularly the acceleration and velocity waves, can be obtained using the LORENTZ transformation equations for the electric and magnetic fields and potentials. However, POINCARÉ assumed that the reader was familiar with LANGEVIN's work and, therefore, gave little background material.[190] Since the details of LANGEVIN's work are of peripheral interest to the problem of the electron's stability, Section 5 will be discussed only briefly.

LANGEVIN demonstrated, utilizing the LIÉNARD-WIECHERT potentials, that the electric and magnetic fields of a moving point charge (a charge whose dimensions are negligible compared to all other distances entering into the problem) are composed of two parts: the velocity wave, dependent on the electron's velocity; the acceleration wave, dependent on the electron's acceleration.[191]

where

$$T' = \begin{pmatrix} \dfrac{k}{l} & 0 & 0 & -\dfrac{k\,\varepsilon}{l} \\ 0 & \dfrac{1}{l} & 0 & 0 \\ 0 & 0 & \dfrac{1}{l} & 0 \\ -\dfrac{k\,\varepsilon}{l} & 0 & 0 & \dfrac{k}{l} \end{pmatrix}.$$

Since $T\,T'=1$, where T is from (184-1), it follows that $T' = T^{-1}$.

[186] POINCARÉ's proof that $l=1$, although more elegant than LORENTZ's is, nevertheless, based on the same assumption, the inequivalence of S and Σ'. This is manifested in an interpretation of the LORENTZ transformation equations which is based solely on mathematics, i.e. group theory. This should be compared with EINSTEIN's proof that $l=1$, a proof based on the equivalence of S and Σ' (see the reference in footnote 2, pp. 46–48).

[187] R. 146–151: O. 515–521. A limited number of copies of the author's translation of Section 5 are available and can be obtained upon request.

[188] R. 146: O. 515.

[189] P. LANGEVIN, "Sur l'origine des radiations et l'inertie électromagnétique," J. de Physique, 4, 165 (1905); reprinted in Œuvres Scientifiques de Paul Langevin (CNRS, 1950) pp. 313–348.

[190] In fact, POINCARÉ concluded this section by referring the reader to LANGEVIN's paper for more detail: "Je me bornerai à renvoyer pour plus détails au Mémoire de M. Langevin dans le Journal de Physique (Année 1905)" (R. 151: O. 521).

[191] For a modern presentation of this calculation see J. D. JACKSON, Classical Electromagnetism (Wiley, 1962), pp. 464–467.

The energy due to the velocity wave depends on the charge distribution: *i. e.* the electromagnetic mass is independent of the acceleration, is concentrated in the neighborhood of the electron and is carried along with it.

The energy due to the acceleration wave is independent of the charge distribution, as well as the position of the observation point, and is directly proportional to the square of the electron's acceleration. Consequently, it is the energy radiated by the electron.

In addition to these two waves there is a third one that results from interference between the velocity and acceleration waves.[192] LANGEVIN defined this as the wave of reorganization (POINCARÉ did not discuss this wave).[193] It has the following properties: it is dependent on the charge distribution; it is directly proportional to the acceleration; it is interpreted as changing the field in the vicinity of the electron when the electron's velocity changes.

POINCARÉ derived the equation for the velocity wave by LORENTZ transformation for the electron from a system of instantaneous rest (S') to a set of axes fixed in the ether (S). He then proved that the properties of the acceleration wave are LORENTZ invariant.[194]

---

[192] LANGEVIN showed that the electric and magnetic fields due to a moving charge can be written as

$$\vec{E} = \vec{E_1} + \vec{E_2}, \qquad \vec{B} = \vec{B_1} + \vec{B_2}$$

where $(\vec{E_1}, \vec{B_1})$ and $(\vec{E_2}, \vec{B_2})$ are the velocity and acceleration waves. The electromagnetic field energy is

$$W = \tfrac{1}{2} \int (E^2 + B^2)\, d\tau = \tfrac{1}{2} \int [(\vec{E_1} + \vec{E_2})^2 + (\vec{B_1} + \vec{B_2})^2]\, d\tau$$

$$= \tfrac{1}{2} \int (E_1^2 + B_1^2)\, d\tau + \tfrac{1}{2} \int (E_2^2 + B_2^2)\, d\tau + \int (\vec{E_1} \cdot \vec{E_2} + \vec{B_1} \cdot \vec{B_2})\, d\tau.$$

The first two terms are the energies due to the velocity and acceleration waves, while the third term is the energy due to the wave of reorganization.

[193] By considering the acceleration wave and the wave of reorganization further light can be thrown on the assumption of quasi-stationary motion (see footnote 28). The work done by the radiation reaction force acting on the electron is equal to the energy arising from these two waves. Therefore, the work done by the radiation reaction force during a displacement δx of the electron is

$$X \delta x = f_1(v)\, a\, \delta x + f_2(v)\, a^2\, \delta x \tag{193-1}$$

where $f_1(v)\, a$ and $f_2(v)\, a^2$ are the hypothetical forces which give rise to waves of reorganization and acceleration respectively. The work done by $\vec{F}_{\text{self}}$ (see (28-1) in footnote 28) for an infinitesimal displacement εdt of the electron in the x-direction has the form of (193-1) if terms proportional to $\ddot{a}, \dddot{a} \ldots$ are neglected. If NEWTON's second law is to be maintained, then the second term in (193-1) must be neglected. Consequently, quasi-stationary motion corresponds to the neglect of the electron's radiation field. It corresponds to almost uniform motion (this argument is due to O. W. RICHARDSON, *The Electron Theory of Matter* (Cambridge, 1916) pp. 260–261.)

[194] "1. The two fields, the electric and magnetic are equal

2. they are mutually perpendicular

3. they are perpendicular to the normal to the wave sphere ..."

(R. 149: O. 519). The second property of the acceleration wave follows from one of the two possible LORENTZ invariants that can be formed from $\vec{E}$ and $\vec{B}$, namely $\vec{E} \cdot \vec{B}$. The second LORENTZ invariant is $E^2 - B^2$ (R. 144: O. 513: K. 169). POINCARÉ proved the LORENTZ invariance of the other two properties of the acceleration wave by making use of these two electromagnetic field invariants.

The next three sections represent the core of "Sur la dynamique ...". They contain the derivation of the mechanism which POINCARÉ found necessary in order to construct a LAGRANGIAN formulation of LORENTZ's theory and, simultaneously, to solve the stability problem raised in Section 1. In addition, they illustrate clearly POINCARÉ's (and LORENTZ's) basic misconceptions concerning the fundamental nature of the LORENTZ transformation and the principle of relativity.

6.7. Contraction of Electrons [§ 6].[195]

Let us consider a single electron put into uniform rectilinear motion. From what we have just seen one can, with the aid of the Lorentz transformation, transfer the study of the field caused by that electron to the case where the electron is at rest; the Lorentz transformation replaces the real moving electron by an imaginary immobile electron.[196]

The final phrase: "... the Lorentz transformation replaces the real moving electron by an imaginary immobile electron," provides evidence for POINCARÉ's purely formal interpretation of the LORENTZ transformation. The imaginary immobile electron is the one in the S' system.

Let α, β, γ; f, g, h be the real field; let α', β', γ'; f', g', h' that which is derived from the Lorentz transformation, so that the imaginary field α', f' corresponds to the case of an electron at rest. We have

$$\alpha' = \beta' = \gamma' = 0, \quad f' = -\frac{\partial \Psi'}{\partial x'}, \quad g' = -\frac{\partial \Psi'}{\partial y'}, \quad h' = -\frac{\partial \Psi'}{\partial z'}$$

and for the real field (from equations 9 of Section 1 [see (67)]):

$$(1) \quad \begin{cases} \alpha = 0, & \beta = \varepsilon h, & \gamma = -\varepsilon g \\ f = l^2 f', & g = k l^2 g', & h = k l^2 h'.^{197} \end{cases}$$

Thus, for POINCARÉ, as for LORENTZ, the instantaneous rest system S' is a fictitious system whose existence is solely to facilitate the calculation of the fields due to the real moving electron. Since, in S', the electron is at rest, only the scalar potential is not zero. Therefore, the components of the electric field in S' are

$$f' = -\frac{\partial \Psi'}{\partial x'}, \quad g' = -\frac{\partial \Psi'}{\partial y'}, \quad h' = -\frac{\partial \Psi'}{\partial z'} \tag{87}$$

(note that here ∂ means partial differentiation). The electric and magnetic fields in the ether fixed system S can then be calculated from (67).

It is now a matter of determining the total energy attributed to the electron's motion, the corresponding action and the electromagnetic field momentum, in order to be able to calculate the electromagnetic mass of the electron. For a distant point, it suffices to consider the electron as reduced to a single point; one can then go back to the equations in (4) of the preceding section which are of general use. But here they do not suffice, because the energy is

---

[195] R. 151–158: O. 521–529. The translation of Sections 6–8 is by the author.
[196] R. 151: O. 521–522.
[197] R. 151–152: O. 522.

principally localized in the parts of the ether in the immediate neighborhood of the electron.[198]

Once the fields are known, the electron's total energy, the LAGRANGIAN (and consequently J) and momentum can be determined. The electromagnetic mass can then be calculated from the field momentum. The LIÉNARD-WIECHERT potentials are insufficient for this calculation because they take into account only observation points far enough from the electron that it appears as a point charge. However, the portion of the electron's energy in its immediate neighborhood now deserves study. This energy, arising from the velocity wave, is assumed to be the source of the electron's mass.

We can make many hypotheses on this subject.

According to Abraham's hypothesis the electrons are rigid spheres. Then, upon applying the Lorentz transformation, since the real electron is spherical, the imaginary electron will become an ellipsoid. From Section 1 the equation of that ellipsoid would be:

$$k^2(x' - \varepsilon t' - \xi t' + \varepsilon \xi x')^2 + (y' - \eta kt' + \eta k \varepsilon x')^2 + (z' - \zeta kt' + \zeta k \varepsilon x')^2 = l^2 r^2.$$

But here we have

$$\xi + \varepsilon = \eta = \zeta = 0, \qquad 1 + \varepsilon \xi = 1 - \varepsilon^2 = \frac{1}{k^2}$$

so that the equation of the ellipsoid becomes:

$$\frac{x'^2}{k^2} + y'^2 + z'^2 = l^2 r^2.$$

If the radius of the real electron is r, then the axes of the imaginary electron would be:

$$klr, lr, lr.$$

On the contrary, according to Lorentz's hypothesis the electron in motion would be deformed, in such a manner that the real electron will become an ellipsoid, whereas the imaginary immobile electron will always be a sphere of radius r; the axes of the real electron will then be:

$$\frac{r}{kl}, \frac{r}{l}, \frac{r}{l}.\text{[199]}$$

It is evident from this passage that the LORENTZ transformation is being interpreted as a mathematical coordinate transformation. As described in Part 3.2, the coordinate transformation employed in the solution of certain potential problems can be interpreted as a transformation to a fictitious system S' wherein the real spherical electron is replaced by an imaginary one that is dilated in the direction of motion. This is the interpretation that LORENTZ used for the equations of transformation according to relativity theory. Similarly, POINCARÉ interpreted $x' = kl(x + \varepsilon t)$ as producing a dilation in the x' direction and showed that the equation for the electron's shape in S' is

$$\frac{x'^2}{k^2} + y'^2 + z'^2 = l^2 r^2. \tag{88}$$

---

[198] R. 152: O. 522.
[199] R. 152: O. 522–523.

Since the axes of the ellipsoid in (88) are

$$klr, lr, lr \qquad (89)$$

the ellipsoid is dilated in the direction of motion. This constitutes ABRAHAM's theory. However, LORENTZ postulated a contraction in the direction of motion since the theorem of corresponding states asserts that the imaginary electron should also be a sphere. Therefore, the axes of the real electron are

$$\frac{r}{kl}, \frac{r}{l}, \frac{r}{l}. \qquad (90)$$

Thus, to LORENTZ *and* POINCARÉ *the contraction hypothesis is independent of the hypothesis of the equations for the* LORENTZ *transformation.* We have seen that this apparent lack of generality of the LORENTZ transformation can be traced back to its purely mathematical description. It was EINSTEIN, in 1905, who discovered the meaning of the LORENTZ transformation by deriving it from the two postulates of his theory.[200] He then demonstrated that one of its consequences is the purely kinematical phenomenon of the contraction of moving objects when viewed from an inertial system which is in relative motion. (More will be said about EINSTEIN's theory in Part 7.3.)

POINCARÉ continued:

Let us denote by

$$A = \tfrac{1}{2} \int f^2 d\tau$$

the *longitudinal electric energy*; by

$$B = \tfrac{1}{2} \int (g^2 + h^2) d\tau$$

the *transverse electric energy*; by

$$C = \tfrac{1}{2} \int (\beta^2 + \gamma^2) d\tau$$

the *transverse magnetic energy*. There is no longitudinal magnetic energy since $\alpha = \alpha' = 0$. Let us designate by A', B', C' the corresponding quantities in the imaginary system. We find immediately

$$C' = 0, \qquad C = \varepsilon^2 B.$$

---

[200] In a letter to CARL SEELIG in 1955 EINSTEIN noted:

There is no doubt, that the special theory of relativity, if we regard its development in retrospect, was ripe for discovery in 1905. LORENTZ had already observed that for the analysis of MAXWELL's equations the transformations which later were known by his name are essential, and POINCARÉ had even penetrated deeper into these connections. Concerning myself, I knew only LORENTZ' important work of 1895 (the two papers quoted above in the German text) but not LORENTZ' later work, nor the consecutive investigations by POINCARÉ. In this sense my work of 1905 was independent. The new feature of it was the realization of the fact that the bearing of the LORENTZ transformation transcended its connection with MAXWELL's equations and was concerned with the nature of space and time in general. A further new result was that the "Lorentz invariance" is a general condition for any physical theory ...

(see M. BORN, *Physics in My Generation* (Springer, 1969), p. 104).

Moreover, we note that the real field depends only on $x + \varepsilon t$, y and z and write

$$d\tau = d(x + \varepsilon t) \, dy \, dz$$
$$d\tau' = dx' \, dy' \, dz' = k l^3 d\tau$$

whence

$$A' = k l^{-1} A, \qquad B' = k^{-1} l^{-1} B, \qquad A = \frac{l A'}{k}, \qquad B = k l B'.$$

According to Lorentz's hypothesis we have $B' = 2A'$, and A', inversely proportional to the radius of the electron, is a constant which is independent of the velocity of the real electron. Thus we find for the total energy

$$A + B + C = A' l k (3 + \varepsilon^2),$$

and for the action (per unit time)

$$A + B - C = \frac{3 A' l}{k}.$$

Let us now calculate the electromagnetic momentum; we shall find

$$D = \int (g\gamma - h\beta) \, d\tau = -\varepsilon \int (g^2 + h^2) \, d\tau = -2\varepsilon B = -4 \varepsilon k l A'.$$

But we must take into account certain relations between the energy $E = A + B + C$, the action per unit time $H = A + B - C$, and the electromagnetic momentum D. The first of these relations is

$$E = H - \varepsilon \frac{dH}{d\varepsilon};$$

the second is

$$\frac{dD}{d\varepsilon} = -\frac{1}{\varepsilon} \frac{dE}{d\varepsilon},$$

whence

(2)
$$D = \frac{dH}{d\varepsilon}, \qquad E = H - \varepsilon D.\text{[201]}$$

POINCARÉ wrote the energy of the electromagnetic field as

$$W = A + B + C$$

where A is referred to as the longitudinal electric energy because it arises from the component of the electric field in the direction of motion (f). For similar reasons B and C are denoted as the transverse electric and magnetic energy.

Since S' is the instantaneous rest system, the magnetic field in it is zero; i.e.

$$\alpha' = \beta' = \gamma' = 0 \tag{91}$$

and

$$C' = \tfrac{1}{2} \int (\beta'^2 + \gamma'^2) \, d\tau' = 0. \tag{92}$$

The magnetic field in S can now be calculated by use of (67):

$$\alpha = 0, \qquad \beta = \varepsilon h, \qquad \gamma = -\varepsilon g. \tag{93}$$

---

[201] R. 152–153: O. 523–524.

By use of (93), it is now possible to relate C and B:

$$C = \frac{1}{2} \int (\beta^2 + \gamma^2) d\tau = \frac{\varepsilon^2}{2} \int (h^2 + g^2) d\tau = \varepsilon^2 B. \tag{94}$$

At this point POINCARÉ used a result from Section 5 to relate $d\tau$ and $d\tau'$ which is [202]

$$d\tau' = kl^3 d\tau. \tag{95}$$

However, he had already implicitly obtained (95) in Section 1 by proving that $l^4 d\tau dt = dt' d\tau'$. Equation (95) can also be obtained from the LORENTZ transformation.[203]

By use of (67) and (95) the longitudinal and transverse electric field energies can be related. In S' the longitudinal electric field energy is

$$A' = \tfrac{1}{2} \int f'^2 d\tau'. \tag{96}$$

Then from (67) and (96)

$$A' = \frac{k}{l} A; \tag{97}$$

similarly

$$B' = \frac{B}{kl}. \tag{98}$$

From (92), (94), (97) and (98) the relation between the field energies in S and S' is

$$A + B + C = \frac{lA'}{k} + klB'(1 + \varepsilon^2). \tag{99}$$

In LORENTZ's theory the electric field is assumed to be spherically symmetric,

$$\frac{E_t^2}{3} = f^2 = g^2 = h^2,$$

where E_t is the resultant electric field; the subscript t on E is necessary because POINCARÉ used E to denote the total field energy. Therefore,

$$B' = \tfrac{1}{2} \int (g^2 + h^2) d\tau = \int f^2 d\tau = 2A'. \tag{100}$$

---

[202] In Section 5 POINCARÉ proved that f was a function of $x + \varepsilon t$.

[203] The JACOBIAN $\partial(x', y', z')/\partial(x, y, z)$ can be obtained by taking advantage of the result of mentioned in footnote 177,

$$d\tau' dt' = l^4 d\tau dt \tag{203-1}$$

and the time transformation

$$t' = kl(t + \varepsilon x). \tag{203-2}$$

Taking the differential of (203-2) yields

$$dt' = kl(1 + \varepsilon \xi) dt, \tag{203-3}$$

and then comparing it with (203-1) gives

$$d\tau' = \frac{l^3}{k(1 + \varepsilon \xi)} d\tau \tag{203-4}$$

or

$$\frac{\partial(x', y', z')}{\partial(x, y, z)} = \frac{l^3}{k(1 + \varepsilon \xi)}. \tag{203-5}$$

In S' (203-3) and (203-4) become

$$dt' = \frac{l}{k} dt, \tag{203-6}$$

$$d\tau' = l^3 k d\tau. \tag{203-7}$$

The total energy (99) can then be written as

$$A + B + C = A'lk(3 + \varepsilon^2).$$ (101)

It is important to note that the terminology "LORENTZ's theory" is sometimes used in a general manner to indicate the class of electron theories whose equations transform according to the LORENTZ transformation (59) with any l.

Similarly, the "action per unit time", which is the LAGRANGIAN, can be expressed as[204]

$$A + B - C = \frac{3A'l}{k}.$$ (102)

In LORENTZ's theory the quantity A' can be calculated as

$$A' = \frac{1}{2}\int f'^2 d\tau' = \frac{1}{6}\int E'_t d\tau' = \frac{e^2}{24\,\pi r}.$$ (103)

Thus, A' is a constant "inversely proportional to the radius of the electron."

The x-component of the momentum of the electromagnetic field,

$$D = \int (g\gamma - h\beta)d\tau,$$ (104)

can be calculated by use of equations (93), the defining equation for B, (98) and (100). The result is

$$D = -\int (g^2 + h^2)d\tau = -2\varepsilon B = -4\varepsilon klA'.$$ (105)

Recalling (103), we have

$$D = -\frac{e^2}{6\,\pi r}kl\varepsilon$$ (106)

which corresponds to (40) with $\varepsilon = -v$.

The important relations between the energy and momentum from classical mechanics are[205]

$$E = H - \varepsilon\frac{\partial H}{\partial\varepsilon},$$ (107a)

---

[204] The validity of (102) is easily shown:

$$J = \frac{1}{2}\int (A + B - C)dt = \frac{1}{2}\int \frac{3A'l}{k}dt.$$

From footnote 203

$$dt = \frac{k}{l}dt';$$

therefore, the equation $J = J'$ is maintained.

[205] See GOLDSTEIN's *Classical Mechanics, op. cit.* (see footnote 166), esp. pp. 53–54 for a detailed exposition of these equations. It is important to note that these equations differ from GOLDSTEIN's because POINCARÉ has defined the electromagnetic field LAGRANGIAN as $U - T$. Consequently, D is always negative in this notation; *e.g.* for a conservative mechanical system $H = U - T$, then $\partial H/\partial\varepsilon = -\partial T/\partial\varepsilon = -m\varepsilon$; but from (107c) $D = \partial H/\partial\varepsilon = -m\varepsilon$ (since $T = \frac{1}{2}m\varepsilon^2$), then

$$\varepsilon\frac{\partial H}{\partial\varepsilon} = -m\varepsilon^2 = -2T.$$

From (107a) (or (107d)) we have

$$E = H - \varepsilon\frac{\partial H}{\partial\varepsilon} = H + 2T = (U - T) + 2T = U + T$$

as expected. Then D calculated from (115) is opposite in sign to ABRAHAM's (20). POINCARÉ was forced to rectify these differences of sign in Section 7 when he discussed $m_\parallel$ and $m_\perp$.

$$\frac{\partial D}{\partial \varepsilon} = -\frac{1}{\varepsilon} \frac{\partial E}{\partial \varepsilon},$$ (107b)

$$D = \frac{\partial H}{\partial \varepsilon},$$ (107c)

$$E = H - \varepsilon D.$$ (107d)

If D as calculated from (104) agrees with (107c), then (107d) is identical with (107a); however, POINCARÉ knew that this was not the case (see Part 4.7). This defect must be remedied if mechanics is to be absorbed by electromagnetism.

The second of equations (2) is always satisfied; but the first is only if $l = (1 - \varepsilon^2)^{\frac{1}{2}} = k^{-\frac{1}{2}}$, that is if the volume of the imaginary electron is equal to that of the real electron — in other words if the electron's volume is constant; this is Langevin's hypothesis.[206]

From (102) and (103), for the class of models that transform according to (59) for any l, the LAGRANGIAN H and the momentum D are

$$H = \frac{3 A' l}{k} = \frac{e^2}{8 \pi r} \frac{l}{k},$$ (108a)

$$D = -\frac{e^2}{6 \pi r} k l \varepsilon.$$ (108b)

The problem is to find the value of l such that D obtained from its electromagnetic definition (104) is equal to the D calculated from H according to LAGRANGIAN mechanics, i.e. $D = \partial H / \partial \varepsilon$. This value of l can be found from the equation

$$\frac{\partial}{\partial \varepsilon} \left(\frac{l}{k} \right) = -\frac{4}{3} k l \varepsilon$$ (109)

which is obtained from comparing the derivative of (108a) with respect to ε with (108b). Only $l = k^{-\frac{1}{2}}$ satisfies (109). The meaning of this restriction (which will soon be discussed in detail) is that the moving electron undergoes a contraction along the direction of motion and a dilation perpendicular to it such that the volumes of the real and imaginary electrons are equal.[207] This is known as LANGEVIN's model. The generality of this proof implies the impossibility of a LAGRANGIAN formulation based solely on the LAGRANGIAN for the electromagnetic field for a model of a deformable electron which transforms according to (59) with $l \neq k^{-\frac{1}{2}}$.[208]

In LORENTZ's theory, wherein $l = 1$, the LAGRANGIAN is

$$H = \frac{e^2}{8 \pi r} (1 - v^2)^{\frac{1}{2}},$$ (110a)

---

[206] R. 153: O. 524.

[207] From the results in footnote 203 the relation between the volume elements in S and S' is

$$d \tau' = k l^3 d \tau.$$ (207-1)

Therefore in LANGEVIN's theory, where $l = k^{-\frac{1}{2}}$, (207-1) becomes $d \tau' = d \tau$.

[208] See the reference in footnote 53; a similar proof can be found on p. 197 of the 1908 edition.

while D, from its electromagnetic definition, is

$$D = - \frac{e^2}{6\pi r} k\varepsilon. \tag{110b}$$

Therefore, only if H is multiplied by $\frac{4}{3}$ will D as calculated from (110a) agree with (110b). This discrepancy is directly related to the appearance of the extra term in (48).

In LANGEVIN's model the analogue of (48) does not contain any troublesome terms because

$$m_\parallel = \frac{e^2}{6\pi r} \frac{\partial}{\partial \varepsilon} (k^{\frac{2}{3}} \varepsilon) = \frac{e^2}{6\pi r} (1-\varepsilon^2)^{-\frac{4}{3}} \left(1 - \frac{1}{3} \varepsilon^2\right). \tag{111}$$

(Adhering to POINCARÉ's notation necessitates the definition $m_\parallel = -\partial G/\partial \varepsilon$ in order to have $m_\parallel > 0$.) The work done by a force acting along the direction of the motion of the electron is

$$m_\parallel \varepsilon \varepsilon' dt = \frac{e^2}{6\pi r} (1-\varepsilon^2)^{-\frac{4}{3}} \left(1 - \frac{1}{3} \varepsilon^2\right). \tag{112}$$

The total electromagnetic energy of the LANGEVIN electron can be calculated from (107a) and (108a) as

$$W = \frac{e^2}{8\pi r} (1-\varepsilon^2)^{-\frac{1}{3}} \left(1 + \frac{1}{3} \varepsilon^2\right). \tag{113}$$

The increment of time in (113) is equal to (112) contrary to the case of $l=1$ (or any $l \neq k^{-\frac{1}{3}}$). Therefore, LANGEVIN's model does not require any non-electromagnetic binding forces.

This [LANGEVIN's model] is in contradiction with the result of Section 4 and with the result that Lorentz obtained by another method. It is necessary to explain this contradiction.[209]

POINCARÉ proved, in Section 4, that only a model with $l=1$ is consonant with the principle of relativity. LORENTZ had also obtained this result by demanding that the force transform in the same manner as the product of mass and acceleration.

POINCARÉ continued by constructing a general LAGRANGIAN formulation for an electromagnetic theory of the electron. This enabled him to investigate simultaneously all possible theories of the electron and then to resolve the discrepancy in LORENTZ's theory.

Before arriving at that explanation, I observe that whatever hypothesis is agreed upon we will have

$$H = A + B - C = \frac{l}{k} (A' + B')$$

or since $C' = 0$

$$H = \frac{l}{k} H'. \tag{3}$$

We can now recover the result $J = J'$ obtained in Section 3. Indeed, we have

$$J = \int H \, dt, \qquad J' = \int H' \, dt'.$$

---

[209] R. 154: O. 524.

We point out that the state of the system depends only on $x + \varepsilon t$, y and z, that is on x', y', z', and that we have:

(4)
$$t' = \frac{l}{k} t + \varepsilon x',$$

$$dt' = \frac{l}{k} dt.$$

Comparing equations (3) and (4) we find that $J = J'$.[210]

From (94), (97) and (98) follows the relativistic transformation equation for the LAGRANGIAN H:

$$H = A + B - C = \frac{l}{k} H'. \tag{114}$$

This equation is valid for any l. Using (114) and $dt' = \frac{l}{k} dt$ (see footnote 203), we have $J = J'$.

POINCARÉ continued:

Let us consider a general hypothesis, which we will be able to suppose to be that of Lorentz, that of Abraham, that of Langevin, or some intermediate hypothesis.

Let r, θr, θr be the three axes of the real electron; those of the imaginary electron will be:

$$klr, \theta lr, \theta lr.$$

Then $A' + B'$ will be the electrostatic energy due to an ellipsoid having as axes klr, θlr, θlr.

Suppose that the electric charge is distributed over the surface of the electron like over a conductor, or uniformly distributed over the electron's interior; the energy will be of the form:

$$A' + B' = \frac{\varphi\left(\frac{\theta}{k}\right)}{klr}$$

where φ is a known function.

Abraham's hypothesis consists in supposing that:

$$r = \text{const.}, \quad \theta = 1.$$

That of Lorentz:

$$l = 1, \quad kr = \text{const.}, \quad \theta = k.$$

That of Langevin:

$$l = k^{-\frac{1}{3}}, \quad k = \theta, \quad klr = \text{const.}$$

We then find:

$$H = \frac{\varphi\left(\frac{\theta}{k}\right)}{k^2 r}.$$

Abraham finds, in a slightly different notation (Göttinger Nachrichten 1902, p. 37):

$$H = \frac{a}{r} \frac{(1 - \varepsilon^2)}{\varepsilon} \log\left(\frac{1 + \varepsilon}{1 - \varepsilon}\right),$$

---

[210] R. 154: O. 524–525.

a is a constant. Thus in Abraham's hypothesis one has $\theta = 1$; then

(5)
$$\varphi\left(\frac{1}{k}\right) = a\,k^2\,\frac{1-\varepsilon^2}{\varepsilon}\,\log\left(\frac{1+\varepsilon}{1-\varepsilon}\right) = \frac{a}{\varepsilon}\,\log\left(\frac{1+\varepsilon}{1-\varepsilon}\right),$$

which defines the function φ.

This being the case, let us imagine that the electron is subject to a constraint, in such a manner that there is a relation between r and θ. In Lorentz's hypothesis that relation would be $\theta r = \text{const.}$, in Langevin's $\theta^2 r^3 = \text{const.}$ We will assume a more general one

$$r = b\,\theta^m,$$

b being a constant; then:

$$H = \frac{1}{b\,k^2}\,\theta^{-m}\,\varphi\left(\frac{\theta}{k}\right).$$

What shape will the electron take when the velocity becomes $-\varepsilon t$, *if one does not assume the intervention of any other forces than those of constraint?* [211]

(Note: the second line from the bottom should read: "... becomes $-\varepsilon$.")

The general model proposed by POINCARÉ enabled him to investigate the manner in which the theories of ABRAHAM, LANGEVIN and LORENTZ evolve from the electromagnetic LAGRANGIAN (19). Since the electron's stability is unaffected by its charge distribution, POINCARÉ generalized (19) to include a charge distribution for both a surface or a volume by replacing $-\dfrac{e^2}{16\,\pi}$ with a positive constant a (recall that POINCARÉ defined the LAGRANGIAN as $E^2 - B^2$ instead of ABRAHAM's $B^2 - E^2$). In addition, v is replaced by ε so that (19) becomes

$$H = \frac{a}{r}\,\frac{(1-\varepsilon^2)}{\varepsilon}\,\log\left(\frac{1+\varepsilon}{1-\varepsilon}\right). \tag{115}$$

The axes of the real and imaginary electrons are a function of a shape parameter θ whose particular values correspond to the three theories in question. The real electron (seen by the observer in S) has axes

$$r,\ \theta r,\ \theta r \tag{116}$$

whereas the imaginary electron has axes

$$k l r,\ \theta l r,\ \theta l r. \tag{117}$$

Then in S' the total energy is due to an ellipsoidal charge distribution with axes (117) and is of the form

$$A' + B' = \frac{\varphi(\theta/k)}{k l r} \tag{118}$$

with

$$H = \frac{\varphi(\theta/k)}{k^2 r}. \tag{119}$$

POINCARÉ arrived at the general equation for H (119) from the relation between H and H':

$$H = \frac{l}{k}\,H' = \frac{l}{k}\,(A' + B') \tag{120}$$

---

[211] R. 154–155: O. 525–526; italics in original.

which is valid for all three models. Substituting (119) into (120) gives

$$H = \frac{l}{k}\left[\frac{\varphi(\theta/k)}{klr}\right] = \frac{\varphi(\theta/k)}{k^2 r};$$

therefore

$$\varphi(\theta/k) = k^2 r H.$$

ABRAHAM's hypothesis was that the real electron is a rigid sphere and the imaginary electron is dilated in the direction of motion. Therefore, (116) and (117) become (121a) and (121b), respectively:

$$(r, r, r),   \tag{121a}$$

$$(kr, r, r).   \tag{121b}$$

This case corresponds to $\theta = 1$.

In LORENTZ's model the real electron undergoes a contraction in the direction of motion while the imaginary electron is a sphere; therefore,

$$\theta = k,$$
$$kr = b \quad (b \text{ is a constant}),   \tag{122}$$
$$l = 1.$$

Equations (116) and (117) become (123a) and (123b), respectively:

$$\left(\frac{b}{k}, b, b\right),   \tag{123a}$$

$$(b, b, b).   \tag{123b}$$

In LANGEVIN's model the deformation occurs in such a way that the volume of the electron remains unchanged; therefore,

$$l = k^{-\frac{1}{3}}, \quad k = \theta$$
$$klr = k^{\frac{2}{3}} r = b,   \tag{124}$$

and (116) and (117) become (125a) and (125b), respectively:

$$\left(\frac{b}{k^{\frac{2}{3}}}, k^{\frac{1}{3}} b, k^{\frac{1}{3}} b\right),   \tag{125a}$$

$$(b, b, b).   \tag{125b}$$

By comparison of (115) with (119) the functional form of φ for $\theta = 1$ (ABRAHAM's theory) can be obtained as

$$\varphi\left(\frac{1}{k}\right) = \frac{a}{\varepsilon}\log\left(\frac{1+\varepsilon}{1-\varepsilon}\right).   \tag{126}$$

It now remains to determine the functional form of φ in the theories of LANGEVIN and LORENTZ. For this purpose POINCARÉ assumed initially that it is the electromagnetic forces which relate r and θ and thereby determine the shape of the electron. In LANGEVIN's theory $\theta l r = k^{\frac{2}{3}} r = b$, while in LORENTZ's theory $\theta r = kr = b$, or in general

$$r = b \theta^m.   \tag{127}$$

Substituting (127) into (119) gives

$$H = \frac{1}{bk^2} \theta^{-m} \varphi(\theta/k), \tag{128}$$

which expresses H as a function of θ thereby providing the equation necessary to determine the electron's shape.

POINCARÉ continued:

That shape will be defined by the equality

(6)
$$\frac{\partial H}{\partial \theta} = 0$$

or

$$-m\theta^{-m-1}\varphi + \theta^{-m}k^{-1}\varphi' = 0$$

or

$$\frac{\varphi'}{\varphi} = \frac{mk}{\theta}.$$

If we want equilibrium to take place in such a manner that $\theta = k$ it is necessary for $\theta/k = 1$, then the logarithmic derivative of φ becomes equal to m.

If we expand $1/k$ and the second member of (5) in powers of ε, equation (5) becomes

$$\varphi\left(1 - \frac{\varepsilon^2}{2}\right) = a\left(1 + \frac{\varepsilon^2}{3}\right),$$

neglecting higher powers of ε.

Upon differentiation, we have

$$-\varepsilon\varphi'\left(1 - \frac{\varepsilon^2}{2}\right) = \frac{2}{3}\varepsilon a.$$

For $\varepsilon = 0$, that is when the argument of φ is equal to one, these equations become:

(7)
$$\varphi = a, \qquad \varphi' = -\frac{2}{3}a, \qquad \frac{\varphi'}{\varphi} = -\frac{2}{3}.$$

We then obtain $m = -\frac{2}{3}$ suitable to Langevin's hypothesis. [So as not to disturb the continuity of this argument, the proof of equation (6) which follows this passage is presented below.[212]]

---

[212] "This result can be traced back to the first of equations (2) from which it really does not differ. In fact, let us suppose that all elements of the electron are subjected to a force $Xd\tau$ parallel to the x-axis, X being the same for all elements: we have then, conforming to the definition of momentum

$$\frac{dD}{dt} = \int Xd\tau.$$

On the other hand the principle of least action gave us

$$\delta J = \int X\,\delta U d\tau dt, \qquad J = \int Hdt, \qquad \delta J = \int D\,\delta U dt$$

δU being the displacement of the center of gravity of the electron, H depends on θ and ε if we suppose that r be linked to θ by the equation of constraint. We then have

$$\delta J = \int\left(\frac{\partial H}{\partial \varepsilon}\,\delta\varepsilon + \frac{\partial H}{\partial \theta}\,\delta\theta\right)dt.$$

On the other hand, $\delta\varepsilon = -d(\delta U)/dt$; whence, integrating by parts:

$$\int\delta\varepsilon dt = \int D\,\delta U dt$$

The conclusion is that if the electron is subject to a constraint between its three axes *and if forces other than those of constraint are excluded*, the shape that the electron will take, when it is placed in motion with a uniform velocity,

or

$$\int \left(\frac{\partial H}{\partial \varepsilon} \, \delta \varepsilon + \frac{\partial H}{\partial \theta} \, \delta \theta \right) dt = \int D \, \delta \varepsilon \, dt$$

whence

$$D = \frac{\partial H}{\partial \varepsilon}, \qquad \frac{\partial H}{\partial \theta} = 0.$$

But the derivative $dH/d\varepsilon$ that appears in the second member of the first of equations (2) is taken by assuming that θ is expressed as a function of ε, so that

$$\frac{dH}{d\varepsilon} = \frac{\partial H}{\partial \varepsilon} + \frac{\partial H}{\partial \theta} \frac{d\theta}{d\varepsilon}.$$

Equation (2) is then equivalent to equation (6)" (R. 156: O. 527). The goal is to prove that the condition

$$\frac{\delta H}{\delta \theta} = 0 \tag{212-1}$$

follows from (107c)

$$D = \frac{\partial H}{\partial \varepsilon}. \tag{107c}$$

First of all, in this case the convective derivative of H with respect to θ (and ε) is equal to the ordinary derivative. The reason can be seen directly from the definition of the convective derivative

$$\frac{\delta}{\delta t} = \frac{\partial}{\partial t} + \vec{v} \cdot \vec{V}. \tag{212-2}$$

If θ or ε replaced t in (212-2) the term with v would obviously become meaningless. Assuming H to be a function of ε and θ and varying J one obtains

$$\delta J = \int \left(\frac{\partial H}{\partial \varepsilon} \, \delta \varepsilon + \frac{\partial H}{\partial \theta} \, \delta \theta \right) dt \tag{212-3}$$

or

$$\delta J = \int \left(D \, \delta \varepsilon + \frac{\partial H}{\partial \theta} \, \delta \theta \right) dt. \tag{212-4}$$

However, we know that

$$\xi = \frac{dU}{dt}$$

(see (80)) and $\xi = -\varepsilon$ in S'. Hence

$$\varepsilon = -\frac{dU}{dt}$$

and consequently

$$\delta \varepsilon = -\frac{d\delta U}{dt} \tag{212-5}$$

(the variational symbol δ commutes with d/dt). By use of (212-5), the first term in (212-4) can be transformed as follows:

$$\int D \, \delta \varepsilon \, dt = -\int D \left(\frac{d\delta U}{dt} \right) dt = \int \frac{dD}{dt} \, \delta U \, dt = \int X \, \delta U \, d\tau \, dt$$

(since $dD/dt = \int X \, d\tau$). Therefore, $\partial H/\partial \theta = 0$ due to the fact that

$$\delta J = \int X \, \delta U \, d\tau \, dt.$$

In other words, if $H = H(\varepsilon(\theta))$ then we have

$$\frac{dH}{d\varepsilon} = \frac{\partial H}{\partial \varepsilon} + \frac{\partial H}{\partial \theta} \frac{d\theta}{d\varepsilon};$$

but $D = \partial H/\partial \varepsilon$. Therefore, $\partial H/\partial \theta = 0$.

will be such that the corresponding imaginary electron will only be a sphere. The case in which the constraint is of constant volume is in accordance with Langevin's hypothesis.[213]

Despite its convective character [§ 6-6] can be interpreted as the usual partial derivative of H.[212] Utilizing (128) for H in [§ 6-6] determines the logarithmic derivative of φ:

$$\frac{\varphi'}{\varphi} = m \frac{k}{\theta}. \tag{129}$$

At this point it can be determined whether LANGEVIN's or LORENTZ's theory can emerge from a purely electromagnetic representation for the electron. In both theories $\theta = k$; therefore,

$$\frac{\varphi'}{\varphi} = m. \tag{130}$$

Furthermore, $\theta = k$ corresponds to $\varphi(\theta/k) = \varphi(1)$, which can be obtained from (126) by expanding $1/k$ in powers of ε and then letting ε pass to zero. The result is

$$\varphi = a, \quad \varphi' = -\frac{2}{3} a, \quad \frac{\varphi'}{\varphi} = -\frac{2}{3} \tag{131}$$

or $m = -\frac{2}{3}$, which implies $\left(\text{see } (127)\right)$ that $r = b\theta^{-\frac{2}{3}} = bk^{-\frac{2}{3}}$. Therefore, only LANGEVIN's model can emerge from the LAGRANGIAN of the electromagnetic field with $\theta = k$. This was to be expected in light of the general proof (stated previously) that only models with $l = k^{-\frac{1}{2}}$ can provide a purely electromagnetic interpretation for a deformable electron. From (131) and (128) the LAGRANGIAN for LANGEVIN's model is[214]

$$H = \frac{a}{b} k^{-\frac{1}{3}}. \tag{132}$$

However, only $l = 1$ is consonant with the principle of relativity. Furthermore, from a restricted viewpoint the LORENTZ transformation equations (59) with $l = k^{-\frac{1}{2}}$ would not produce the correct compensatory terms to account for the MICHELSON-MORLEY experiment.

---

[213] R. 155–157: O. 527–528; italics in original.

[214] LANGEVIN's prediction for $m_{\parallel}$ can be calculated from this H as

$$D = \frac{\partial H}{\partial \varepsilon} = -\frac{4}{3} \frac{a}{b} \varepsilon k^{\frac{2}{3}} \tag{214-1}$$

and

$$m_{\parallel} = -\frac{\partial D}{\partial \varepsilon} = \frac{4}{3} \frac{a}{b} \frac{\partial}{\partial \varepsilon} \left(\varepsilon k^{\frac{2}{3}} \right) \tag{214-2}$$

(the minus sign is put in so as to have $m_{\parallel} > 0$; recall that POINCARÉ defined L as $U - T$). Equation (214-2) agrees with (111) if

$$\frac{a}{b} = \frac{e^2}{8\pi r}. \tag{214-3}$$

Equation (214-3), which asserts that a/b is the electron's electrostatic (or rest) mass, will turn out to be a non-fortuitous result, the relevance of which POINCARÉ seems not to have perceived.

The problem is how to produce a factor of $\frac{4}{3}$ in H and at the same time have H proportional to $k^{-\frac{1}{2}}$. Poincaré's statement of this problem was as follows:

Let us pose the following problem: what supplementary forces other than those of constraint would it be necessary to introduce in order to calculate Lorentz's law or more generally any law other than that of Langevin ?

The simplest hypothesis, and the first one that we shall examine, is that the supplementary forces can be derived from a special potential obtainable from the three axes of the ellipsoid, and consequently from θ and r. Let $F(\theta, r)$ be the potential; in that case we have for the action:

$$J = \int [H + F(\theta, r)]\, dt,$$

and the equilibrium conditions can be written as

(8)
$$\frac{dH}{d\theta} + \frac{dF}{dr} = 0, \qquad \frac{dH}{dr} + \frac{dF}{dr} = 0.$$

If we suppose that r and θ are related by the connection $r = b\theta^m$, then we can regard r as a function of θ, consider F as if dependent only on θ, and maintain only the first of equations (8) with

$$H = \frac{\varphi}{bk^2\theta^m}, \qquad \frac{dH}{d\theta} = -\frac{m\,\varphi}{bk^2\theta^{m+1}} + \frac{\varphi'}{bk^3\theta^m}.$$

It is necessary that for $k = 0$ equation (8) should be satisfied, which gives, when the calculation in equation (7) is taken into account,

$$\frac{dF}{d\theta} = \frac{ma}{b\theta^{m+3}} + \frac{2}{3}\frac{a}{b\theta^{m+3}}$$

whence

$$F = -\frac{a}{b\theta^{m+2}}\frac{m + \frac{2}{3}}{m + 2},$$

and in Lorentz's hypothesis, where $m = -1$,

$$F = \frac{a}{3b\theta}.$$

Suppose now that there is no connection and consider r and θ as two independent variables. We keep the two equations (8), from which we obtain

$$H = \frac{\varphi}{k^2 r}, \qquad \frac{dH}{d\theta} = \frac{\varphi'}{k^3 r}, \qquad \frac{dH}{dr} = -\frac{\varphi}{k^2 r^2}.$$

Equations (8) have to be satisfied for $k = 0$, $r = b\theta^m$; this gives

(9)
$$\frac{dF}{dr} = \frac{a}{b^2\theta^{2m+2}}, \qquad \frac{dF}{d\theta} = \frac{2}{3}\frac{a}{b\theta^{m+3}}.$$

One way of satisfying these conditions is to put

(10)
$$F = A r^\alpha \theta^\beta,$$

A, α and β being constants. The equations (9) have to be satisfied for $k = 0$, and $r = b\theta^m$. The result is

$$A\alpha b^{\alpha-1}\theta^{m\alpha-m+\beta} = \frac{a}{b^2\theta^{2m+2}}, \qquad A\beta b^\alpha \theta^{m\alpha+\beta-1} = \frac{2}{3}\frac{a}{b\theta^{m+3}}.$$

Identifying like terms, we find

(11) $\alpha = 3\gamma$, $\beta = 2\gamma$, $\gamma = -\dfrac{m+2}{3m+2}$, $A = \dfrac{a}{\alpha b^{\alpha+1}}$.

But the volume of the ellipsoid is proportional to $r^3\theta^2$; therefore, the supplementary potential is proportional to the power γ of the electron's volume.

According to Lorentz's hypothesis we have $m = -1$, $\gamma = 1$.

Thus we recover Lorentz's hypothesis provided that we add a supplementary potential proportional to the volume of the electron.

Langevin's hypothesis corresponds to $\gamma = \infty$.[215]

It is necessary to calculate the terms that must be combined with the electromagnetic LAGRANGIAN (115) to produce LORENTZ's theory, namely, the additional energy of the LORENTZ electron due to the mechanical cohesive forces acting on it. Furthermore, these forces account for the connection $kr = b$ when $\theta = k$. Electromagnetic forces could account only for the connection $k^{\frac{3}{2}}r = b$ when $\theta = k$.

The simplest assumption is to add one term, which is a function of r and θ, to the LAGRANGIAN (115). The action is then

$$J = \int [H(r, \theta) + F(r, \theta)]\,dt. \tag{133}$$

For J to be stationary under variation of r and θ it is necessary that[216]

$$\frac{\partial H}{\partial r} + \frac{\partial F}{\partial r} = 0, \tag{134a}$$

$$\frac{\partial H}{\partial \theta} + \frac{\partial F}{\partial \theta} = 0. \tag{134b}$$

First, consider the case in which r is a function of θ, namely $r = b\theta$. (The general case of r and θ as independent variables follows.) From (128) and (134b) we have

$$\frac{\partial F}{\partial \theta} = \frac{m\,\varphi}{b\,k^2\,\theta^{m+1}} - \frac{\varphi'}{b\,k^3\,\theta^m}. \tag{135}$$

But from (131) $k = \theta$ necessitates that $\varphi = a$ and $\varphi' = -\frac{2}{3}a$; however, since $\partial F/\partial\theta \neq 0$, the supplementary potential F constrains m from equalling $-\frac{2}{3}$. The result of setting $k = \theta$, $\varphi = a$ and $\varphi' = -\frac{2}{3}a$ in (135), and then integrating it, is

$$F = -\frac{a}{b\,\theta^{m+2}} \cdot \frac{m+\frac{2}{3}}{m+2} \tag{136}$$

(note that $m = -\frac{2}{3}$ gives $F = 0$). In LORENTZ's theory $m = -1$; therefore, we have

$$F = \frac{a}{3b\theta} = \frac{a}{3bk}. \tag{137}$$

---

[215] R. 157–158: O. 528–529; italics in original.
[216] In order that the variation of J shall vanish, *i.e.*

$$\delta J = \int \left[\left(\frac{\partial H}{\partial r} + \frac{\partial F}{\partial r} \right) \delta r + \left(\frac{\partial H}{\partial \theta} + \frac{\partial F}{\partial \theta} \right) \delta\theta \right] dt = 0,$$

the coefficient of δr and $\delta\theta$ must vanish (r and θ are considered to be independent coordinates).

In general H and F are functions of both r and θ, and therefore (134a, b) must be used. Using these two equations and assuming a separable form for F in order to facilitate integration, *i.e.*

$$F = A r^\alpha \theta^\beta \qquad (138)$$

(A is a constant and should not be confused with $A = \frac{1}{2}\int f^2 d\tau$)[217] as well as $\varphi = a$, $\varphi' = -\frac{2}{3}a$, we easily show that

$$\alpha = 3\gamma, \qquad \beta = 2\gamma, \qquad \gamma = -\frac{m+2}{3m+2}, \qquad A = \frac{a}{\alpha b^{\alpha+1}} \qquad [\S\,6\text{-}11].$$

In LORENTZ's model $m = -1$; therefore,

$$F = A r^3 \theta^2 = \frac{a}{3 b^4}\, k^2 r^3 \qquad (139)$$

(this is equivalent to (137) because $kr = b$). In LANGEVIN's model $\gamma = \infty$ but $F = 0$.

From (62) (with $l = 1$ and $\xi = -\varepsilon$) it is obvious that F is proportional to the volume of the electron. We shall see that this is crucial for ensuring that F has suitable properties for a LORENTZ transformation. Therefore, if and only if F is present is it possible to obtain LORENTZ's theory, because F acts as a device which makes possible values of m other than $-\frac{2}{3}$. Consequently, one can now investigate the class of models of a deformable electron wherein $l \neq k^{-\frac{1}{3}}$.

From (128), (133) and (137) the LAGRANGIAN for LORENTZ's theory is

$$H_L = H + F = \frac{4}{3}\, \frac{a}{b}\, (1 - \varepsilon^2)^{\frac{1}{2}}. \qquad (140)$$

As we shall see, an unfortunate error prevented POINCARÉ from drawing the proper conclusions from this result. The electron's momentum calculated from (140) by using (107c) agrees with the one calculated from its electromagnetic definition (108b) (with $l = 1$) if $a/b = e^2/8\pi r$ (see also footnote 214). A detailed discussion of this result as well as the interpretation of the non-electromagnetic energy F will be given in Section 8.

POINCARÉ then set out to prove that $m = -1, l = 1$ (LORENTZ's theory) is the only value of m and l which prevent the detection of absolute motion.

6.8. Quasi-Stationary Motion [§ 7].[218]

It remains to be seen whether the hypothesis of the contraction of electrons accounts for the impossibility of detecting absolute motion. I shall begin by studying the quasi-stationary motion of an isolated electron, or one subject only to the influence of other distant electrons.

We define a motion as a quasi-stationary motion when the variation of the velocity is very slow so that the electric and magnetic energies due to the electron's motion differs little from what they would be for uniform motion.

---

[217] The interchangeable use of symbols, *e.g.*, A as the constant in (138) as well as $A = \frac{1}{2}\int f^2 d\tau$, F representing the supplementary potential as well as the x-component of the vector potential, H representing the LAGRANGIAN as well as the z-component of the vector potential, and the carelessness in regard to minus signs, suggest that this work is incomplete.

[218] R. 158–163: O. 529–536.

Moreover, we know that Abraham proceeding from the notion of quasi-stationary motion arrived at the transverse and longitudinal electromagnetic masses.[219]

The problem is to examine the equation for the quasi-stationary motion of either an electron or one which is acted upon by the fields arising from other distant electrons. POINCARÉ noted that the quasi-stationary approximation corresponds to neglecting the radiation field of the electron, taking into account only the energy arising from the velocity wave and the wave of reorganization. This is the approximation in which "the electric and magnetic energy due to the electron's motion differs little from what it would be for uniform motion." This approximation, which was originally used by ABRAHAM (and, as has been noted, is essential for an unambiguous definition of mass) is now examined in detail.

I think I should explain this work precisely. Let H be the action per unit time:

$$H = \tfrac{1}{2} \int \left(\Sigma f^2 - \Sigma \alpha^2 \right) d\tau,$$

where we consider for the moment that the electric and magnetic fields are due only to an isolated electron. In the preceding section, the motion was considered as uniform, and we regarded H as dependent on the velocity ξ, η, ζ of the center of gravity of the electron (these three components in the preceding section had the values $-\varepsilon, 0, 0$) and on the parameters r and θ which define the shape of the electron.[220]

POINCARÉ was careful to point out that the fields contained in H are the self-fields of the electron in question. These fields generate the electron's mass and therefore the electron itself.

But if the motion is not uniform H depends not only on the value of $\xi, \eta, \zeta, r, \theta$ at the instant under consideration, but the values of these same quantities at other instants which differ by quantities of the same order as the time taken by light to go from one point of the electron to another; in other words, H depends not only on $\xi, \eta, \zeta, r, \theta$, but also on their derivatives to all orders with respect to time.[221]

The scalar and vector potentials for a moving electron (the LIÉNARD-WIECHERT potentials) can be developed as a sum of terms depending on the time derivatives to all orders in the particle's acceleration. The reaction force of the radiation is calculated from this series. Then, truncating this series at the second term gives (see footnote 28)

$$\vec{F}_{\text{self}} = -m_e \vec{a} + \alpha_2 \dot{\vec{a}}.$$

The statement: "... which differ by quantities of the same order as the time taken by light to go from one point on the electron to another," can be verified by an order of magnitude calculation. In footnote 28 it was stated that

$$\alpha_2 \sim \frac{e^2}{c^3};$$

[219] R. 158: O. 529–530.
[220] R. 158: O. 530.
[221] R. 158–159: O. 530.

then

$$\alpha_2 \dot{a} \sim \frac{e^2}{c^3} \frac{L}{T^3} \quad \text{where} \quad \dot{a} \sim \frac{L}{T^2} \frac{1}{T}$$

and T is the time for an electron to move a distance L equal to its diameter. (For the purpose of this calculation take $c \neq 1$ in order to keep track of dimensions.) This calculation can be rewritten as

$$\alpha_2 \dot{a} \sim \frac{e^2}{c^2 T^2} \frac{L}{c T}$$

(note that $e^2/c^2 T^2$ has the dimensions of force). In footnote 28 it was also pointed out that $m_e \sim \dfrac{e^2}{r c^2}$ (this is the dimensions of $m_\parallel$ or $m_\perp$). Then the condition that

$$m_e a \gg \alpha_2 \dot{a}$$

becomes

$$\frac{e^2}{r c^2} \frac{L}{T^2} \gg \left(\frac{e^2}{c^2 T^2} \right) \frac{L}{c T}$$

or

$$\frac{r}{c T} \ll 1.$$

Therefore, the condition that α_2 can be neglected is that the time during which the electron changes its speed from v_1 to v_2 must be such that $r/cT \ll 1$. This demonstration can be extended easily to the higher order time derivatives. Although this argument was directed towards the self-force, it leads to the same result for the fields.

POINCARÉ continued:

> Now, the motion will be called quasi-stationary whenever the partial derivatives of H with respect to the successive derivatives of $\xi, \eta, \zeta, r, \theta$ will be negligible compared to the partial derivatives of H with respect to the quantities $\xi, \eta, \zeta, r, \theta$ themselves.[222]

This condition is a restatement of the assertion that the series for the radiation reaction force must be truncated at $\vec{a}$ to maintain NEWTON's second law in electron physics. This is impossible if derivatives of H with respect to $\xi, \eta, \zeta, r, \theta$, of order higher than the first, are admitted. However, an electromagnetic world-picture which can only account for quasi-stationary motion is at best an approximate one.

POINCARÉ then wrote the equations for quasi-stationary motion:

> The equations for such a motion can be written as:

$$\frac{dH}{d\theta} + \frac{dF}{d\theta} = \frac{dH}{dr} + \frac{dF}{dr} = 0$$

(1)

$$\frac{d}{dt} \frac{dH}{d\xi} = - \int X d\tau, \quad \frac{d}{dt} \frac{dH}{d\eta} = - \int Y d\tau, \quad \frac{d}{dt} \frac{dH}{d\zeta} = - \int Z d\tau.$$

> In these equations F has the same meaning as in the preceding section; X, Y, Z are the components of the force which acts on the electron. This force is due solely to the electric and magnetic fields produced by the other electrons.[223]

---

[222] R. 159: O. 530.
[223] Loc. cit.

The set of equations

$$\frac{d}{dt}\frac{\partial H}{\partial \xi} = -\int X d\tau,$$

$$\frac{d}{dt}\frac{\partial H}{\partial \eta} = -\int Y d\tau, \qquad (141)$$

$$\frac{d}{dt}\frac{\partial H}{\partial \zeta} = -\int Z d\tau$$

are a consequence of the quasi-stationary approximation and represent NEWTON's second law, $F = ma$. As noted previously, the minus sign arises from POINCARÉ's definition of H as $U - T$, making $D = -\partial H/\partial \xi$, etc. The electron's momentum is derived from its self-field, and the fields of other electrons give rise to the components of force X, Y, Z. That (141) follows from the assumption of quasi-stationary motion can be seen from the results mentioned in footnote 28. Truncating the expansion of $\vec{F}_{self}$ at $\vec{a}$ gives

$$\vec{F}_{ext} - m_e \vec{a} = \vec{0}$$

(the mechanical mass m_0 is non-existent) or more generally

$$\vec{F}_{ext} - \frac{d\vec{D}}{dt} = \vec{0}.$$

For POINCARÉ, the inexactness of NEWTON's second law in electron physics (in mechanics it was a convention) represented another instance wherein NEWTONIAN mechanics had to be altered in the realm of the physical sciences. See the discussion in Part 5.2 for the status that POINCARÉ assigned to the first and third laws.

POINCARÉ continued in "Sur la dynamique ...":

> Observe that H depends on ξ, η, ζ only through the combination
>
> $$V = \sqrt{\xi^2 + \eta^2 + \zeta^2},$$
>
> that is through the magnitude of the velocity; we have, therefore, again calling D the momentum,
>
> $$\frac{dH}{d\xi} = \frac{dH}{dV}\frac{\xi}{V} = -D\frac{\xi}{V},$$
>
> whence
>
> (2) $\qquad -\frac{d}{dt}\frac{dH}{d\xi} = \frac{D}{V}\frac{d\xi}{dt} - D\frac{D\xi}{V^2}\frac{dV}{dt} + \frac{dD}{dV}\frac{\xi}{V}\frac{dV}{dt}$
>
> (2′) $\qquad -\frac{d}{dt}\frac{dH}{d\eta} = \frac{D}{V}\frac{d\eta}{dt} - D\frac{\eta}{V^2}\frac{dV}{dt} + \frac{dD}{dV}\frac{\eta}{V}\frac{dV}{dt}$
>
> with
>
> (3) $\qquad V\frac{dV}{dt} = \Sigma\xi\frac{d\xi}{dt}$. [224]

In general H is a function of

$$V = \sqrt{\xi^2 + \eta^2 + \zeta^2}. \qquad (142)$$

---

[224] R. 159–160: O. 530–531.

The chain rule for differentiation yields

$$\frac{\partial H}{\partial \xi} = \frac{\partial H}{\partial V} \frac{\partial V}{\partial \xi} = - D \frac{\xi}{V}$$

where POINCARÉ sets $D = -\partial H/\partial V$ (the reason for the minus sign will soon be explained). The x-component of the time rate of change of the electron's momentum is then

$$- \frac{d}{dt}\left(\frac{\partial H}{\partial \xi}\right) = \frac{d}{dt}\left(D\frac{\xi}{V}\right) = \frac{\xi}{V}\frac{dD}{dV}\frac{dV}{dt} + \frac{D}{V}\frac{d\xi}{dt} - D\frac{\xi}{V^2}\frac{dV}{dt}. \quad (143)$$

The equation for the y-component is obtained in a similar manner as

$$- \frac{d}{dt}\left(\frac{\partial H}{\partial \eta}\right) = \frac{d}{dt}\left(D\frac{\eta}{V}\right) = \frac{\eta}{V}\frac{dD}{dV}\frac{dV}{dt} + \frac{D}{V}\frac{d\eta}{dt} - D\frac{\eta}{V^2}\frac{dV}{dt}. \quad (144)$$

Utilizing the time derivative of (142) in the form

$$V \frac{dV}{dt} = \Sigma\xi\frac{d\xi}{dt} \quad (145)$$

and then specializing to motion in the x-direction, $i.e.$ $\xi = V, \eta = \zeta = 0$, we may reduce (143) and (144) to

$$- \frac{d}{dt}\frac{\partial H}{\partial \xi} = \frac{dD}{dV}\frac{d\xi}{dt} \quad (146)$$

(the second and third terms in (143) cancel),

$$- \frac{d}{dt}\frac{\partial H}{\partial \eta} = \frac{D}{V}\frac{d\eta}{dt} \quad (147)$$

(the first and third terms in (144) vanish since $\eta = \zeta = 0$). An equation similar to (147) holds for motion in the z-direction. Comparing (146) and (147) with (141) gives

$$\frac{dD}{dV}\frac{d\xi}{dt} = \int X d\tau, \quad (148a)$$

$$\frac{D}{V}\frac{d\eta}{dt} = \int Y d\tau, \quad (148b)$$

$$\frac{D}{V}\frac{d\zeta}{dt} = \int Z d\tau. \quad (148c)$$

The longitudinal and transverse masses of the electron can be obtained from (148) according to the "usual" definition of mass as the ratio of force to acceleration:

$$m_\parallel = \frac{dD}{dV}, \quad (149a)$$

$$m_\perp = \frac{D}{V}. \quad (149b)$$

At this point POINCARÉ chose to define D as [225]

$$D = - \frac{\partial H}{\partial V} \quad (150)$$

---

[225] Note that POINCARÉ writes

$$D = - \frac{\partial H}{\partial V} = - \frac{\delta H}{\delta V}$$

as proven in footnote 212.

even though he had already used this equation several times. Replacing r and θ "as functions of V" means that $r = b/k$ and $\theta = k$. Particularizing (128) to LORENTZ's theory ($\varphi = a$, $kr = b$, $\theta = k$ and $m = -1$), we obtain

$$H = A\,(1 - V^2)^{\frac{1}{2}} \tag{151}$$

where $A = a/b$.[226]

POINCARÉ continued:

We shall choose the units in such a manner that the constant factor A is equal to one, and I put $\sqrt{1 - V^2} = H$, whence

$$H = +h, \quad D = \frac{V}{h}, \quad \frac{dD}{dV} = \frac{1}{h^3}, \quad \frac{dD}{dV}\frac{1}{V^2} - \frac{D}{V^3} = \frac{1}{h^3}.$$

Let us put

$$M = V\frac{dV}{dt} = \Sigma\xi\frac{d\xi}{dt}, \quad X_1 = \int X d\tau,$$

and we find for the equation of quasi-stationary motion,

$$(5) \qquad\qquad h^{-1}\frac{d\xi}{dt} + h^{-3}\xi M = X_1.\text{[227]}$$

POINCARÉ, in the elegant style in which he wrote scientific articles, dispensed with the constant A by setting it equal to one, leaving to others the task of keeping track of constants like 4π, c and e.

The electron's momentum can now be calculated by use of (151)

$$D = -\frac{\partial H}{\partial V} = \frac{V}{\sqrt{1 - V^2}} = \frac{V}{h}, \tag{152}$$

where $H = h$. In my analysis of Section 8 I shall argue that POINCARÉ should have used H_L from (140) in these equations instead of H. Consequently, D in (152) is incorrect because it differs from its electromagnetic definition by a factor of $\frac{4}{3}$. However, this is the contradiction that POINCARÉ was trying to resolve.

The reason for the minus sign in (150), compared to the definition of D on (107c), now becomes clear: it exists because D/V is $m_\perp$ (see (149b)) which must be positive. The time rate of change on D can now be calculated as

$$\frac{dD}{dt} = \frac{dD}{dV}\frac{d\xi}{dt} = \left(\frac{1}{h} + \frac{V^2}{h^3}\right)\frac{d\xi}{dt} = \frac{1}{h}\frac{d\xi}{dt} + \frac{V}{h^3}\left(V\frac{d\xi}{dt}\right), \tag{153}$$

but $\dfrac{d\xi}{dt} = \dfrac{dV}{dt}$, so

$$X_1 = \int X d\tau = \frac{1}{h}\frac{d\xi}{dt} + \frac{V}{h^3}M \tag{154}$$

where $M = V\dfrac{dV}{dt}$.

Next POINCARÉ calculated how the equations for quasi-stationary motion behave according to the LORENTZ transformation:

---

[226] Yet another meaning for A.
[227] R. 160: O. 531–532.

If we take the actual direction of the velocity as the x-axis, they become:

$$\xi = V, \quad \eta = \zeta = 0, \quad \frac{d\xi}{dt} = \frac{dV}{dt};$$

equations (2) and (2') become

$$-\frac{d}{dt}\frac{dH}{d\xi} = \frac{dD}{dV}\frac{d\xi}{dt}, \quad -\frac{d}{dt}\frac{dH}{d\eta} = \frac{D}{V}\frac{d\eta}{dt}$$

and the last three equations (1) become

(4) $\qquad \dfrac{dD}{dV}\dfrac{d\xi}{dt} = \int X\,d\tau, \quad \dfrac{D}{V}\dfrac{d\eta}{dt} = \int Y\,d\tau, \quad \dfrac{D}{V}\dfrac{d\zeta}{dt} = \int Z\,dt.$

This is why Abraham has given to dD/dV the name *longitudinal mass* and to D/V the name *transverse mass*; let us recall $D = dH/dV$.

According to Lorentz's hypothesis, we have

$$D = -\frac{dH}{dV} = -\frac{\partial H}{\partial V},$$

$\dfrac{\partial H}{\partial V}$ represents the derivative with respect to V, after r and θ have been replaced by their values as functions of V as obtained from the first two equations (1); we shall have therefore, after that substitution

$$H = +A\sqrt{1-V^2}.$$

Let us see how these equations change under the Lorentz transformation. Let us put $1 + \xi\varepsilon = \mu$, and we shall have

$$\mu\xi' = \xi + \varepsilon, \quad \mu\eta' = \frac{\eta}{k}, \quad \mu\zeta' = \frac{\zeta}{k},$$

whence we easily obtain

$$\mu h' = \frac{h}{k}.$$

We have

$$dt' = k\mu\,dt,$$

whence

$$\frac{d\xi'}{dt'} = \frac{d\xi}{dt}\frac{1}{k^3\mu^3}, \quad \frac{d\eta'}{dt'} = \frac{d\eta}{dt}\frac{1}{k^2\mu^2} - \frac{d\xi}{dt}\frac{\eta\varepsilon}{k^2\mu^3},$$

$$\frac{d\zeta'}{dt'} = \frac{d\zeta}{dt}\frac{1}{k^2\mu^2} - \frac{d\xi}{dt}\frac{\zeta\varepsilon}{k^2\mu^3}.$$

Also

$$M' = \frac{d\xi}{dt}\frac{\varepsilon h^2}{k^3\mu^4} + \frac{M}{k^3\mu^3},$$

and

(6) $\qquad h'^{-1}\dfrac{d\xi'}{dt'} + h'^{-3}\xi'M' = \left[h^{-1}\dfrac{d\xi}{dt} + h^{-3}(\xi+\varepsilon)M\right]\mu^{-1}$

(7) $\qquad h'^{-1}\dfrac{d\eta'}{dt'} + h'^{-3}\eta'M' = \left(h^{-1}\dfrac{d\eta}{dt} + h^{-3}\eta M\right)\mu^{-1}h^{-1}.$

Let us go back to equations (11′) of Section 1; we can consider X_1, Y_1, Z_1 as having the same meaning as that in equations (5). On the other hand $l=1$ and $\dfrac{\varrho'}{\varrho}=k\mu$; these equations then become

(8)
$$\begin{cases} X_1'=\mu^{-1}(X_1+\varepsilon\,\varSigma X_1\,\xi), \\ Y_1'=k^{-1}\mu^{-1}Y_1. \end{cases}$$

Let us calculate $\varSigma X_1\,\xi$ with the aid of equation (5). We shall find

$$\varSigma X_1\,\xi=h^{-3}M,$$

whence

(9)
$$\begin{cases} X_1'=\mu^{-1}(X_1+\varepsilon h^{-3}M), \\ Y_1'=k^{-1}\mu^{-1}Y_1. \end{cases}$$

On comparing equations (5), (6), (7) and (9) we obtain

(10)
$$\begin{cases} h'^{-1}\dfrac{d\xi'}{dt'}+h'^{-3}\xi'M'=X_1', \\ h'^{-1}\dfrac{d\eta'}{dt'}+h'^{-3}\eta'M'=Y_1', \end{cases}$$

which demonstrates that the equations for quasi-stationary motion are not changed by the Lorentz transformation; but that does not prove that Lorentz's hypothesis is the only one which leads to this result.[228]

The problem is to determine how the equation for quasi-stationary motion (154) and

$$Y_1=\frac{1}{h}\frac{d\eta}{dt},\tag{155}$$

$$Z_1=\frac{1}{h}\frac{d\zeta}{dt}\tag{156}$$

behave under the LORENTZ transformation. The transformation equations for (154—6) are derived by assuming, on the basis of the principle of relativity, that they are valid in the $\varSigma'$ system. Then they will be transformed to S according to LORENTZ's equations in order to determine whether the relation between X_1', Y_1', Z_1' and X_1, Y_1, Z_1 is in agreement with [§ 1-11′]. POINCARÉ (see footnote 155), uncharacteristically, presented this derivation in full detail, leaving very few steps to be filled in by the reader. The probable reason for this is that, having LORENTZ's incorrect result in mind, he wanted to point out explicitly the steps necessary to derive the transformation equations from $\varSigma'$ to S for the electron's acceleration. POINCARÉ's general equations of transformation for the acceleration reduce to those of LORENTZ when $\varSigma'$ becomes S'. POINCARÉ's step-by-step presentation of this result makes it unnecessary to provide additional detail here. The result is that the equations of motion in the quasi-stationary approximation have the correct properties for a LORENTZ transformation; consequently, these equations cannot reveal absolute motion.

However, can other hypotheses besides LORENTZ's also lead to this result? This question is now answered by POINCARÉ in full generality:

---

[228] R. 160–161: O. 532–533.

In order to establish this point, we are going to restrict ourselves, in the same way that Lorentz has done, to certain particular cases, which would clearly be sufficient for us to demonstrate a negative proposition.

How shall we at first sight go about extending the hypotheses on which the preceding calculation rests ?

1. Instead of supposing that $l = 1$ in the Lorentz transformation let us assume some arbitrary l.

2. Instead of supposing that F is proportional to the volume and consequently that H is proportional to h, let us assume that F is an arbitrary function of θ and r, such that $\left(\text{after } \theta \text{ and } r \text{ have been replaced by their values as functions of } V, \text{ taken from the first two equations (1)}\right)$ H becomes an arbitrary function of V.

I observe first that if we suppose $H = h$, then we will be obliged to have $l = 1$; thus equations (6) and (7) hold, except that the second members will be multiplied by $1/l$; similarly for equation (9), except that the second members will be multiplied by $1/l^2$; and finally equations (10), except that the second members will be multiplied by $1/l$. If one does not want the equations of motion altered by the Lorentz transformation, that is that equations (10) are not to differ from equation (5) except for accents on the letters, then it is necessary to suppose that $l = 1$.[229]

POINCARÉ's program was to investigate the properties of the equations for quasi-stationary motion under LORENTZ transformations. In LORENTZ's model $l = 1$, and in LANGEVIN's model $l = k^{-\frac{1}{4}}$. Whereas in LORENTZ's model F is proportional to the electron's volume, in LANGEVIN's model $F = 0$. In general the form of F is determined by the value of l, characteristic of the model in question.

POINCARÉ then pointed to a subtlety in the derivation of $H = h$; namely that $H = h$ was derived from (128) in which l did not appear because it cancelled when H' from (118) was substituted into (120). Therefore, no restriction was placed on l because the addition of F to H served to only allow m to equal -1 instead of $-\frac{2}{3}$. Consequently, factors of l in the transformation equations (59d), (63) and [§ 1-11'] (see footnote 155) must be taken into account when X_1', Y_1', Z_1' are transformed to S according to LORENTZ's equations. These factors of l appear in

$$M' = \Sigma \, \xi' \, \frac{d\xi'}{dt'} = \frac{1}{l} \left[\frac{d\xi}{dt} \, \frac{\varepsilon h^2}{k^3 \mu^4} + \frac{M}{k^3 \mu^3} \right]$$

(see the equation directly above [§ 7-6]). Then the right-hand side of [§ 7-6] (and [§ 7-7] as well) become multiplied by $1/l$. From [§ 1-11'] (see footnote 155) the right-hand side of the first of equations [§ 7-8] is multiplied by $(k/l^5) \, \varrho/\varrho'$ and the second by $(1/l^5) \, \varrho/\varrho'$. Now, from (63) $\varrho'/\varrho = k\mu/l^3$ therefore, the right-hand side of [§ 7-8] is multiplied by $1/l^2$. However, according to the principle of relativity the equations of motion should be left unaltered. Thus $l = 1$. It is important to note that the velocity transformation equations are independent of l.

Taking into account the entire class of theories which admit of a real deformable electron that transforms according to (59), POINCARÉ went on to demonstrate

---

[229] R. 161–162: O. 533.

that the equations for quasi-stationary motion are consonant with the principle of relativity only if $l = 1$:

Suppose now that we have $\eta = \zeta = 0$, whence $\xi = V$, $\dfrac{d\xi}{dt} = \dfrac{dV}{dt}$; equations (5) will take the form

(5') $-\dfrac{d}{dt}\dfrac{dH}{d\xi} = \dfrac{dD}{dV}\dfrac{d\zeta}{dt} = X_1$, $-\dfrac{d}{dt}\dfrac{dH}{d\eta} = \dfrac{D}{V}\dfrac{d\eta}{dt} = Y_1$.

Let us set

$$\dfrac{dD}{dV} = f(V) = f(\xi), \qquad \dfrac{D}{V} = \varphi(V) = \varphi(\xi).$$

If the equations of motion are not altered by the Lorentz transformation, we will have obtained

$$f(\xi)\dfrac{d\xi}{dt} = X_1,$$

$$\varphi(\xi)\dfrac{d\eta}{dt} = Y_1,$$

$$f(\xi')\dfrac{d\xi'}{dt'} = X_1' = l^{-2}\mu^{-1}(X_1 + \varepsilon\Sigma X_1\xi) = l^{-2}\mu^{-1}X_1(1+\varepsilon\xi) = l^{-2}X_1,$$

$$\varphi(\xi')\dfrac{d\eta'}{dt'} = Y_1' = l^{-2}k^{-1}\mu^{-1}Y_1$$

and consequently

(11) $\begin{cases} f(\xi)\dfrac{d\xi}{dt} = l^2 f(\xi')\dfrac{d\xi'}{dt'}, \\[2mm] \varphi(\xi)\dfrac{d\eta}{dt} = l^2 k\mu\,\varphi(\xi')\dfrac{d\eta'}{dt'}. \end{cases}$

But we have

$$\dfrac{d\xi'}{dt'} = \dfrac{d\xi}{dt}\dfrac{1}{k^3\mu^3}, \qquad \dfrac{d\eta'}{dt'} = \dfrac{d\eta}{dt}\dfrac{1}{k^2\mu^2},$$

whence

$$f(\xi') = f\left(\dfrac{\xi+\varepsilon}{1+\varepsilon\xi}\right) = f(\xi)\dfrac{k^3\mu^3}{l^2},$$

$$\varphi(\xi') = \varphi\left(\dfrac{\xi+\varepsilon}{1+\varepsilon\xi}\right) = \varphi(\xi)\dfrac{k^2\mu^2}{l^2},$$

whence upon eliminating l^2 we find the functional equation

$$k^2\mu^2\,\dfrac{\varphi\left(\dfrac{\xi+\varepsilon}{1+\varepsilon\xi}\right)}{\varphi(\xi)} = \dfrac{f\left(\dfrac{\xi+\varepsilon}{1+\varepsilon\xi}\right)}{f(\xi)}$$

or, upon setting

$$\dfrac{\varphi(\xi)}{f(\xi)} = \Omega(\xi) = \dfrac{D}{V\dfrac{dD}{dV}},$$

then

$$\Omega\left(\dfrac{\xi+\varepsilon}{1+\varepsilon\xi}\right) = \Omega(\xi)\dfrac{1+\varepsilon^2}{(1+\xi\varepsilon)^2}.$$

This equation must be satisfied for all values of ξ and ε. For $\zeta = 0$ we find

$$\Omega(\varepsilon) = \Omega(0)(1-\varepsilon^2),$$

19*

whence
$$D = A \left(\frac{V}{\sqrt{1 - V^2}} \right)^m,$$

A being a constant and where I have made $\Omega(0) = 1/m$.

We then find
$$\varphi(\xi) = \frac{A}{\xi} \left(\frac{\xi}{\sqrt{1 - \xi^2}} \right)^m, \qquad \varphi(\xi') = \frac{A\mu}{\xi + \varepsilon} \left(\frac{\xi + \varepsilon}{\sqrt{1 - \xi^2}\sqrt{1 - \varepsilon^2}} \right)^m.$$

Since $\varphi(\xi') = \varphi(\xi) \frac{k\mu}{l^2}$, we have
$$(\xi + \varepsilon)^{m-1} (1 - \varepsilon^2)^{-m/2} = - \xi^{m-1} (1 - \varepsilon^2)^{-\frac{1}{2}} l^{-2}.$$

As l does not depend on ε (since, if there are many electrons, l remains the same for all the electrons of which the velocities can differ), this identity cannot be satisfied unless
$$m = 1, \qquad l = 1.^{230}$$

POINCARÉ's general proof is obtained by comparing the x and y-components of the force acting on the electron in the S and Σ' frames. The longitudinal and transverse masses are taken as general functions of V:

$$\frac{\partial D}{\partial V} = f(V) = f(\xi), \tag{157a}$$

$$\frac{D}{V} = \varphi(V) = \varphi(\xi) \tag{157b}$$

where $V = \xi$. Then the equation of motion in S becomes

$$-\frac{d}{dt} \frac{\partial H}{\partial \xi} = \frac{\partial D}{\partial V} \frac{d\xi}{dt} = f(\xi) \frac{d\xi}{dt} = X_1, \tag{158a}$$

$$-\frac{d}{dt} \frac{\partial H}{\partial \eta} = \frac{D}{V} \frac{d\eta}{dt} = \varphi(\xi) \frac{d\eta}{dt} = Y_1. \tag{158b}$$

Equation (158a) written in S' can be related to S by taking into account that^{231}

$$\frac{d\xi'}{dt'} = k^3 \frac{d\xi}{dt}. \tag{159}$$

---

^{230} R. 162–163: O. 534–535.

^{231} The equation of transformation of the acceleration from Σ' to S are

$$\frac{d\xi'}{dt'} = \frac{d\xi}{dt} \frac{1}{k^2 \mu^3},$$

$$\frac{d\eta'}{dt'} = \frac{d\eta}{dt} \frac{1}{k^2 \mu^2} - \frac{d\xi}{dt} \frac{\eta\xi}{k^2 \mu^2},$$

$$\frac{d\zeta'}{dt'} = \frac{d\zeta}{dt} \frac{1}{k^2 \mu^2} - \frac{d\xi}{dt} \frac{\zeta\xi}{k^2 \mu^2}.$$

However, in S', $\xi = -\varepsilon$, $\eta = \zeta = 0$, and the above equations become

$$\frac{d\xi'}{dt'} = \frac{d\xi}{dt} k^3,$$

$$\frac{d\eta'}{dt'} = \frac{d\eta}{dt} k^2,$$

$$\frac{d\zeta'}{dt'} = \frac{d\zeta}{dt} k^2$$

which are the same as (44) for $l = 1$ (see also footnote 39).

Then

$$X_1' = f(\xi') \frac{d\xi'}{dt'} = k^3 f(\xi') \frac{d\xi}{dt}. \tag{160}$$

In a similar manner (158b) becomes

$$Y_1' = \varphi(\xi') \frac{d\eta'}{dt'} = \varphi(\xi') k^2 \frac{d\eta}{dt}, \tag{161}$$

since

$$\frac{d\eta'}{dt'} = k^2 \frac{d\eta}{dt}. \tag{162}$$

Comparing (160) and (161) with (158) and utilizing the equations [232]

$$X_1' = l^{-2} X_1,$$

$$Y_1' = l^{-2} k Y_1$$

makes it possible to relate f and φ in S and S' as follows:

$$f(\xi) = f(\xi') k^3 l^2, \tag{163a}$$

$$\varphi(\xi) = \varphi(\xi') k l^2. \tag{163b}$$

The result of dividing (163a) by (163b) and recalling that $\xi' = 0$ in S' is

$$\Omega(\xi) = \Omega(0)(1 - V^2) \tag{164}$$

where $\Omega(\xi) = \varphi(\xi)/f(\xi)$. There is an obvious misprint in POINCARÉ's equation

$$\Omega\left(\frac{\xi + \varepsilon}{1 + \varepsilon\xi}\right) = \Omega(\xi) \frac{1 + \varepsilon^2}{(1 + \xi\varepsilon)^2}:$$

the factor $1 + \varepsilon^2$ should be $1 - \varepsilon^2$.

By use of (157) equation (164) can be rewritten as the differential equation

$$\frac{D}{V \frac{dD}{dV}} = \Omega(0)(1 - V^2) \tag{165}$$

where $\Omega(0)$ is a constant which POINCARÉ, with no loss of generality, took to be $1/m$ (this m is not the one in $r = b\theta^m$). Equation (165) is easily integrated to give

$$D = A \left(\frac{V}{\sqrt{1 - V^2}}\right)^m. \tag{166}$$

---

[232] Substituting (63) into [11'] in footnote 155 gives

$$X_1' = \frac{1}{l^2 \mu}(X_1 + \varepsilon \Sigma X_1 \xi), \tag{232-1a}$$

$$Y_1' = \frac{1}{l^2 k \mu} Y_1, \tag{232-1b}$$

$$Z_1' = \frac{1}{l^2 k \mu} Z_1. \tag{232-1c}$$

Since the motion is in the x-direction, $\Sigma X_1 \xi = X_1 \xi$ and (231-1a) becomes

$$X_1' = \frac{X_1}{l^2}.$$

From (166) the equation for $\varphi(\xi)$ is

$$\varphi(\xi) = \frac{D}{V} = \frac{A}{\xi}\left(\frac{\xi}{\sqrt{1-\xi^2}}\right)^m \tag{167}$$

(since $V = \xi$). However, (166) and consequently (167) should be valid for all values of ξ and ε, *i.e.* when S' is not the frame of the particle at rest; therefore,

$$\varphi(\xi') = \frac{A}{\xi'}\left(\frac{\xi'}{\sqrt{1-\xi'^2}}\right)^m. \tag{168}$$

The relation between ξ' and ξ is

$$\xi' = \frac{\xi+\varepsilon}{1+\varepsilon\xi} = \frac{\xi+\varepsilon}{\mu}. \tag{169}$$

Then

$$\sqrt{1-\xi'^2} = \sqrt{1-\left(\frac{\xi+\varepsilon}{1+\varepsilon\xi}\right)^2} = \frac{1}{\mu}\sqrt{1-\varepsilon^2}\sqrt{1-\xi^2}. \tag{170}$$

By means of (169) and (170) the general result (168) becomes

$$\varphi(\xi') = \frac{A\mu}{\xi+\varepsilon}\left(\frac{\xi+\varepsilon}{\sqrt{1-\varepsilon^2}\sqrt{1-\xi^2}}\right)^m. \tag{171}$$

The generalized version of (163 b) is

$$\varphi(\xi') = \frac{k\mu}{l^2}\varphi(\xi)$$

(now the factor l reappears), which, after use of (167), (169) and (171), becomes

$$\frac{A\mu}{\xi+\varepsilon}\left(\frac{\xi+\varepsilon}{\sqrt{1-\varepsilon^2}\sqrt{1-\xi^2}}\right)^m = \frac{k\mu A}{l^2\xi}\left(\frac{\xi}{\sqrt{1-\xi^2}}\right)^m, \tag{172}$$

or

$$(\xi+\varepsilon)^{m-1}(1-\varepsilon^2)^{-m/2} = \xi^{m-1}(1-\varepsilon^2)^{-\frac{1}{2}}l^{-2}. \tag{173}$$

(Note: There is a misprint in POINCARÉ's version of (173) which is

$$(\xi+\varepsilon)^{m-1}(1-\varepsilon^2)^{-m/2} = -\xi^{m-1}(1-\varepsilon^2)^{-\frac{1}{2}}l^{-2}.$$

The minus sign on the right-hand side should be a plus sign.) In order that (173) can be satisfied it is necessary that $m = 1$ and $l = 1$.

POINCARÉ continued:

> Thus Lorentz's hypothesis is the only one that is compatible with the impossibility of detecting absolute motion. If we admit that impossibility, then it is necessary to admit that the electron in motion undergoes a contraction in such a manner as to become an ellipsoid of revolution two of whose axes remain constant; it is also necessary to admit, as we have demonstrated in the previous section, the existence of a supplementary potential proportional to the electron's volume.[233]

Therefore, only if $l = 1$ can the equations of motion transform under (59) in such a way as to exclude the possibility of detecting motion with respect to the ether. Since $l = 1$ corresponds to LORENTZ's model, the electron's dimensions per-

---

[233] R. 163: O. 535.

pendicular to the direction of motion remain constant while those along the direction of motion are shortened by an amount $\sqrt{1-V^2}$. However, in order to obtain a LAGRANGIAN formulation for $l=1$ it is necessary to assume the existence of non-electromagnetic forces which enter the theory through a supplementary potential.

POINCARÉ then concluded Section 7:

> Lorentz's analysis turns out to be fully confirmed, but we are now better able to return and account for the real motive which occupied us in our work; that reason should be sought in the considerations of Section 4. *The transformations which do not change the equations of motion have to form a group, and that does not happen unless $l=1$.* Since we are not able to recognize if an electron is at rest or in absolute motion, it is necessary that when it is in motion it undergoes a deformation which is precisely that which is imposed by the transformation corresponding to the group.[234]

Therefore, the major result of this section is in agreement with the one deduced from group-theoretical considerations in Section 4.

By means of the principle of least action POINCARÉ went on to prove in Section 8 that the form for F, derived for quasi-stationary motion, was correct for any type of motion.

6.9. Arbitrary Motion [§ 8].[235]

The preceding results only apply to the case of quasi-stationary motion, but it is easy to extend them to the general case; it suffices to apply the principles of Section 3, that is to start with the principle of least action.

The expression for the action is

$$J = \int dt\, d\tau \left(\frac{\Sigma f^2}{2} - \frac{\Sigma \alpha^2}{2} \right).$$

It is advisable to add on a term, representing the supplementary potential F from Section 6; this term will certainly take the form

$$J_1 = \int \Sigma (F)\, dt,$$

where $\Sigma(F)$ represents the sum of the supplementary potentials attributed to the different electrons, each of them being proportional to the volume of the corresponding electron.

I write (F) between parentheses to prevent confusing it with the vector F, G, H.[236]

The action J is now considered to be the sum of terms from the self-fields of many electrons. To obtain LORENTZ's theory ($l=1$) necessitates the addition of a term $\int F\, dt$ to J. Since many electrons are being taken into account, the additional term must be written as

$$J_1 = \int \Sigma (F)\, dt \tag{174}$$

---

[234] R. 163: O. 535–536; italics in original.
[235] R. 164–166: O. 536–538.
[236] R. 164: O. 536.

where $\Sigma(F)$ is the sum of the supplementary potentials from each electron whose self-field is contained in J.

Poincaré finally expresses concern over his repeated use of symbols by emphasizing that F should not be confused with the x-component of the vector potential (see also footnotes 217, 226). Indeed, it seems as if " Sur la dynamique..." were somewhat hastily put together to furnish support for Lorentz's theory. Additional proof for this conjecture will be presented shortly.

It is assumed implicitly that the self-fields arising from each electron in the assembly do not interfere so that J can be written as $J = J(\text{electron}_1) + J(\text{electron}_2) + \cdots$.

Poincaré continued:

The total action is then $J + J_1$. We have seen in Section 3 that J is not changed by the Lorentz transformation; it is now necessary to show that it is the same for J_1.

We have for one of the electrons

$$(F) = \omega_0 \tau,$$

ω_0 being a special coefficient of the electron and τ its volume. I can then write

$$\Sigma(F) = \int \omega_0 d\tau,$$

where the integral is to be taken over all space, but in such a way that the coefficient becomes zero outside of the electron, and in the interior of each electron it becomes equal to a coefficient special to that electron. We then have

$$J_1 = \int \omega_0 d\tau\, dt,$$

and after the Lorentz transformation

$$J_1' = \int \omega_0' d\tau'\, dt'.$$

Thus we have $\omega_0 = \omega_0'$. This is so because if a point belongs to an electron, the corresponding point after the Lorentz transformation once again belongs to the same electron. Previously we have found in Section 3 that

$$d\tau'\, dt' = l^4 d\tau\, dt$$

and since we now suppose that $l = 1$,

$$d\tau'\, dt' = d\tau\, dt,$$

we then have

$$J_1 = J_1' \quad \text{Q.E.D.}[237]$$

For Lorentz's theory it has already been shown that F is proportional to the electron's volume $\big(\text{see } (139)\big)$:

$$F = A r^3 k^2 \tag{175}$$

(where $A = a/3\,b^4$) or, equivalently,

$$F = \frac{a}{3bk}. \tag{176}$$

---

[237] R. 164–164: O. 536–537.

In general (175) can be written as

$$F = \omega_0 \tau \tag{177}$$

where ω_0 is a coefficient characteristic of the electron in question. The entire assembly of electrons can be taken into account by writing

$$\Sigma(F) = \int \omega_0 d\tau \tag{178}$$

where the integral is taken over all space and ω_0, and therefore $\Sigma(F)$, are non-zero inside each electron. The total action $J + J_1$ written in the primed frame is $J' + J_1'$. Since $J = J'$ (from Section 3), it is necessary only to show that $J_1 = J_1'$, so that the principle of least action remains LORENTZ invariant. In the primed system J_1 is

$$J_1' = \int \omega_0' dt' d\tau_0'. \tag{179}$$

It was shown (see footnote 121) that

$$dt' d\tau' = l^4 dt d\tau; \tag{180}$$

however, in LORENTZ's theory, where $l = 1$, $dt d\tau$ is a LORENTZ invariant volume element. Furthermore, $\omega_0 = \omega_0'$ because ω_0' is a function of a point on the electron which must remain unchanged by the LORENTZ transformation; therefore, $J_1 = J_1'$.

POINCARÉ continued:

The theorem is thus a general one. It has given us at the same time a solution to the question that we posed at the end of Section 1: to find the complementary forces not altered by the Lorentz transformation. The supplementary potential (F) satisfies that condition.

We are therefore able to generalize the result stated at the end of Section 1 and write:

If the electron's inertia is exclusively of electromagnetic origin, if it is subjected to no forces other than those of electromagnetic origin, or to the forces due to the supplementary potential (F), then no experiment can detect evidence of absolute motion.[238]

In Section 1 POINCARÉ had demonstrated that a deformable electron could not be in equilibrium solely under the action of electromagnetic forces. The problem that was posed was to derive the form of the additional forces necessary to maintain equilibrium and have the proper properties under LORENTZ transformation. By means of a covariant principle of least action POINCARÉ proved that LORENTZ's theory could not be derived from the purely electromagnetic LAGRANGIAN (19). Rather, it was necessary to postulate an additional term to (19). The form of the non-electromagnetic energy F was obtained by demanding that the electron be stable under a variation of its action $\big($see (134)$\big)$. Assuming a separable form for F, POINCARÉ then demonstrated that in LORENTZ's theory F was proportional to the electron's volume. In this Section he proved that F had the desired properties under LORENTZ transformation.

---

[238] R. 165: O. 537; italics in original.

POINCARÉ then discussed the mechanical properties of F:

What are the forces which generate the potential (F)? They evidently can be compared with a pressure that exists in the interior of the electron; all of this implies that the electron is hollow and subjected to a constant internal pressure (independent of the volume); the work done by a similar pressure would evidentally be proportional to the variation of the volume.

I want to point out, however, that this pressure is negative. Recall equation (10) of Section 6, which according to Lorentz's hypothesis can be written

$$F = A r^3 \theta^2.$$

Equation (11) of Section 6 gives us

$$A = \frac{a}{3 b^4}.$$

Our pressure is equal to A, a coefficient taken as a constant, which, moreover, is negative.

Let us now evaluate the mass of the electron, by which I mean the "experimental mass", that is to say, the mass for small velocities. We have (cf. § 6)

$$H = \frac{\varphi\left(\frac{\theta}{k}\right)}{k^2 r}, \quad \theta = k, \quad \varphi = a, \quad \theta r = b,$$

whence

$$H = \frac{a}{bk} = \frac{a}{b} \sqrt{1 - V^2}.$$

For V very small I can write

$$H = \frac{a}{b}\left(1 - \frac{V^2}{2}\right)$$

which implies that the mass, both longitudinal and transverse, will be a/b.

Now, a is a numerical constant, which shows that *the pressure due to our supplementary potential is proportional to the fourth power of the experimental mass of the electron.*

Since the Newtonian attraction is proportional to the experimental mass, one is tempted to infer that there exists a general relation between the causes giving rise to gravitation and those which give rise to the supplementary potential.[239]

An understanding of the physical meaning of F would have been facilitated had POINCARÉ first evaluated the electron's mass in terms of a and b and then associated a pressure with F. The explicit form of F is required in order to understand what stresses it is cancelling. Towards this end, let us work backward through the passage above.

POINCARÉ had shown in Section 6, the necessity of adding the supplementary potential F to the electromagnetic LAGRANGIAN (115) to obtain the proper

---

[239] R. 165–166: O. 537–538; italics in original.

LAGRANGIAN H_L for LORENTZ's theory:

$$H_L = H + F = \frac{4}{3}\frac{a}{b}\sqrt{1-V^2}. \tag{140}$$

The electron's momentum is then

$$D = -\frac{\partial H_L}{\partial V} = \frac{4}{3}\frac{a}{b}\frac{V}{\sqrt{1-V^2}} \tag{181}$$

which agrees with LORENTZ's result (40) calculated from the electromagnetic definition of the momentum if $a/b = e^2/8\pi r$,[240] which is the electron's electrostatic mass.

However, according to POINCARÉ, a/b is "the experimental mass, that is to say, the mass for small velocities". Hence, it is the value for both the transverse and longitudinal masses in the limit $V \ll 1$. This statement is a result of noting how H behaves for $V \ll 1$. It is a peculiar statement because according to LORENTZ's theory a/b is *not* the correct limiting value for $m_\parallel$ and $m_\perp$. Rather, according to LORENTZ's theory, the limit $V \ll 1$ of $m_\parallel$ and $m_\perp$ is

$$m_\parallel = m_\perp = \frac{e^2}{6\pi r} = \frac{4}{3}m_e' \tag{182}$$

where $m_e' = a/b = \dfrac{e^2}{8\pi r}$.

Two conjectures can be made at this point about what happened:

(1) POINCARÉ made a mistake and should have looked at the limit of $H_L = \frac{4}{3}a/b\sqrt{1-V^2}$ instead of H.

(2) POINCARÉ realized that there was something wrong with the equation for the electron's momentum, which instead of $D = \frac{4}{3}m_e'Vk$ should be $D = m_e'Vk$.

I am inclined towards the first conjecture because in the days of pre-EINSTEINIAN relativity there was no reason to worry about whether the electrostatic mass (rest mass) or some multiple of it occurred in the momentum. It was assumed that the symbol m in the momentum represented the "experimental mass." Furthermore, it was POINCARÉ's goal to construct a LAGRANGIAN formalism for LORENTZ's theory wherein the value of both $m_\parallel$ and $m_\perp$ in the $V \ll 1$ limit was $e^2/6\pi r$ and not $e^2/8\pi r$. Moreover, it was well known that $e^2/6\pi r$ was the mass limit at low velocity for ABRAHAM's theory and in general for all electron theories which used the electromagnetic field momentum. Consequently, I shall continue the discussion of POINCARÉ's work assuming that he meant H_L and not H and, therefore, that $\frac{4}{3}a/b$ and not a/b is the experimental mass for low velocities. It is important to note that this error was also reflected in Section 7 where H and not H_L was used in the equations of motion. Additional evidence for the validity of the first conjecture will be presented when the explicit form of the stress associated with F is calculated.

POINCARÉ went on to note that since the additional *negative* energy is proportional to the electron's volume it can be represented as a pressure multiplied by a volume:

$$F = T_p \cdot \text{Volume}$$

---

[240] The same result for a/b as in LANGEVIN's model (see footnote 214).

where T_P is usually referred to as the POINCARÉ pressure.[241] Consequently, T_P can be obtained directly from (175) written as

$$F = \frac{A}{\frac{4}{3}\pi} \cdot \left(\frac{4}{3}\pi r^3 k^2 \right). \tag{183}$$

Since F represents the negative energy due to the mechanical stress acting on the electron, the stress T_P can be identified as

$$T_P = - \frac{|A|}{\frac{4}{3}\pi} \tag{184}$$

where $|A| = |a/3\,b^4|$. T_P is negative because it balances the repulsive electromagnetic stresses which tend to expand the electron's volume, i.e. T_P acts on the electron's elastic surface causing its volume to decrease. Thus, T_P is a constant negative cohesive pressure that acts on the inner surface of LORENTZ's deformable electron causing it to assume a spherical shape of radius r when at rest and to undergo a contraction when in motion.

We use $|A|$ in (184) to emphasize yet another anomaly in POINCARÉ's notation. Namely that since A turned out to be a negative quantity, then from (176) the "experimental mass" a/b should also be negative, i.e. $a/b = - e^2/(8\pi r)$. Clearly, this was not what POINCARÉ meant and was merely an oversight. However, it does reflect back to (140) which, to conform with POINCARÉ's sudden shift in notation, should be $- H_L$.

POINCARÉ's assumption that the electron at rest is a hollow deformable sphere with a uniform surface charge distribution is made purely on the grounds of simplicity. There is no loss of generality because the stability problem is independent of the charge distribution. This assumption enabled him to split the stability problem into two parts: the POINCARÉ stress on the interior balancing the repulsive COULOMB forces on the exterior (there is no electromagnetic field inside this electron).

In the brief version of "Sur la dynamique ...", published in the *Comptes Rendus*, POINCARÉ made the misleading comment that T_P is an external pressure. This statement was corrected in the version in the *Rendiconti* "... the electron is hollow and subjected to a constant internal pressure."[239] Then in 1912 POINCARÉ presented again a detailed exposition of T_P.[242] Using a hydrostatic analogy, he

[241] T_P is the quantity referred to in all discussions of POINCARÉ's work on the self-stress problem. See, for example, footnote 30; L. DE BROGLIE, *New Perspectives in Physics*, translated by A. J. POMERANS (Basic Books, 1962), esp. pp. 41, 45 and 48; S. SCHWEBER, *An Introduction to Relativistic Quantum Field Theory* (Row, Peterson and Company, 1961), esp. p. 514.

[242] H. POINCARÉ, *La dynamique de l'électron* (Dumas, 1913), esp. pp. 58–60. This is an extremely interesting work because it shows that POINCARÉ's views on relativity remained unchanged to the end of his life.

LOUIS DE BROGLIE, in his popular exposition *Matter and Light*, propagated POINCARÉ's misleading statement in the Comptes rendus version of "Sur la dynamique ..." to the effect that T_P is an external pressure: "... the great difficulty is to understand how such a sphere, containing electricity all of the same sign, can exist in a stable manner, since its constituent parts ought to repel each other. We are driven to imagine, with Henri Poincaré, that there is a pressure at the surface of the electron coming from outside and preventing the particle exploding; but we remain wholly at a loss to explain the origin of such a pressure." (L. DE BROGLIE, *Matter and Light: The New Physics*, translated by W. H. JOHNSTON (Norton, 1939; reprinted by Dover, n.d.), p. 108.)

demonstrated that the force associated with T_P is equal and opposite to the body force (the sum of the self force plus the external force) exerted on the deformable electron. He also reaffirmed, by means of this technique, that T_P has the proper properties under Lorentz transformation.

It can hardly be overemphasized that this was as far as POINCARÉ went in 1905. His crucial result is the impossibility of constructing an electron theory that is consonant with the principle of relativity, using only the LAGRANGIAN of the electromagnetic field. Rather, it is necessary to add a supplementary and non-electromagnetic energy to H. These remarks were aptly summarized by POINCARÉ in " Sur la dynamique ...": "*If the electron's inertia is exclusively of electromagnetic origin, if it is subjected to no forces other than those of electromagnetic origin, or to the forces due to the supplementary potential (F), then no experiment can detect evidence of absolute motion.*"[238] The generality of this result demonstrated the impossibility of formulating a relativistic pre-EINSTEINIAN electromagnetic world-picture.

7. The Relevance of Poincaré's Results to Twentieth-Century Physics

> "Fait incroyable, le Mémoire de Poincaré est à peu près inconnu et n'est presque jamais cité. Il est introuvable en librairie. C'est là une lacune regrettable, qu'il était urgent de combler."
>
> From ÉDOUARD GUILLAUME's introduction (1924) to POINCARÉ's *La Mécanique nouvelle*.

7.1. General Perspective. "Sur la dynamique ..." was not "almost unknown and ... hardly ever cited", as ÉDOUARD GUILLAUME asserted in the introduction to the reprint volume of POINCARÉ's *La Mécanique nouvelle*,[243] published in 1924.

---

[243] H. POINCARÉ, *La Mécanique nouvelle* (Gauthier-Villars, 1924). This is a reprint volume containing three of POINCARÉ's works: a lecture entitled "La Mécanique nouvelle" delivered to the Association française pour l'Avancement des Sciences at the Congrès de Lille in 1909; the Comptes rendus and Rend. di Palermo versions of "Sur la dynamique ..." In his introduction ÉDOUARD GUILLAUME argued that POINCARÉ should share in the acclaim that was accorded only to EINSTEIN. Some reasons for his opinion are: POINCARÉ also demonstrated that the LORENTZ transformations form a group (p. vii); POINCARÉ also derived the relativistic addition law for velocities (p. vii); POINCARÉ obtained the same results as the proponents of "*L'École relativiste*" (p. viii). In addition, GUILLAUME implied that EINSTEIN's notion of the relativity of time was a generalization of LORENTZ's local time, which was given a physical meaning by POINCARÉ (pp. vi–vii).
A detailed study of the response to relativity theory has been made by S. GOLDBERG, "The Early Response to Einstein's Special Theory of Relativity, 1905–1911: A Case Study of National Differences", unpublished Ph. D. Thesis, Harvard University (1968). Papers based on portions of this case study have already appeared in print:
  a) S. GOLDBERG's "Henri Poincaré and Einstein's ...", *op. cit.* (footnote 8(c)),
  b) "Poincaré's Silence and Einstein's ...", *op. cit.* (footnote 8(d)),
  c) "The Lorentz Theory of Electrons ...", *op. cit.* (footnote 22(c)),
  d) "The Abraham Theory of the Electron ...", *op. cit.* (footnote 22(c)).
  e) "In Defense of the Ether: The British Response to Einstein's Special Theory of Relativity; 1905–1911" *Historical Studies in the Physical Sciences: Second Annual Volume* 1970, R. MCCORMMACH ed. (University of Pennsylvania Press, 1970), pp. 89–125. With respect to the reception of EINSTEINIAN relativity in France, GOLDBERG notes

In fact, starting in 1907, it was cited in many important studies; however, the primary emphasis in them was already on EINSTEIN's paper of 1905, which could then be regarded as an important generalization of LORENTZ's theory.[244]

MAX PLANCK, in the first complete exposition of EINSTEINIAN kinematics written in 1907, referred to POINCARÉ's research on the electron's structure.[245]

HERMANN MINKOWSKI, in his seminal work of 1907 on a four-dimensional formulation of relativity entitled "Das Relativitätsprinzip", cited POINCARÉ's work in "Sur la dynamique ..." at least four times.[246] What especially impressed him was POINCARÉ's emphasis on how electrodynamical quantities transform according to the LORENTZ group, LORENTZ invariants and, more generally, the symmetries inherent in the equations of electromagnetism.

MAX ABRAHAM, in the second edition of his book *Theorie der Elektrizität* (1908) cited "Sur la dynamique ..." as the place where the name "LORENTZ transformations" first appeared.[247]

H. A. LORENTZ, in 1909, published *The Theory of Electrons*, which is an expanded version of a series of lectures he had given at Columbia University in the spring of 1906. In it can be found the first complete published account of the application of the POINCARÉ stress to LORENTZ's theory.[248]

In 1911 MAX VON LAUE's widely read monograph *Das Relativitätsprinzip*[249] appeared. In it POINCARÉ's work on the stability of the electron is discussed within the context of PLANCK's and MINKOWSKI's results. VON LAUE showed what is

---

that: "Apparently Poincaré's influence was responsible for the fact that almost no one else in France had much to say about Einstein's theory prior to Poincaré's death ..." ("In Defense ...", p. 97).

I am grateful to Professor GERALD HOLTON for making a copy of this thesis available to me.

A good illustration of GOLDBERG's conclusion is that PAUL LANGEVIN, a student and friend of POINCARÉ, did not publish anything on EINSTEINIAN relativity until after he had obtained a secure position at the Collège de France in 1909. In 1911 LANGEVIN published an important exposition of EINSTEIN's theory entitled "L'évolution de l'espace et du temps", Scientia, **10**, 31–54 (1911). This article brought LANGEVIN to EINSTEIN's attention, thereby serving as the catalyst for a lifelong friendship.

[244] For an enlightening discussion of the period of transition, 1905–1910, see T. HIROSIGE, "Theory of Relativity and the Ether," Japanese Studies in the History of Science, **7**, 37–53 (1968). During this period the crucial question was whether the velocity dependence of the transverse mass conformed to the theory of ABRAHAM or LORENTZ-EINSTEIN.

[245] M. PLANCK, "Zur Dynamik bewegter Systeme," Sitzungsber. d. k. preuss. Akad. Wiss., **13**, 542–570 (1907), esp. p. 544. This paper was also published in Ann. d. Phys. **26**, 1–34 (1908). Although PLANCK did not explicitly cite "Sur la dynamique ...", he must have had it in mind since it was POINCARÉ's only technical paper on this subject. A previous important paper by PLANCK on the LORENTZ-EINSTEIN theory, in which POINCARÉ is not mentioned, is "Das Prinzip der Relativität und die Grundgleichungen der Mechanik," Verh. d. p. Ges., **6**, 136–141 (1906).

[246] H. MINKOWSKI, "Das Relativitätsprinzip," *op. cit.* (footnote 151(a)), esp. pp. 928, 929, 931 and 938.

[247] M. ABRAHAM, *Theorie der Elektrizität*, Volume 2 (Teubner, 1908), esp. p. 370.

[248] H. A. LORENTZ, *T. E.*, esp. pp. 213–215. See also H. A. LORENTZ, *Lectures on Theoretical Physics: Volume III* (Macmillan, 1931), translated by L. SILBERSTEIN & A. P. H. TRIVELLI, esp. pp. 261–265. This section is from a set of lectures given by LORENTZ in 1910–1912 at Leiden.

[249] M. VON LAUE, *Das Relativitätsprinzip* (Friedr. Vieweg und Sohn, 1911).

required of *any* field-theoretical description of an extended elementary particle: the self-stress must vanish in the particle's rest frame so that its energy and momentum transform as what are now referred to as four-vectors.

Indeed, a complete study of "Sur la dynamique …" cannot end with the year 1906; rather it necessitates tracing the impact of POINCARÉ's results through more than one-half century of developments in physical theory to the present-day quantum theory of the electron. This requires first a discussion of LORENTZ's application of the POINCARÉ stress. Then a comparison of the relativity theories of EINSTEIN and LORENTZ will be followed by an account of MINKOWSKI's work. Next a study of VON LAUE's exposition of POINCARÉ's results will serve as the background for a discussion of the elementary particle theories of GUSTAV MIE (1912), HERMANN WEYL (1918) and EINSTEIN (1919). Then subsequent work on classical electron theory by ENRICO FERMI (1922), P. A. M. DIRAC (1938), BERNARD KWAL (1949) and FRITZ ROHRLICH (1960) will be presented. Part 7 will conclude with a discussion of the relevance of "Sur la dynamique …" to the quantum theory of the electron, quantum electrodynamics.

7.2. Lorentz's Application of the Poincaré Stress. The first application of POINCARÉ's results to LORENTZ's theory was made by LORENTZ himself. Let us reconstruct his arguments from POINCARÉ's equations.[248]

The instability of the deformable electron in S' is due to the stress that arises from the electron's COULOMB field. This can be calculated from the MAXWELL stress tensor as[250]

$$T' = \frac{e^2}{32\,\pi^2 r^4}. \tag{185}$$

---

[250] The MAXWELL stress tensor written in dyadic form is

$$\overset{\leftrightarrow}{T} = \vec{E}\vec{E} + \vec{B}\vec{B} - \tfrac{1}{2}\overset{\leftrightarrow}{II}(E^2 + B^2)$$

(see JACKSON's *Classical Electrodynamics*, *op. cit.* (footnote 191), p. 149). Taking the z'-axis along the outward normal to the spherical electron (in S') gives

$$T' = \frac{E'^2}{2} = \frac{e^2}{32\,\pi^2 r^4}.$$

By use of the MAXWELL-LORENTZ equations to eliminate ϱ and $\varrho\vec{v}$ the LORENTZ force equation can be written as

$$\vec{V} \cdot \overset{\leftrightarrow}{T} - \frac{\partial \vec{g}_e}{\partial t} = \varrho\vec{E} + \varrho\vec{v} \times \vec{B}$$

where $\vec{g}_e = \vec{E} \times \vec{B}$. (See JACKSON's *Classical Electrodynamics*, *op. cit.* (footnote 191), pp. 190–194.) In the electron's rest frame $\vec{B}' = 0$; therefore,

$$\vec{V}' \cdot \overset{\leftrightarrow}{T'} = \varrho'\vec{E}'.$$

This is the mathematical expression of the statement that in S' the non-vanishing electrostatic field gives rise to a stress which tends to tear the electron apart. The terminology stress tensor arose in connection with the concept put forth by MAXWELL of an ether which actually exerts stresses on charged bodies. However, if stresses are transmitted by the ether, then it cannot be at rest. The reason is that forces are being exerted by one part of the ether on another part. This point, which caused HELMHOLTZ to admit a moving ether into his theory, was of great concern to LORENTZ. He finally chose to deny the existence of MAXWELL stresses but pointed out that the mathematical formalism of the stress tensor could still be utilized. His reason was

Since F transforms like H, its form in S' can be calculated directly from (176) as

$$F' = \frac{a}{3b}. \tag{186}$$

Dividing F' by the volume of the electron in S' gives the stress T'_P:

$$T'_P = \frac{a}{3b} \cdot \frac{1}{(\frac{4}{3}\pi r^3)} = -\frac{e^2}{32\pi^2 r^4}, \tag{187}$$

where according to POINCARE's notation $a/b = -e^2/(8\pi r)$. Thus T'_P cancels the self-stress (185) thereby ensuring the electron's stability in S' and resolving the inconsistency raised in Section 1 of "Sur la dynamique ...". Note that if a/b instead of $\frac{4}{3}a/b$ is taken as the "experimental mass", then T'_P from (187) does not cancel the self-stress T' in (185). That the resolution of this problem is directly linked to the disappearance of the extra term in (48) can be shown in the following way. The work W_T done by the POINCARÉ stress T_P in deforming the electron is $-F$:

$$W_T = -F \tag{188}$$

or

$$W_T = \frac{e^2}{24\pi r}(1 - v^2)^{\frac{1}{2}}. \tag{189}$$

The total energy W of a LORENTZ electron is the sum of the portion calculated from electromagnetic field theory W_{elm}, from (48) and W_T:

$$W = W_{\text{elm}} + W_T = \frac{e^2}{6\pi r}(1 - v^2)^{-\frac{1}{2}}. \tag{190}$$

Therefore,

$$\frac{1}{v}\frac{dW}{dv} = m_{\parallel} = \frac{dD}{dv} \tag{191}$$

instead of the result in (50). Equation (191) can also be obtained by using H_L (see (140)) as the LAGRANGIAN for LORENTZ's model instead of H, as POINCARÉ did in Section 8 of "Sur la dynamique ...".

However, LORENTZ proved that although the electron in his theory is stable in S' because of the POINCARÉ stress, i.e. it does not explode in S', nevertheless it is unstable with respect to changes of shape.[251] This necessitates, for theories of a deformable electron, the ad hoc postulation of additional stresses that vanish in S'. ABRAHAM's theory was saved from this dilemma because its electron was postulated to be a rigid sphere; however, this theory was not consonant with the prin-

---

that "the ether is undoubtedly different from ordinary matter" (LORENTZ, T.E., p. 30). See T.E. (esp. pp. 30–32) for LORENTZ's discussion of the properties that must be attributed to the fixed ether. These characteristics serve to distinguish LORENTZ's ether from any elastic body that is encountered in continuum mechanics. That the ether is an anomalous elastic substance was already known from the ether theories of FRESNEL, CAUCHY, STOKES, MacCULLAGH and KELVIN.

[251] H. A. LORENTZ, T.E. (Dover, 1952), pp. 215, 335–335. See also LORENTZ, "The Connection Between Momentum and the Flow of Energy. Remarks Concerning the Structure of Electrons and Atoms," Versl. Kon. Akad. Wetensch. Amsterdam, 26, 981 (1917), in Collected Papers, 5, pp. 314–329, esp. p. 328.

ciple of relativity. However, in 1908, BUCHERER[252] confirmed the LORENTZ-EINSTEIN theory on the basis of a set of experiments that avoided KAUFMANN's technical difficulties.

Thus, LORENTZ's electron theory, despite its remaining internal problems, was in agreement with the experimental data. With respect to the remaining problems and the future of his theory *circa* 1909 LORENTZ noted:

> Notwithstanding all this, it would, in my opinion, be quite legitimate to maintain the hypothesis of the contracting electrons, if by its means we could really make some progress in the understanding of phenomena. In speculating on the structure of these minute particles we must not forget that there may be many possibilities not dreamt of at present; it may very well be that other internal forces serve to ensure the stability of the system, and perhaps, after all we are wholly on the wrong track when we apply to the parts of an electron our ordinary notion of force.[253]

---

[252] A. H. BUCHERER, "Messungen an Becquerelstrahlen. Die experimentelle Bestätigung der Lorentz-Einsteinschen Theorie," Phys. Zeit., 9, 755–762 (1908). However, there was some hesitation to accept BUCHERER's work due mainly to the criticism of A. BESTELMEYER, "Bemerkung zu der Abhandlung Herrn A. H. Bucherer's 'Die experimentelle Bestätigung des Relativitätsprinzips'," Ann. d. Phys., 30, 166–174 (1909). The publication of BUCHERER to which BESTELMEYER referred was published in Ann. d. Phys., 28, 513–536 (1909). For further references to this controversy see JAMMER's *Concepts of Mass, op. cit.* (footnote 14), esp. pp. 166–167. The more accurate experiments of G. NEUMANN ("Die träge Masse schnell bewegter Elektronen," Ann. d. Phys., 45, 529–579 (1914)) and C. E. GUYE & CH. LAVANCHY ("Vérification expérimentale de la formule de Lorentz-Einstein par les rayons cathodiques de grande vitesse," Arch. Sci. phys. nat., 41, 286, 353, 441 (1916)) were considered as decisive in favor of the prediction of LORENTZ and EINSTEIN. BUCHERER's experiment of 1908 is mentioned by LORENTZ in *T.E.* in a note added to the 1909 edition when it was reprinted in 1915 (p. 339): "Later experiments by Bucherer ... have confirmed the formula for the transverse electromagnetic mass, so that, in all probability, the only objection that could be raised against the hypothesis of the deformable electron and the principle of relativity has now been removed." On the other hand KAUFMANN's unfavorable results were discussed briefly on pp. 42–43. However, in passages which appeared in the edition of 1909 of *T.E.*, LORENTZ presented convincing evidence for his theory over those of ABRAHAM and LANGEVIN on the basis of the precise optical experiments of RAYLEIGH and BRACE (pp. 17–20). LORENTZ proved that the double refraction predicted by ABRAHAM or LANGEVIN should have been observed by RAYLEIGH and BRACE, while his model accounted for their null result.
It is of interest to note, as JAMMER points out on p. 167 of the work cited in footnote 14, that BUCHERER's experimental methods were analyzed in 1938 by C. T. ZAHN & A. A. SPEES ("A critical analysis of the classical experiments on the variation of electron mass," Phys. Rev. 53, 511–521 (1938)), who discovered that due to a defect in his velocity filters BUCHERER's resolution was not good enough to decide between the predictions from the theories of ABRAHAM or LORENTZ-EINSTEIN for the transverse mass. For a critical discussion of the current state of this problem see JAMMER, pp. 168–171.

[253] H. A. LORENTZ, *T.E.*, p. 215. Even as late as 1917 LORENTZ was prepared, if necessary, to introduce additional terms into his theory so as to maintain the extended electron's stability because "as it is, quantities have already been introduced of a nonelectromagnetic nature by the supplementary stress $[T_P]$ and the energy corresponding to it." (H. A. LORENTZ, "The Connection Between ..." *op. cit.* (footnote 251) p. 329.)

20 Arch. Hist. Exact Sci., Vol. 10

Lorentz went on to suggest an alternative to the postulation of additional mechanical stresses: "... we may rest content with simply admitting for the moving electron, without any further discussion, the ellipsoidal form with the smaller axis in the line of translation."[254]

Therefore, the stability problem was never completely resolved within the framework of Lorentz's theory of the electron. In fact, Poincaré, in his last major paper on electron theory, avoided the problem by noting: "Is its [the Lorentz electron] equilibrium stable? That is a very difficult question and we make no response here."[255]

7.3. The Theories of Einstein and Lorentz. Einstein asserted almost immediately in "Zur Elektrodynamik bewegter Körper" that "[t]he introduction of a 'luminiferous ether' will prove to be superfluous inasmuch as the view here to be developed will not require an 'absolute stationary space' provided with special properties ..."[256] Thus, in Einstein's theory of relativity the space-time coordinates in *all* inertial frames have a physical interpretation: all inertial reference systems are equivalent. Then by means of a critical analysis of the manner in which space and time are measured Einstein demonstrated that they are relativistic concepts. Towards a deductive theory of relativity he then raised the principle of relativity (applied now to all of physics and not just to electromagnetism) and the constancy of the speed of light in vacuum to postulates. This enabled him to derive the equations of relativistic transformation that Lorentz and Poincaré previously had to postulate. Einstein then demonstrated that a contraction in length and a dilation in time followed from these equations. He attributed a kinematical interpretation to these consequences of the two postulates of relativity. Lorentz and Poincaré, on the other hand, had to postulate the contraction of length and then, consonant with an electromagnetic world-picture, gave to it a dynamical interpretation, *i.e.* it was a real and not an apparent effect. Furthermore, they attributed no reality to the effect of dilation of time. For Poincaré it was merely a shift in the origin of the time variable. For Einstein an understanding of the nature of time was the key to a truly universal principle of relativity. Einstein then derived on an exclusively kinematical basis the relativistic law for the addition of velocities. Poincaré had obtained this, within the confines of the electromagnetic world-picture, by demanding that the Maxwell-Lorentz equations be Lorentz-covariant.

Einstein then turned to electrodynamics and demonstrated that (59), written in the conventional manner as

$$x' = k(x - vt),$$
$$y' = y,$$
$$z' = z, \tag{59'}$$
$$t' = k(t - vx),$$

---

[254] *Ibid.*, p. 215.

[255] H. Poincaré, *La dynamique de l'électron* (Dumas, 1913), p. 60.

[256] A. Einstein, "Zur Elektrodynamik bewegter Körper," *op. cit.* (footnote 2), p. 38.

constitute the set of space-time transformations which leave the MAXWELL-LORENTZ equations unchanged.

EINSTEIN then showed that LORENTZ's theory is consonant with the principle of relativity: if the fields transform as in (67) (with ε replaced by $-v$); if the vector v on the right hand side of (1 d) is the velocity of the moving charge, since it transforms like a velocity; and if the charge density transforms like (63) (once again with ε replaced by $-v$). Consequently, "on the basis of our kinematical principles, the electrodynamic foundation of LORENTZ's theory of the electrodynamics of moving bodies is in agreement with the principle of relativity."[257]

EINSTEIN's universal interpretation of the principle of relativity manifested itself in the section entitled "Dynamics of the Slowly Accelerated Electron" wherein he avoided all of the basic problems that confronted the proponents of an electromagnetic world-picture. He treated the problem of an electron acted upon by an external electromagnetic field by simply stating $F = ma$ for the particle instantaneously at rest and then transforming this equation according to (67) and (59') to a moving frame of reference. Ignoring the origin of the electron's rest mass, he defined it simply as some number m', the same kind of term that would be inserted into $F = ma$ for a block sliding down an inclined plane.

Thus, EINSTEIN was able to dodge problems concerning the nature of matter: his approach to the electron, in this paper, was a kinematical one. This was contrary to the reductionist goal of the electromagnetic world-picture. For EINSTEIN, electromagnetism and mechanics were two branches of physics which conformed to the principle of relativity; he made no attempt to reduce one to the other. Thus, the laws of mechanics could be extended, in a relativized form, into the domain of microscopic physics.

That this was an attractive proposal was evidenced by the fact that MAX PLANCK immediately began a series of articles in which he constructed a relativistic kinematics.[245, 258] Problems concerning the structure of the electron could now

---

[257] *Ibid.*, p. 60.

[258] It is important to note that EINSTEIN's unfortunate choice for the definition of force in his paper of 1905 (see the reference in footnote 2) led him to the famous incorrect equation for the transverse mass,

$$m_\perp = \frac{m'}{1 - \dfrac{v^2}{c^2}}$$

instead of

$$m_\perp = \frac{m'}{\sqrt{1 - \dfrac{v^2}{c^2}}}.$$

Indeed, this constitutes another piece of convincing evidence that he was not aware of LORENTZ's paper of 1904 which contains the correct result for the transverse mass. EINSTEIN would very probably have commented on this discrepancy.

EINSTEIN went on to assert, in the reference in footnote 2, that the equations for $m_\parallel$ and $m_\perp$ should be valid for all bodies. However, it should be possible to derive a result of such universal importance from a nonelectromagnetic argument. This objection was raised by G. N. LEWIS & R. C. TOLMAN in 1909 ("The Principle of Relativity and Non-Newtonian Mechanics," Phil. Mag., **18**, 510–523 (1909)). They derived the mass equation (192), which is the correct equation to put into the rate of change of momentum, from an argument based on conservation of momentum and the LORENTZ

20*

be circumvented by simply defining m' as the rest mass of a point particle. The disagreeable restriction to quasi-stationary motion and its implication that the laws of mechanics were of only approximate validity for microscopic physics was obviated, thus perhaps accounting for PLANCK's not referring to POINCARÉ's exposition of relativistic mechanics although it was correct (albeit within the confines of the quasi-stationary approximation). In EINSTEIN's framework the relativistic version of NEWTON's second law is exact for bodies of all dimensions. Consequently, the inertia of all bodies increases from the value of m' in its rest frame in accordance with the equation

$$m = \frac{m'}{\sqrt{1 - v^2/c^2}} \,. \tag{192}$$

Although EINSTEIN's functional form for the velocity dependence of the mass is the same as the form in LORENTZ's theory, the conceptual framework from which it evolved, and, therefore, the meaning of the terms, are entirely different. Thus, the EINSTEIN contraction is not the same as the LORENTZ contraction which asserts the existence of absolute lengths. The differences between LORENTZ's and EINSTEIN's theories were not appreciated by most of those concerned with relativity theory (including PLANCK) until the concept of the ether rapidly began to disappear (c. 1910).[244]

7.4. Hermann Minkowski's Four-Dimensional Formulation of Relativity. It is reasonable to conjecture that what especially impressed MINKOWSKI in "Sur la dynamique ..." was POINCARÉ's realization that the charge density (ϱ) and the components of the convection current $(\varrho\,\xi,\ \varrho\eta,\ \varrho\zeta)$ transform according to the LO-RENTZ transformation in the same way as do the coordinates t and $x,\ y,\ z$; similarly for $\Sigma X\xi$ and X, Y, Z as well as for Ψ and F, G, H. Indeed, in "Sur la dynamique ..." POINCARÉ had discovered that certain combinations of these quantities, as well as the electromagnetic field components, are LORENTZ invariants, $e.g.\ x^2 + y^2 + z^2 - c^2 t^2$ and $E^2 - B^2$. Furthermore, in Section 9, POINCARÉ had presented a four-dimensional formulation of a LORENTZ-covariant gravitational theory in which "the Lorentz transformation is simply a rotation of this [four-dimensional] space about a fixed origin."[259]

MINKOWSKI's mathematical insight, coupled with an appreciation of EIN-STEIN's formulation of the principle of relativity, enabled him to discover the meaning of the symmetries uncovered by POINCARÉ. In particular, the LORENTZ invariant $x^2 + y^2 + z^2 - c^2 t^2$ implied the inseparability of space and time. Thus, the covariance of the equations of physics is displayed by using not a three-dimensional vector formalism, but rather one in a four-dimensional space.

---

transformation (see JAMMER's *Concepts of Mass, op. cit.* (footnote 14), esp. pp. 158–165 for a discussion of this episode; note that JAMMER, in describing EINSTEIN's paper of 1905, on p. 160 writes EINSTEIN's result for the transverse mass as

$$m_\perp = \frac{m'}{\sqrt{1 - \dfrac{v^2}{c^2}}}$$

thereby inadvertently correcting his error).

[259] R. 168: O. 542: K. 176.

MINKOWSKI first stated his formulation of physics in a four-dimensional vector space in a lecture delivered at Göttingen on 5 November, 1907, entitled "Das Relativitätsprinzip."[151(a)] He referred to POINCARÉ's work in "Sur la dynamique ..." at least four times in the course of this lecture. In particular, he emphasized that it was his goal to exhibit those symmetries in the equations of physics which "had not occurred to any of the previously mentioned authors, not even POINCARÉ himself [the others were EINSTEIN, LORENTZ and PLANCK] . . ."[260]

On 21 December, 1907 MINKOWSKI submitted for publication a detailed account of his four-vector formalism, entitled "Die Grundgleichungen für die elektromagnetischen Vorgänge in bewegten Körpern."[151(b)] "Sur la dynamique ..." is explicitly cited twice in this paper.[261]

Then, at Cologne on 21 September, 1908, MINKOWSKI delivered a lecture, entitled "Raum und Zeit".[262] He began by emphasizing the philosophical and physical implications of the fact that the spatial and time intervals are not separately LORENTZ invariant; rather the LORENTZ invariant is the combination $x^2 + y^2 + z^2 - c^2 t^2$:

> Henceforth, space by itself, and time by itself, are doomed to fade away into mere shadows, and only a kind of union of the two will preserve an independent reality.[263]

7.5. Max von Laue's Explanation of the Poincaré Stress within the Context of Einsteinian Relativity. By 1911 it had become apparent that a theory of a stable deformable electron whose mass is derived from its self-field could not be constructed using pre-EINSTEINIAN definitions of kinematical quantities. Therefore, the fundamental goal of the electromagnetic world-picture of ABRAHAM and LORENTZ was unattainable.

---

[260] H. MINKOWSKI, "Das Relativitätsprinzip," *op. cit.* (footnote 151 (a)), p. 129. MAX BORN recalled that while at Göttingen "in a seminar on the theory of electrons [which began in 1905], held not by a physicist but by a mathematician, Hermann Minkowski ... We studied papers by Hertz, Fitzgerald, Larmor, Lorentz, Poincaré, and others but also got an inkling of Minkowski's own ideas which were published only two years later" (MAX BORN, *Physics in My Generation* (Springer, 1969), p. 101).

[261] See reference in footnote 151 (b), pp. 54 and 129.

[262] H. MINKOWSKI, "Raum und Zeit," *op. cit.* (footnote 151 (c)). POINCARÉ is not mentioned in this work.

[263] H. MINKOWSKI, "Raum und Zeit," *op. cit.* (footnote 151 (c)), p. 75 of the Dover edition. EINSTEIN's opinion on MINKOWSKI's contribution to the theory of relativity is recorded in his "Autobiographical Notes":

> Minkowski's important contribution to the theory lies in the following: Before Minkowski's investigation it was necessary to carry out a Lorentz-transformation on a law in order to test its invariance under such transformations; he, on the other hand, succeeded in introducing a formalism such that the mathematical form of the law itself guarantees its invariance under Lorentz-transformations. By creating a four-dimensional tensor-calculus he achieved the same thing for the four-dimensional space which the ordinary vector-calculus achieves for the three spatial dimensions. He also showed that the Lorentz-transformation (apart from a different algebraic sign due to the special character of time) is nothing but a rotation of the coordinate system in the four-dimensional space.

(P. SCHILPP, ed., *Albert Einstein: Philosopher-Scientist* (The Library of Living Philosophers, 1949), p. 59).

In order to explain this statement fully it is necessary to begin with a discussion of one of the crucial results of EINSTEIN's theory — the mass-energy equivalence[264]

$$E = mc^2 \tag{193}$$

where m is the relativistic mass (192) and E is the particle's total energy. This equation is a purely kinematical result and therefore no assumptions have been made concerning the origin of the rest mass m'. In the particle's rest frame (193) becomes

$$E' = m'c^2. \tag{194}$$

The analogue of (193) in LORENTZ's theory is obtained by writing (190) as

$$E_e = \frac{4}{3} \frac{m'_e c^2}{\sqrt{1 - v^2/c^2}} \tag{195}$$

where $m'_e = a/b = e^2/(8\pi r)$ is the electron's rest mass. In the particle's rest frame, (195) becomes

$$E'_e = \tfrac{4}{3} m'_e c^2. \tag{196}$$

The factor of $\tfrac{4}{3}$ in (196) also appears in the momentum of the LORENTZ electron

$$\vec{P}_e = \frac{4}{3} \frac{m'_e \vec{v}}{\sqrt{1 - v^2/c^2}} \tag{197}$$

instead of

$$\vec{P} = \frac{m' \vec{v}}{\sqrt{1 - v^2/c^2}} \tag{198}$$

as in EINSTEIN's theory. If, for the moment, the difference in interpretation of m'_e and m' is set aside, the $\tfrac{4}{3}$ factor seems to be what distinguishes LORENTZ's kinematics from EINSTEIN's. Recall, however, that this factor results from the construction of a consistent LAGRANGIAN formulation for LORENTZ's theory. The understanding of the occurrence of the $\tfrac{4}{3}$ factor (and the extra term in (48)) in LORENTZ's theory resulted from the work of PLANCK[245] and MINKOWSKI.[151] Their papers provided a clear understanding of the relativistic transformation properties of physical quantities.

The historical discussion of this factor began with MAX VON LAUE's influential book *Das Relativitätsprinzip*,[249] wherein he related POINCARÉ's work to the results of PLANCK and MINKOWSKI. VON LAUE's discussion of an extended, deformable electron rests on his calculation of the transformation properties of the energy-momentum tensor $T_{\mu\nu}$. The spatial portions of this tensor are the MAXWELL stresses,[250] the time part T_{44} is the electromagnetic energy density $u = \tfrac{1}{2}(E^2 + B^2)$ and the mixed parts are $-ic\vec{g}_e$ where $\vec{g}_e$ is the electromagnetic field momentum density. VON LAUE showed that the relation between $T_{\mu\nu}$ in S and S' for a spherically symmetric charge distribution is[265]

$$T_{44} = k^2 \left[T'_{44} - \left(\frac{v}{c} \right)^2 T'_{xx} \right], \tag{199a}$$

$$(g_e)_x = \frac{v}{c^2} k^2 \left[T'_{44} - T'_{xx} \right] \tag{199b}$$

---

[264] See Chapter 13 of the reference in footnote 14 for a history of this equation.

[265] Equation (199) corresponds to equation (162) in the work mentioned in footnote 249 which is written in the Σ' system. Note that my notation for $T_{\mu\nu}$ is the negative of VON LAUE's.

(where (59′) has been used). Integrating (199) over all space in S, with $k d\tau = d\tau'$, gives

$$E_e = k \int d\tau' \left[T'_{44} - \left(\frac{v}{c}\right)^2 T'_{xx} \right], \tag{200a}$$

$$P_e = \frac{v}{c^2} k \int d\tau' \left[T'_{44} - T'_{xx} \right], \tag{200b}$$

where

$$P_e = \int (g_e)_x \, d\tau, \qquad E_e = \int T_{44} d\tau = \tfrac{1}{2} \int (E^2 + B^2) \, d\tau$$

and as usual the subscript e denotes kinematical quantities derived from the self-fields. Since $T_{\mu\nu}$ is a traceless tensor, $T'_{xx} = -\tfrac{1}{3} u' = -\tfrac{1}{3} T'_{44}$ and therefore (200) becomes

$$E_e = k \left[\int d\tau' \, T'_{44} \right] \left[1 + \frac{1}{3} \left(\frac{v}{c}\right)^2 \right], \tag{201a}$$

$$P_e = \frac{4}{3} k \frac{v}{c^2} \int T'_{44} \, d\tau' \tag{201b}$$

or [266]

$$E_e = \frac{m'_e c^2}{\sqrt{1 - v^2/c^2}} \left[1 + \frac{1}{3} \left(\frac{v}{c}\right)^2 \right], \tag{202a}$$

$$P_e = \frac{4}{3} \frac{m'_e v}{\sqrt{1 - v^2/c^2}} . \tag{202b}$$

Von Laue noted that this equation could have been derived alternatively by assuming quasi-stationary motion.[267] In fact, (202a) is another form of (48), and (202b) is (197). This was one of the reasons why Einstein's theory was considered initially to be a generalization of Lorentz's. Von Laue recognized that (202a) does not have the proper Lorentz transformation properties for the energy of a particle because of the term $\frac{1}{3} \left(\frac{v}{c}\right)^2$; the forces holding the deformable electron together have not been considered. The electron has been treated as an open system.

This situation is analogous to studying the static equilibrium of a fluid without assessing the forces exerted by the walls of the container. In fact, von Laue presented Poincaré's work by using the relativistic equations for the energy and momentum of a continuous medium. He considered the electron as an elastic sphere with a uniform surface charge distribution. Moreover, this system is considered to be closed, i.e. in equilibrium. Therefore, the energy and momentum of the electron will transform like four-vectors because the contributions from both the self-field and the mechanical force of constraint (from the Poincaré stress) are taken into account. The energy and momentum of the electron are

$$E = E_{\text{mech}} + E_e, \tag{203a}$$

$$P = G_{\text{mech}} + P_e. \tag{203b}$$

The mechanical energy E_{mech} and the mechanical momentum G_{mech} are:[268]

$$E_{\text{mech}} = k \left[E'_{\text{mech}} + \left(\frac{v}{c}\right)^2 p' V' \right] \tag{204a}$$

$$G_{\text{mech}} = \frac{v}{c^2} k [E'_{\text{mech}} + p' V'] \tag{204b}$$

---

[266] This corresponds to equation (113) on p. 98 of von Laue's *Relativitätsprinzip*.
[267] Cf. ibid., p. 98.
[268] Equations (204) are the relativistic transformation equations for the mechanical energy and momentum relations for an elastic body.

where p' is an invariant pressure, which in this case is T'_P, and V' is the electron's volume in S'. Consequently, the quantity $p'V'$ is

$$p'V' = T'_P V' = -\frac{m'_e c^2}{3} = -\frac{E'_e}{3}. \tag{205}$$

Thus (203) becomes

$$E = \frac{(E'_{\text{mech}} + m'_e c^2)}{\sqrt{1 - \dfrac{v^2}{c^2}}} \tag{206a}$$

$$P = \frac{(E'_{\text{mech}} + m'_e c^2)}{\sqrt{1 - \dfrac{v^2}{c^2}}}. \tag{206b}$$

Under LORENTZ transformation (206) has the desired properties of an energy and momentum (even though E_{mech}, G_{mech}, E_e and p_e do not) if the rest mass m is taken to be

$$M' = \frac{E'_{\text{mech}} + m'_e c^2}{c^2}. \tag{207}$$

For a purely electromagnetic model, $G_{\text{mech}} = 0$, which implies that

$$E'_{\text{mech}} = -T'_p V' = \tfrac{1}{3} m'_e c^2 \tag{208}$$

(from (204b)), so that [269]

$$M' = \frac{4}{3} m'_e = \frac{e^2}{6\pi r c^2}. \tag{209}$$

The result of substituting (205) and (208) into (204b) is [270]

$$E_{\text{mech}} = \frac{m'_e c^2}{3k} \tag{210a}$$

----

[269] Therefore, statements to the effect that the $\tfrac{4}{3}$ factor spoils the LORENTZ invariance of the LORENTZ-EINSTEIN theory are incorrect, e.g. "An additional $\tfrac{4}{3}$ factor is present [in the LORENTZ theory]. This factor was to plague the theory for a long time, especially when Lorentz-invariant formulations were being attempted: it is inconsistent with this invariance property." (F. ROHRLICH, Classical Charged Particles (Addison-Wesley, 1965), p. 13.) As I have shown this problem did not "plague" VON LAUE, and we shall see that it lay dormant until 1922 and then had to be rediscovered two more times in a span of 38 years. See also PANOFSKY & PHILLIPS, Classical Electricity and Magnetism (Addison-Wesley, 1961), p. 385, where it is asserted that POINCARÉ reduced the $\tfrac{4}{3}$ factor to unity by postulating the existence of mechanical stresses. Consequently, the observed mass "obeys the relativistically correct equations," i.e. no $\tfrac{4}{3}$ factor in the momentum. This statement implies that POINCARÉ knew of EINSTEIN's paper of 1905, or, better yet, of PLANCK's and MINKOWSKI's work. Moreover, additional proof that POINCARÉ's work should have produced the $\tfrac{4}{3}$ factor had he considered H_L instead of H is found in an exposition of his work by PAUL LANGEVIN, "L'inertie de l'énergie et ses conséquences" (Lecture delivered before the French Society of Physics on 26 March, 1913), Œuvres Scientifiques de Paul Langevin (CNRS, 1950), pp. 397–426, esp. pp. 412–414. On p. 414 LANGEVIN noted that the rest energy of the deformable electron is the sum of its electrostatic energy and the energy due to the POINCARÉ stress, i.e.

$$E' = \frac{e^2}{8\pi r} + \frac{e^2}{24\pi r} = \frac{e^2}{6\pi r} = \frac{4}{3} m'_e$$

(this corresponds to equation (19) on p. 414).
[270] Cf. VON LAUE, pp. 166–167.

which is identical to (176) (recall that $m'_e = a/b$). This provides proof that POINCARÉ erroneously inferred, by using H instead of H_L, that a/b is the "experimental mass". From (207), for a purely mechanical model we have

$$M' = \frac{E'_{\text{mech}}}{c^2}.$$ (210b)

Before discussing this result, VON LAUE showed that it is connected with a general theorem. Namely, that if the self-stress of a system vanishes in its rest frame then it is in static equilibrium and acts like a particle whose energy and momentum are:[271]

$$E = \frac{m'c^2}{\sqrt{1 - V^2/c^2}}$$ (211a)

$$p = \frac{m'v}{\sqrt{1 - V^2/c^2}}.$$ (211b)

This result follows immediately from a generalization of (200) to any system which can be described by an energy-momentum tensor. Therefore, as a result of VON LAUE's theorem E and p in (211) transform like a four-vector and can be associated with a particle of rest mass m':

$$E^2 - c^2 p^2 = m'^2 c^4.$$

In this manner POINCARÉ unknowingly maintained the relativistic invariance of LORENTZ's theory. By emphasizing the dualistic nature of POINCARÉ's work, i.e. both mechanical and electromagnetic forces were used, VON LAUE has accounted for the appearance of the $\frac{4}{3}$ factor in (195), (196) and (197). The electron's energy and momentum in a dualistic theory are given by (206).

VON LAUE concluded his discussion by noting that nothing definite about the electron's structure could be inferred from a dualistic theory since it leads to a two-term expression for M' (207). This implies an infinity of choices between the two extremes of a purely mechanical model and a purely electromagnetic model. Nevertheless, no matter which model is chosen, knowledge of the electron in 1911 allowed the inference that its radius could not be smaller than 10^{-13} cm.[272]

VON LAUE's theorem clearly demonstrates that the electromagnetic world-picture of ABRAHAM and LORENTZ could not possibly succeed. ABRAHAM used the field momentum and energy to describe an electron in analogy to the case of radiation. This was a natural choice for a description of an electron within a pre-EINSTEINIAN electromagnetic world-picture. LORENTZ used the field momentum for the same reason. Therefore, the stability problem, characterized by the extra term in (48), is non-existent in a dualistic electron theory having the proper LORENTZ

---

[271] Cf. VON LAUE, pp. 168–170. GUSTAV MIE, in "Grundlagen einer Theorie der Materie," Ann. d. Phys., **40**, 1–66 (1913), referred to this result as VON LAUE's theorem (p. 7).

[272] Cf. VON LAUE, p. 170. The radius 10^{-13} cm for a purely electromagnetic model is arrived at from the equation

$$M'c^2 = \frac{e^2}{6\pi r}$$

by use of measurements of e/M' (at very low velocities) and the value of e from electrochemical experiments.

transformation properties. POINCARÉ realized this but failed to see fully its consequences because his concept of a principle of relativity was inextricably linked
specifically to a theory of the electron and more globally to an electromagnetic
world-picture. The mass-energy equivalence from EINSTEIN's theory must have
seemed quite strange to proponents of the electromagnetic world-picture; although
energy is equivalent to mass, the converse is not true in the theories of ABRAHAM
and LORENTZ.[273]

Although VON LAUE succeeded in clarifying the essence of POINCARÉ's work
in the light of EINSTEINIAN relativity, his procedure is nevertheless aesthetically
displeasing. The inclusion of mechanical stresses of unknown origin to cancel the
self-stress, thereby closing the system, brings in an aura of the *ad hoc*. Also, the
duality of the theory brings in E'_{mech}, which represents another adjustable parameter in addition to the radius of the electron. Most important is that (211),
which has the proper LORENTZ transformation properties, was obtained by
combining E_{mech}, G_{mech}, E_e and p_e which do not transform separately like four-
vectors. This is an unsatisfactory state of affairs because MAXWELL-LORENTZ
electrodynamics is a properly covariant theory.

7.6. Post-Einsteinian Classical Field Theories of the Electron.
Post-EINSTEIN-
IAN[274] theories of an extended electron were formulated in such a manner as to
exclude mechanical forces. Theories of this type will be referred to as unitary

---

[273] Even in 1917 LORENTZ still clung to a dualistic electron theory. He noted that
by means of the POINCARÉ stresses the electron's energy and momentum could be
written as

$$E = \frac{M'c^2}{\sqrt{1 - \dfrac{v^2}{c^2}}}, \tag{30}$$

$$G = \frac{M'v}{\sqrt{1 - \dfrac{v^2}{c^2}}} \tag{31}$$

(H. A. LORENTZ, "The Connection Between...", *op. cit.* (footnote 251), p. 322, equations
(30) and (31)) where $M' = \frac{4}{3} m'_e$. He went on to assert that: "The electromagnetic
field of the electron which moves with the velocity v and of which the shape is modified
into an ellipsoid with semi-axes R, R and $R\sqrt{1-v^2/c^2}$ can easily be determined, so that
one can calculate the momentum in that field. This turns out to have exactly the value
(31) from which one may conclude that the momentum is wholly of electromagnetic
nature. On the other hand one does not obtain for the electromagnetic energy the
value (30) but an amount $\frac{1}{4} mc^2\sqrt{1-v^2/c^2}$ [this is the term $(e^2/(24\pi r))\sqrt{1-v^2}$ in (48)] less.
One must therefore imagine that energy of a nonelectromagnetic nature exists to
that same amount" (H. A. LORENTZ, "The Connection..." *op. cit.* (footnote 251),
pp. 322–323). This passage implies that the factor of $\frac{4}{3}$ did not "plague" LORENTZ
and that he had not yet understood the full implications of EINSTEIN's theory. However,
by 1921, when EINSTEIN's theory had been firmly entrenched for eleven years, LORENTZ
began to have doubts about the electromagnetic world-picture. At the Solvay conference
of 1921, he noted that due to relativity theory his model of the electron might not be
correct (see *Atoms and Electrons* (Gauthier-Villars, 1923) esp. p. 23).

[274] See R. McCORMMACH, "Einstein, Lorentz and the Electron Theory," *Historical
Studies in the Physical Sciences: Second Annual Volume 1970*, R. McCORMMACH ed.
(University of Pennsylvania Press, 1970), pp. 41–87, for a discussion of EINSTEIN's
unpublished work on a semi-classical non-linear theory of electrons and light quanta.

theories. A first attempt in this direction was made by GUSTAV MIE in 1912.[275] At the expense of altering the MAXWELL-LORENTZ equations MIE was able to construct a formalism wherein the repulsive self-stress in the electron's rest frame was balanced by other electromagnetic forces. However, it proved to be impossible to find a LAGRANGIAN that satisfied certain conditions imposed by the theory and simultaneously corresponded to a real electron.

The next notable attempt was made in HERMANN WEYL's theory (1918) based on the premise that all processes are reducible to electromagnetic and gravitational interactions.[276] However, the field equations were extremely complicated, having been constructed on the basis of WEYL's generalization of RIEMANNIAN geometry, and could not be integrated.

An interesting attempt was made by EINSTEIN in 1919 in which the factor $\frac{4}{3}$ appeared.[277] In response to the theories of MIE and WEYL, which EINSTEIN considered unsatisfactory, he decided to investigate the possibility of drawing conclusions as to the structure of matter from the general theory of relativity. EINSTEIN assumed that the extended electron is held together by gravitational forces (instead of the POINCARÉ stresses which would have introduced an undesired duality). The mathematical expression of this assumption is an alteration of the general relativistic field equations inside the electron in such a way that the four divergence of $T_{\mu\nu}$ vanishes in this region.[278] This serves to replace the cohesive

---

[275] G. MIE, "Grundlagen einer Theorie der Materie," Ann. d. Phys., **37**, 511–534 (1912); *ibid.*, **39**, 1–40 (1912); *ibid.*, **40**, 1–66 (1913). For an explicit discussion of this work see W. PAULI, *Theory of Relativity*, translated by G. FIELD (Pergamon, 1958), esp. pp. 188–192. This is a translation of PAULI's article "Relativitätstheorie" in Encykl. math. Wiss. Vol. V19 (Teubner, 1921).

[276] H. WEYL, "Gravitation und Elektrizität," Sitzungsber. d. k. preuss. Akad. Wiss., 465–480 (1918). Reprinted in *P.R.C.*, pp. 201–216. See also PAULI's *Theory of Relativity, op. cit.* (footnote 275), esp. pp. 192–202, and H. WEYL, *Space-Time-Matter*, translated by HENRY L. BROSE (Dover, 1952), esp. pp. 282–312 (this is an unabridged reproduction of the fourth edition of this book published by Methuen in 1922).

[277] A. EINSTEIN, "Spielen Gravitationsfelder im Aufbau der materiellen Elementarteilchen eine wesentliche Rolle?" Sitzungsber. d. k. preuss. Akad. Wissensch. 349–356 (1919). Reprinted in *P.R.C.*, pp. 191–198.

[278] It is possible to write the LORENTZ force equation and the energy equation in terms of $T_{\mu\nu}$ (see JACKSON, *Classical* ..., *op. cit.* (footnote 191), esp. pp. 383–386) as

$$f_\mu = \frac{\partial T_{\mu\nu}}{\partial x_\nu}. \tag{278-1}$$

The spatial part of this equation is

$$\vec{F} = \vec{V} \cdot \overset{\leftrightarrow}{T} - \frac{\partial \vec{g}_e}{\partial t} \tag{278-2}$$

(see footnote 250) and the temporal part is

$$\frac{\partial \mathscr{E}_{\text{mech}}}{\partial t} = -\vec{V} \cdot \vec{g}_e - \frac{\partial u}{\partial t}, \tag{278-3}$$

where $\mathscr{E}_{\text{mech}}$ is the mechanical energy density. The result of integrating (278-2) and (278-3) over a volume V and then rearranging terms is

$$\frac{d}{dt} [\vec{G}_{\text{mech}} + \vec{p}_e] = \int \vec{V} \cdot \overset{\leftrightarrow}{T} d\tau, \tag{278-4a}$$

$$\frac{d}{dt} [E_{\text{mech}} + E_e] = -\int \vec{V} \cdot \vec{g}_e d\tau. \tag{278-4b}$$

These equations are the conservation laws for momentum and energy for the system of particles and the electromagnetic field. Equations (278-4) can be simplified if a

mechanical POINCARÉ pressure by a gravitational pressure.[279] EINSTEIN then demonstrated that $\frac{3}{4}$ of the electron's energy can be ascribed to the electromag-

volume of integration is chosen large enough that the integrals on the right-hand side vanish. In the case in question only the velocity fields are contained in $T_{\mu\nu}$; thus, these integrals become negligible over an integration volume only slightly larger than the electron's volume (recall the discussion of the velocity fields in Part 6.6). As was discussed in footnote 26, POINCARÉ arrived at (278-4a) by substituting for ϱ and $\varrho\vec{v}$ from the MAXWELL-LORENTZ equations. Then by integrating over all space (or, as previously noted, over a volume sufficiently large compared to the electron's volume) he found that

$$\frac{d}{dt}[\vec{G}_{\text{mech}} + \vec{p}_e] = \vec{0} \tag{278-5}$$

where $\vec{p}_e$ was the $\vec{G}$ mentioned in footnote 26. Consequently, it was necessary to ascribe a momentum to the electromagnetic field in order to maintain the law of conservation of momentum. However, as pointed out, $\vec{p}_e$ can not in general be associated with a particle. On the other hand, if $f_\mu = 0$, i.e., there are no sources, then

$$\frac{\partial T_{\mu\nu}}{\partial x_\nu} = 0, \tag{278-6}$$

and it is possible to define an energy-momentum four-vector

$$p_\mu = \frac{i}{c}\int T_{\mu 4} d\tau \tag{278-7}$$

where $p_\mu = \left(\vec{p}, i\frac{E}{c}\right)$. The proof that p_μ has the desired properties under the LORENTZ transformation is due to FELIX KLEIN in "Über die Integralform der Erhaltungssätze und die Theorie der räumlich geschlossenen Welt," Nachr. Ges. Wiss. Göttingen, 394–423 (1918). However, the procedure to obtain p_μ, i.e. (278-7), is due to H. WEYL (see H. WEYL, Space-Time-Matter, op. cit. (footnote 276), esp. p. 201). Consequently, the energy and momentum of the electromagnetic field can be associated with a particle only in the absence of sources. It can be shown that the rest mass energy of this particle is zero (see W. HEITLER, The Quantum Theory of Radiation (Oxford U. Press, 1954), esp. pp. 17–18).

The formal definition of a general static system in equilibrium, i.e. a closed system, can be obtained from (278-1). If f_μ is written as the four divergence of a second rank tensor $\Theta_{\mu\nu}$ (referred to as the kinetic energy-momentum tensor in PAULI's Theory of Relativity, op. cit. (footnote 275), esp. p. 117), then (278-1) becomes

$$\frac{\partial R_{\mu\nu}}{\partial x_\nu} = 0 \tag{278-8}$$

where $R_{\mu\nu} = \Theta_{\mu\nu} - T_{\mu\nu}$. Equation (278-8) is equivalent to (278-4) and therefore contains the conservation laws of energy and momentum for the system of particle and electromagnetic field. From (278-7) an energy-momentum four-vector can be constructed from $R_{\mu\nu}$. Consequently, this system is closed if both electrical and mechanical forces are taken into account. However, the separate systems do not have the desired properties under the LORENTZ transformation and are therefore not closed. For a detailed relativistic discussion of closed systems see C. MØLLER, The Theory of Relativity (Oxford U. Press, 1952), esp. pp. 163–217.

[279] This would correspond to a sweeping generalization of POINCARÉ's statement in the paper of 1906: "Since the Newtonian attraction is proportional to the experimental mass, one can attempt to infer that there exists a general relation between the causes giving rise to gravitation and those which give rise to the supplementary potential" (R. 166: O. 538). I am of course not implying that EINSTEIN was in any way influenced by POINCARÉ's gravitational theory, which was formulated along entirely dissimilar lines. However, he was aware by 1921 of at least some of the content of POINCARÉ's paper of 1906. In a set of lectures given at Princeton University in 1921 EINSTEIN pointed to the analogy between his gravitational stresses in the paper of

netic field and $\frac{1}{4}$ to the gravitational field.[280] However, there is one equation too few for a complete determination of both the gravitational and electromagnetic fields for a static spherically symmetric electron.

7.7. The Elusive 4/3 Factor. The first attempt to dispose of the $\frac{4}{3}$ factor was made by ENRICO FERMI in 1922.[281] FERMI was concerned that this factor contradicts the general theory of relativity which asserts the equality of gravitational and inertial masses (assuming that the gravitational mass is purely mechanical and the inertial mass is purely electromagnetic in origin). Moreover, a purely mechanical interpretation of (193) yields a result from LORENTZ's theory (195) that is at variance with the special theory of relativity. Unaware of VON LAUE's discussion of the origin of this factor, FERMI characteristically found one on his own.[282] He demonstrated that the $\frac{4}{3}$ factor in the electron's energy (see (195)) was reduced to unity by demanding that the variation of the action containing only the electromagnetic field LAGRANGIAN produces an energy that has the proper LORENTZ transformation properties. However, he ignored the stability problem.

FERMI's procedure is similar to that of BERNARD KWAL[283] who, in 1949, rediscovered and clarified FERMI's procedure. KWAL's basic premise was that it

1919 (*op. cit.*, footnote 277) and the POINCARÉ stresses: "In order to be consistent with the facts [that electrons are stable] it is necessary to introduce energy terms, not contained in Maxwell's theory, so that the single electric particles may hold together in spite of the mutual repulsions between their elements. For the sake of consistency with the fact, Poincaré has assumed a pressure to exist inside these particles which balances the electrostatic repulsion." (A. EINSTEIN, *The Meaning of Relativity*, translated by E. P. ADAMS (5th ed., Princeton University Press, 1970), p. 106.)

I am grateful to Professor ABRAHAM PAIS for permission to publish the following conversation he had with EINSTEIN at the Institute for Advanced Study in 1954.

PAIS: "What was the impact of Poincaré's paper ['Sur la dynamique ...'] on your thinking?"

EINSTEIN: "I never read it."

[280] Equation (196) can be written as

$$\tfrac{3}{4} E' = m_e' c^2,$$

which implies that $\frac{3}{4}$ of the electron's rest mass is of electromagnetic origin.

[281] E. FERMI, "Correzione di una grave discrepanza tra la teoria delle masse elettromagnetiche e la teoria della relativitá," Rendiconti dei Lincei, **31**, 184–187, 306–309 (1922); reprinted in *Collected Papers: Vol. I, Italy 1921–1938*, E. SEGRÉ, ed. (U. of Chicago Press, 1962), pp. 24–32.

[282] E. PERSICO on p. 24 of FERMI's *Collected Papers* ..., *loc. cit.* states that FERMI was unaware of VON LAUE's explanation of the appearance of the $\frac{4}{3}$ factor on p. 218 of the third edition of VON LAUE's work cited in footnote 249 (this description is the same as the one in the first edition).

[283] B. KWAL, "Les expressions et de l'impulsion du champ électromagnétique propre de l'électron en mouvement," J. Phys. Rad., 193–194 (1949). JACKSON (p. 597 of the work cited in footnote 191), ROHRLICH (on p. 17 of the work cited in footnote 269) and JAMMER (p. 183 of the work cited in footnote 14) assert incorrectly that FERMI's reduction of the $\frac{4}{3}$ factor to unity was rediscovered, although in a different manner, by W. WILSON. A close examination of this paper reveals that WILSON rediscovered VON LAUE's explanation of 1911 of how the $\frac{4}{3}$ factor arises and not a method to eliminate it! This is seen by comparing equation (10) on p. 737 of WILSON's paper which (rewritten in my notation) is

$$M' = k E / c^2$$

where $E = E'_{\text{mech}} + \dfrac{e^2}{8\,\pi r}$ with (207) in this paper.

should be possible to construct an energy-momentum four-vector from $T_{\mu\nu}$ since MAXWELL-LORENTZ electrodynamics is properly covariant (this was the second objection to VON LAUE's exposition raised in Part 7.5). This can be accomplished by taking the scalar product of $T_{\mu\nu}$ with the velocity four-vector V_ν and then integrating over the invariant volume element $kd\tau$. By construction the resulting vector p_μ is a four-vector:

$$p_\mu = \frac{k}{c} \int T_{\mu\nu} V_\nu \, d\tau \tag{212}$$

where the components of the velocity four-vector are $k\vec{v}/c$ and ik and repeated Greek indices are summed from one to four (this is the EINSTEIN summation convention). It is important to note that (212) reduces to the electrostatic energy in the electron's rest frame, as it should. By use of the definition of V_ν (212) can be rewritten as

$$E_e = k^2 \int d\tau \, [u - \vec{v} \cdot \vec{g}_e], \tag{213a}$$

$$p_e = k^2 \int d\tau \left[\vec{g}_e + \frac{\vec{v} \cdot \overleftrightarrow{T}}{c^2} \right]. \tag{213b}$$

It can be shown easily that the last two terms on the right-hand side of (213), which distinguish it from its analogue in the LORENTZ-POINCARÉ theory, aside from the factors of k^2, reduce the $\frac{4}{3}$ factor to unity.[284] Again, the stability question has been ignored.

KWAL's result was rediscovered by FRITZ ROHRLICH in 1960.[285] ROHRLICH made the important observation that the "stability problem is not related to the transformation properties of the theory".[286] Consequently, it is possible by means of (213) to have a unitary, relativistically covariant theory of an unstable extended electron.

In 1938, interest was again aroused in the classical theory of the electron through the work of P. A. M. DIRAC.[287] The main thrust of this work was to search for techniques that would be useful in the resolution of certain problems in quantum electrodynamics. However, no suitable theory of the extended electron has been found; only point-charge theories have had any success. In fact, in DIRAC's point-charge theory the observed mass is taken to be the sum of the mechanical mass and the electromagnetic mass. Thus the infinity arising from the electromagnetic mass in the limit as $r \to 0$ is ignored.

7.8. A Quantum Field Theory of the Electron: Quantum Electrodynamics. The rapid advances in the quantum theory of the atom, starting in 1913, which promised to shed new light on the concept of an elementary particle, resulted for a period of time in an almost complete neglect of the problem of the classical electron.

[284] Cf. J. D. JACKSON, Classical ..., op. cit. (footnote 191), esp. p. 596 and the reference in footnote 285 (below), esp. p. 162.
[285] F. ROHRLICH, "Self-Energy and Stability of the Classical Electron," Am. J. Phys., **28**, 639–643 (1960).
[286] Ibid., p. 643.
[287] P. A. M. DIRAC, "Classical theory of radiating electrons," Proc. Roy. Soc. (London) A **167**, 148–169 (1938). For a detailed description of this work and of recent developments in the classical electron problem see F. ROHRLICH's Classical Charged Particles, op. cit. (footnote 269).

Research on this problem, as we have seen, was conducted primarily by those interested in the general theory of relativity.

Indeed, there were unexpected results from quantum theory such as the fact that the electron has a fourth degree of freedom, namely, a spin which has no counterpart in a classical theory. Moreover, the wave-particle duality renders unclear the concept of an electron with a well defined radius.

In quantum electrodynamics, as in the theories of ABRAHAM and LORENTZ, the electron is represented within the framework of a field theory. However, in this theory it is represented by two fields: the quantized field of the bare electron and the quantized electromagnetic field which accompanies it. Yet in this case it is possible to make the volume integral of the self-stress (which is the sum of terms due to the quantized field of the electron and its quantized self-field) vanish in the electron's rest frame.[288]

Consequently, in quantum theory, the electron accompanied by its own electromagnetic field behaves like a particle. Thus, the stability problem can be dealt with in quantum field theory; however, charge and self-energy (mass) divergences have to be overcome. These divergences are studied within the so-called renormalization program. Although charge divergences have no classical counterpart, the divergence of the electron's self-energy existed in the theories of ABRAHAM and LORENTZ. In these theories the self-energy, *i.e.* the energy of an electron at rest, varies as $1/r$ and therefore diverges as r goes to zero. In quantum electrodynamics the divergence is a logarithmic one which is less severe.

8. Concluding Remarks

More than six decades of EINSTEINIAN relativity have tended to obscure the historical significance of the efforts of LORENTZ and POINCARÉ. One sometimes forgets that their efforts represented what was very probably the limit to which the techniques and concepts available in 1905 could be pushed towards an electromagnetic world-picture. But it is sobering to think that the basic problem with which POINCARÉ dealt still exists and indeed has not yet been resolved adequately within the framework of the new world-picture, a world-picture in which the notion of what constitutes physical reality is not yet clear.

LORENTZ's paper of 1904 was an attempt to make his electromagnetic theory *the* basis of physics. However, LORENTZ was unsuccessful because of his effort to maintain a continuous link from the work of 1892 to 1895 to 1904. LORENTZ's ultimate failure to accept relativity theory fully is related to the fact that he was basically a nineteenth-century materialist.

For a scientist of POINCARÉ's talents the awareness of LORENTZ's theory should have been the impetus for the discovery of a theory of relativity. POINCARÉ seemed to have all the requisite concepts for a relativity theory: a discussion of the various null experiments to first and second order accuracy in v/c; a discussion of the role of the speed of light in length measurements; the correct relativistic transformation equations for the electromagnetic field and the charge density; a relativistically invariant action principle; the correct relativistic equation for the addition of velocities; the concept of the LORENTZ group; a rudimentary notion

---

[288] See W. HEITLER, *The Quantum Theory of Radiation* (Oxford U. Press, 1954). This book contains an excellent discussion of quantum electrodynamics.

of the four-vector formalism and of four-dimensional space; a correct relativistic kinematics (in the quasi-stationary approximation); and a LORENTZ covariant gravitational theory in which gravity is propagated through the ether with the speed of light. His relativity theory was to be an inductive one with the laws of electromagnetism as the basis for all of physics. However, a strict adherence to this type of research program prevented him from understanding the universal applicability of the principle of relativity and therefore the importance of the constancy of the velocity of light in all inertial frames. That this was a reflection of POINCARÉ's philosophy will be discussed in the sequel to this article.

POINCARÉ's "Sur la dynamique ..." essentially heralded the end of the search for a relativistic electromagnetic theory of an extended deformable electron based strictly on LORENTZ's theory. Yet, he asserted that an electromagnetic world-picture was still possible if only certain cohesive mechanical stresses were admitted. The probable reason for this statement is that the POINCARÉ stresses were derived from a relativistically invariant action principle. These stresses were necessary for a theory which conformed to the principle of relativity to be obtained from the electromagnetic field LAGRANGIAN.

VON LAUE's exposition of POINCARÉ's results served not only to present them in the light of the techniques of EINSTEINIAN relativity (as developed by PLANCK and MINKOWSKI), but, in addition, pointed out the pitfalls that LORENTZ could have avoided had he constructed an electron theory with the tools that were available four years later. In EINSTEIN's theory of relativity, energy and momentum transform in such a manner that the extra term in (48) is non-existent. These transformation properties cannot be deduced from LORENTZ's theory without the POINCARÉ stresses. Had POINCARÉ investigated the manner in which the field momentum density transforms he would have found that it is not a vector. However, he did not realize that in a universal relativity theory the basic role is played by the energy and momentum instead of the force. In an electromagnetic world-picture, on the other hand, force is the basic factor because it is a function of the fields that are determined by the charge (which is assumed to be given) through the MAXWELL-LORENTZ equations. In this way the massive electron is constructed or generated from its self-fields.

POINCARÉ's adherence to the electromagnetic world-picture in which the principle of relativity was a law open to experimental verification remained strong to the end of his life. The importance which POINCARÉ attached to experimental facts (in accordance with conventionalism) is illustrated graphically by the uncertain tenor of "Sur la dynamique ...":

> Perhaps the abandonment of this definition [two lengths are equal if they are traversed by light in the same time] would suffice to overthrow Lorentz's theory ...[289]
>
> . . . .
>
> I have therefore not hesitated to publish these incomplete results, even though at the present time the entire theory may seem to be threatened by the discovery of magneto-cathode rays.[290]

[289] R. 132: O. 498: K. 149.
[290] *Loc. cit.*, the word "magneto" was omitted from KILMISTER's translation. It is not clear what POINCARÉ meant by a "magneto-cathode ray."

However, by 1908 his mood had changed to one of optimism. He questioned the validity of KAUFMANN's experiment: "The question [whether or not the principle of relativity is valid] is of such importance that one would wish to see Kaufmann's experiment repeated by another experimenter."[291] A footnote to this quotation showed that POINCARÉ's optimism was justified: "At the moment of going to press we learn that M. Bucherer has repeated the experiment, surrounding it with new precautions, and that, unlike Kaufmann, he has obtained results confirming Lorentz's views."[292] In addition, POINCARÉ asserted that the mass of the negative electron was of purely electromagnetic origin:

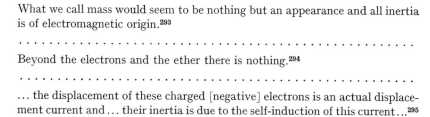

> What we call mass would seem to be nothing but an appearance and all inertia is of electromagnetic origin.[293]
>
> . . . .
>
> Beyond the electrons and the ether there is nothing.[294]
>
> . . . .
>
> ... the displacement of these charged [negative] electrons is an actual displacement current and ... their inertia is due to the self-induction of this current...[295]

Moreover, he was willing to accept the consequences of the new mechanics: "... there is no more matter, since positive electrons have no longer any real mass, or at least no constant real mass [whereas negative electrons definitely have no "real" mass]. The actual principles of our mechanics based on the constancy of mass must accordingly be modified."[296] Thus, NEWTON's laws lose their status as conventions in the physical sciences where the characteristic length is of the order of the electron's diameter. POINCARÉ's views on relativity theory remained unchanged in other works in this area, "La Mécanique nouvelle"[297(a)] and "La dynamique de l'électron."[297(b)]

In his essay entitled "Space and Time", in the last volume of his collected philosophical works, POINCARÉ carefully differentiated the principle of physical relativity referred to in 1900 as the principle of relative motion and the principle

---

[291] H. POINCARÉ, *S.M.*, *op. cit.* (footnote 117(a)), p. 228.

[292] *Ibid.*, p. 228.

[293] *Ibid.*, p. 207.

[294] *Ibid.*, p. 209.

[295] *Ibid.*, p. 226.

[296] *Ibid.*, p. 227.

[297] (a) This constitutes the sixth lecture of the POINCARÉ "Festspiele" at Göttingen 22–28 April, 1909 (H. POINCARÉ, *Sechs Vorträge über ausgewählte Gegenstände aus der reinen Mathematik und mathematischen Physik* (Teubner, 1910)). It is a qualitative account of relativity theory and contains no reference to EINSTEIN's work. POINCARÉ referred to the relativity theory as LORENTZ's theory. This is consistent with his narrow interpretation of the principle of relativity. The following year, at the LORENTZ "Festspiele", LORENTZ spoke about EINSTEIN's theory of relativity: "To discuss Einstein's principle of relativity here in Göttingen where Minkowski has taught seems to me a particularly welcome task" (H. A. LORENTZ, "Alte und neue Fragen der Physik," Phys. Zeit., **11**, 1234–1257, (1910), esp. p. 1236, in *Collected Papers*, 7, 205–257).

In 1909 POINCARÉ delivered a similar lecture at Lille (see the reference in footnote 243) with the same title.

(b) H. POINCARÉ, *La dynamique de l'électron*, *op. cit.* (footnote 242).

of relativity[298], in the physical sciences. The former "can serve to define space"[299] as we know it in our everyday movements about the world. It is linked to the three-dimensional group of displacements and rotations by means of which we construct our notions of EUCLIDEAN geometry. It is therefore a convention like EUCLIDEAN geometry. However, POINCARÉ went on to note: "What, then, is the revolution which is due to the recent progress in physics? The principle of relativity, in its former respect [as the principle of physical relativity] has had to be abandoned; it is replaced by the principle of relativity according to LORENTZ. It is the transformations of 'the group of Lorentz' that do not falsify the differential equations of dynamics."[300] He went on to note that according to this group objects in motion are deformed, which implies that in LORENTZ's principle of relativity there are no rigid bodies. Furthermore, POINCARÉ did not ascribe the status of a convention to LORENTZ's principle of relativity.

That the ether and the electromagnetic world-picture remained of the utmost importance to POINCARÉ is illustrated by the title of a lecture he delivered on 11 April, 1912, three months before his death on 17 July, 1912, entitled "The Relations Between Matter and the Ether".[301] What is so peculiar about this title is that POINCARÉ never explicitly discussed the ether. This lecture was essentially about the state of quantum theory and atomic physics circa 1912, i.e. he discussed PERRIN's experiments; radioactivity; the electromagnetic world-picture with respect to atomic physics ("It is to the self-induction of the convection currents produced by the movements of these electrons [in the RUTHERFORD atom] that the atom which is made up of them owes its apparent inertia, which we call its mass.");[302] spectroscopy, in particular, the work of BALMER, RUNGE, KAISER and RYDBERG; WEISS' magneton; PLANCK's quantum theory, and his own brilliant work on quantum theory entitled, "Sur la théorie des Quanta";[303] and finally a brief mention (hostile) of EINSTEIN's work on the *nature of light*, followed by a reference to "*Lorentz's principle of relativity*".[304]

Thus, POINCARÉ's firm attachment to an inductive theory of relativity was consonant with the fundamental tenet of his philosophy of science — "Experiment is the sole source of truth"[67] — because "it is not merely a principle which it is a question of saving, it is the indubitable results of the experiments of Michelson."[122]

---

[298] H. POINCARÉ, *Mathematics and Science: Last Essays*, translated by JOHN BOLDUC (Dover, 1963), esp. pp. 21–24 (Chapter II, pp. 15–24 of *Last Essays...*, is a translation of POINCARÉ's lecture "L'Espace et le Temps" given on 4 May, 1912 at the University of London).

[299] *Ibid.*, p. 22.

[300] *Ibid.*, p. 23.

[301] *Ibid.*, pp. 89–101 (Chapter VII, pp. 89–101 of *Last Essays...* is a translation of POINCARÉ's lecture "Les rapports de la matière et de l'éther" given on 11 April, 1912 before the French Society of Physics and published in Journal de Physique théorique et appliquée, **2**, 347 (1912)).

[302] *Ibid.*, p. 91.

[303] H. POINCARÉ, "Sur la théorie des Quanta," Comptes rendus de l'Académie des Sciences, **153**, 1103–1108 (1911); reprinted in *Oeuvres*, pp. 620–625. A detailed version of this paper with the same title appeared in Journal de Physique théorique et appliquée, **2**, 5–34 (1912); reprinted in *Oeuvres*, pp. 625–653. For a survey of this work see R. MCCORMMACH, "Henri Poincaré and Quantum Theory," Isis, **58**, 37–55 (1967).

[304] H. POINCARÉ, *Last Essays...*, *op. cit.* (footnote 298), p. 99; italics added.

This interaction between POINCARÉ's epistemology and his physics, however, is the theme of the next paper.

Acknowledgment. I benefited greatly from Professor ERWIN HIEBERT's insights into POINCARÉ and from discussions in his seminar at Harvard University in the spring of 1971. It is also a pleasure to thank Professor C. TRUESDELL, O. COSTA DE BEAUREGARD and R. McCORMMACH for their comments on earlier versions of this paper. Above all I am grateful to Professor GERALD HOLTON for the advice, criticism and encouragement that he offered me during the many discussions we had not only on this work, but on the entire subject of the history and philosophy of physics. Needless to say, certain conclusions in this paper are not necessarily endorsed by these scholars and I alone am responsible for them.

This paper was written during the summer of 1971 at Harvard University while I was on leave from the Lowell Technological Institute. I thank Professor HOLTON for the hospitality that he extended to me during this period at Harvard. I am on sabbatical leave at Harvard University for 1972–1973 as a National Endowment for the Humanities Fellow.

Bibliography

ABRAHAM, M., "Dynamik des Elektrons," Nachr. Ges. Wiss. Göttingen, 20–41 (1902).
—, Prinzipien der Dynamik des Elektrons," Ann. d. Phys., **10**, 105–179 (1903).
—, "Die Grundhypothesen der Elektronentheorie," Phys. Zeit., **5**, 576–579 (1904).
—, *Theorie der Elektrizität* (1st ed., Volume 2; Leipzig: Teubner, 1905; 2nd ed., Volume 2, 1908).
BECKER, R., & F. SAUTER, *Electromagnetic Fields and Interactions* (2 vols.; New York: Blaisdell 1964).
BELL, E. T., *The Development of Mathematics* (New York: McGraw-Hill 1945).
BESTELMEYER, A., "Bemerkung zu der Abhandlung Herrn. A. H. Bucherer's 'Die experimentelle Bestätigung des Relativitätsprinzips'," Ann. d. Phys., **30**, 166–174 (1909).
BORN, M., "Die träge Masse und das Relativitätsprinzip," Ann. d. Phys., **28**, 571–584 (1909).
—, *Physics in My Generation* (New York: Springer 1969).
BORK, A. F., "The 'Fitzgerald' Contraction" Isis, **57**, 199–207 (1966).
BRUSH, S. G., "Note on the History of the Fitzgerald-Lorentz Contraction", Isis, **58**, 230–232 (1967).
BUCHERER, A. H., *Mathematische Einführung in die Elektronentheorie* (Leipzig: Teubner 1904).
—, "Messungen an Becquerelstrahlen. Die experimentelle Bestätigung der Lorentz-Einsteinschen Theorie," Phys. Zeit., **9**, 755–762 (1908).
—, "Die experimentelle Bestätigung des Relativitätsprinzips," Ann. d. Phys., **28**, 513–536 (1909).
CRÉMIEU, V., "Sur les expériences de M. Rowland relatives à l'effet magnétique de la 'convection électrique'," Comptes rendus de l'Académie des Sciences, **132**, 797–800 (1901).
— & H. PENDER, "On the Magnetic Effect of Electrical Convection," Phil. Mag., **6**, 442–464 (1903).
— & H. PENDER, "Nouvelles recherches sur la convection électrique," Comptes rendus de l'Académie des Sciences, **136**, 548–550 (1903).
— & H. Pender, "Recherches Contradictoires sur l'effet Magnétique de la Convection Électrique." Journal de physique théorique et appliquée, 2, 641–666 (1903).
DANTZIG, T., *Henri Poincaré: Critic of Crisis* (New York: Scribner 1954).
DE BROGLIE, L., *New Perspectives in Physics*, translated by A. J. POMERANS (New York: Basic Books 1962).
DIRAC, P. A. M., "Classical theory of radiating electrons," Proc. Roy. Soc. (London) A **167**, 148–169 (1938).

21*

Duhem, P., *The Aim and Structure of Physical Theory*, translated by P. P. Wiener (New York: Atheneum 1962).

Einstein, A., "Zur Elektrodynamik bewegter Körper," Ann. d. Phys., **17**, 891–921 (1905). (Reprinted in *The Principle of Relativity: A Collection of Original Memoirs on the Special and General Theory of Relativity by H. A. Lorentz, A. Einstein, H. Minkowski and H. Weyl*, translated by W. Perrett & G. B. Jeffery (New York: Dover, n.d.), pp. 37–65. This reprint volume is referred to as *P.R.C.*)

—, "Spielen Gravitationsfelder im Aufbau der materiellen Elementarteilchen eine wesentliche Rolle?" Sitzungsber. d. k. preuss. Akad. Wiss., 349–356 (1919). (Reprinted in *P.R.C.*, pp. 191–198.)

—, *The Meaning of Relativity*, translated E. P. Adams (5th ed.; Princeton: Princeton University Press 1970).

—, "Autobiographical Notes," in *Albert Einstein: Philosopher-Scientist*, P. A. Schilpp ed. (Evanston: The Library of Living Philosophers, 1949).

Fermi, E., "Correzione di una grave discrepanza tra la teoria delle masse elettromagnetiche e la teoria della relativitá," Rendiconti dei Lincei, **31**, 184–187, 306–309 (1922). (Reprinted in *Collected Papers: Vol. I, Italy 1921–1938*, E. Segré ed. (2 vols.; Chicago: University of Chicago Press 1962), pp. 24–32.)

Goldberg, S., "The Early Response to Einstein's Special Theory of Relativity, 1905–1911: A Case Study of National Differences," Ph. D. Thesis (unpublished), Harvard University (1968).

—, "Henri Poincaré and Einstein's Theory of Relativity," Am. J. Phys., **35**, 934–944 (1967).

—, "Poincaré's Silence and Einstein's Relativity: The Role of Theory and Experiment in Poincaré's Physics," Brit. J. Hist. Sci., **5**, 73–84 (1970).

—, "The Abraham Theory of the Electron: The Symbiosis of Experiment and Theory," Archive for History of Exact Sciences, **7**, 7–25 (1970).

—, "The Lorentz Theory of Electrons and Einstein's Theory of Relativity," Am. J. Phys., **37**, 498–513 (1969).

—, "In Defense of the Ether: The British Response to Einstein's Special Theory of Relativity: 1905–1911," *Historical Studies in the Physical Sciences: Second Annual Volume 1970*, R. McCormmach ed. (Philadelphia: University of Pennsylvania Press, 1970), pp. 89–125.

Goldstein, H., *Classical Mechanics* (Reading: Addison-Wesley, 1965).

Guye, C. E., & Ch. Lavanchy, "Vérification expérimentale de la formula de Lorentz-Einstein par les rayons cathodiques de grande vitesse," Arch. Sci. phys. nat., **41**, 286, 353, 441 (1916).

Heaviside, O., "On the electromagnetic effects due to the motion of electrification through a dielectric," Phil. Mag., **27**, 324–339 (1889).

Heitler, W., *The Quantum Theory of Radiation* (London: Oxford University Press 1954).

Hirosige, T., "Origins of Lorentz' Theory of Electrons and the Concept of the Electromagnetic Field," *Historical Studies in the Physical Sciences: Volume I*, R. McCormmach ed. (Philadelphia; University of Pennsylvania Press 1969), pp. 151–209.

—, "Electrodynamics before the Theory of Relativity, 1890–1905," Japanese Studies in the History of Science, **5**, 1–49 (1966).

—, "Theory of Relativity and the Ether," *ibid.*, **7**, 37–53 (1968).

Holton, G., "On the Origins of the Special Theory of Relativity," Am. J. Phys., **28**, 627–636 (1960).

—, "On the Thematic Analysis of Science: The Case of Poincaré and Relativity," Mélanges Alexandre Koyré (Paris: Hermann 1964), pp. 257–268.

—, "Einstein, Michelson and the 'Crucial' Experiment," Isis, **60**, 133–197 (1969).

Jackson, J. D., *Classical Electrodynamics* (New York: Wiley 1962).

Jammer, M., *Concepts of Mass* (Cambridge: Harvard University Press 1961).

Kahan, T., "Sur les Origines de la théorie de la relativité restreinte," Revue d'Histoire des Sciences, *XIII*, 159–165 (1959).

KAUFMANN, W., "Die magnetische und elektrische Ablenkbarkeit der Becquerel-strahlen und die scheinbare Masse der Elektronen," Nachr. Ges. Wiss. Göttingen, 143–155 (1901).

—, "Die elektromagnetische Masse des Elektrons," Phys. Zeit., 4, 54–57 (1902).

—, "Über die 'Elektromagnetische Masse' der Elektronen," Nachr. Ges. Wiss. Göttingen, 90–103 (1903).

KESWANI, G. H., "Origin and Concept of Relativity," Brit. J. Phil. Sci., 15, 286–306; 16, 19–32 (1965).

KILMISTER, C. W. ed., Special Theory of Relativity (New York: Pergamon 1970).

KLEIN, F., "Über die Integralform der Erhaltungssätze und die Theorie der räumlich geschlossenen Welt," Nachr. Ges. Wiss. Göttingen, 394–423 (1918).

KWAL, B., "Les expression et de l'impulsion du champ électromagnétique propre de l'électron en mouvement," J. Phys. Rad., 193–194 (1949).

LANGEVIN, P., "La physique des électrons," Revue générale des Sciences pures et appliquées, 16, 257–276 (1905).

—, "Sur l'origine des radiations et l'inertie électromagnétique," J. de Physique, 4, 165 (1905). (Reprinted in Œuvres Scientifiques de Paul Langevin (Paris: C.N.R.S., 1950), pp. 313–348.)

—, "L'inertie de l'énergie et ses conséquences," in Œuvres Scientifique ..., pp. 397–426.

LARMOR, J., Aether and Matter (Cambridge: Cambridge U. Press 1900).

LAUE, M. VON, Das Relativitätsprinzip (1st ed.; Braunschweig; Friedr. Vieweg und Sohn 1911: 2nd ed.: 1913).

LEBON, ERNEST, Savants du Jour: Henri Poincaré Biographie, Bibliographie analytique des écrits (2nd ed.; Paris: Gauthier-Villars 1912).

LEWIS, G. N., & R. C. TOLMAN, "The Principle of Relativity and Non-Newtonian Mechanics," Phil. Mag., 18, 510–523 (1909).

LORENTZ, H. A., "La théorie électromagnétique de Maxwell et son application aux corps mouvants," Arch. Néerl., 25, 363 (1892). (Reprinted in Collected Papers (9 vols.; The Hague: Nijhoff, 1935–1939), Vol. 2, 164–343.)

—, Versuch einer Theorie der elektrischen und optischen Erscheinungen in bewegten Körpern (Leiden: Brill 1895). (Reprinted in Collected Papers, Vol. 5, 1–137.)

—, "Théorie simplifiée des phénomènes électriques et optiques dans des corps en mouvement," Versl. Kon. Akad. Wetensch. Amsterdam, 7, 507 (1899). (Reprinted in Collected Papers, Vol. 5, 139–155.)

—, "The Theory of Radiation and the Second Law of Thermodynamics," ibid., 9, 418 (1900). (Reprinted in Collected Papers, Vol. 6, 265–279.)

—, "Considérations sur la pesanteur," Ibid., 8, 603 (1900). (Reprinted in Collected Papers, Vol. 5, 198–215.)

—, Boltzmann's and Wien's laws of Radiation," ibid., 9, 572 (1901). (Reprinted in Collected Papers, Vol. 6, 280–292.)

—, "Elektromagnetische Theorien physikalischer Erscheinungen," Phys. Zeit. 1, 498, 514 (1900). (Reprinted in Collected Papers, Vol. 8, 333–352.)

—, "Über die scheinbare Masse der Ionen," ibid., 2, 78 (1901). (Reprinted in Collected Papers, Vol. 3, 113–116.)

—, "Contributions to the Theory of Electrons," Proc. Roy. Acad. Amsterdam, 5, 608 (1903). (Reprinted in Collected Papers, Vol. 3, 132–154.)

—, "Weiterbildung der Maxwellschen Theorie. Elektronentheorie," Encykl. math. Wiss. Vol. V2, Art. 14 (Leipzig: Teubner 1904).

—, "Electromagnetic Phenomena in a System Moving with any Velocity Less than that of Light," Proc. Roy. Acad. Amsterdam, 6, 809 (1904). (Reprinted in P.R.C., pp. 11–34.)

—, "Ergebnisse und Probleme der Elektrontheorie," Elektotechn. Verein zu Berlin, 1904 (Berlin; Springer 1905). (Reprinted in Collected Papers, Vol. 8, 79–124.)

—, The Theory of Electrons (Leiden; Brill 1909; rev. ed., 1915; New York: Dover, 1952). This book is referred to as T.E.

—, "Alte und neue Fragen der Physik," Phys. Zeit., 11, 1234–1257 (1910). (Reprinted in Collected Papers, Vol. 7, 205–257.)

LORENTZ, H. A., "Deux mémoires de Henri Poincaré sur la physique mathématique," Acta mathematica, **38**, 293 (1914). (Reprinted in *Collected Papers*, Vol. **7**, 258–273.)

—, "The Connection Between Momentum and the Flow of Energy. Remarks Concering the Structure of Electrons and Atoms," Versl. Kon. Akad. Wetensch. Amsterdam, **26**, 981 (1917). (Reprinted in *Collected Papers*, Vol. **5**, 314–329.)

—, *Lectures on Theoretical Physics: Volume III*, translated by L. SILBERSTEIN & A. P. H. TRIVELLI (London: MacMillan, 1931).

McCORMMACH, R., "Henri Poincaré and Quantum Theory," Isis, **58**, 37–55 (1967).

—, "H. A. Lorentz and the Electromagnetic View of Nature," Isis, **61**, 459–497 (1970).

—, "Einstein, Lorentz and the Electron Theory," *Historical Studies in the Physical Sciences: Second Annual Volume 1970*, R. McCORMMACH ed. (Philadelphia: University of Pennsylvania Press 1970).

MIE, G., "Grundlagen einer Theorie der Materie," Ann. d. Phys., **37**, 511–534: *ibid.*, **39**, 1–40 (1912); *ibid.*, **40**, 1–66 (1913).

MILLER, A. I., "Comment on: Poincaré's Rendiconti Paper on Relativity: Part I," Am. J. Phys. **40**, 923 (1972).

MILLER, J. D., "Rowland and the Nature of Electric Currents," Isis, **63**, 5–27 (1972).

MINKOWSKI, H., "Das Relativitätsprinzip," lecture delivered before the Math. Ges. Göttingen on 5 Nov. 1907 published in Ann. d. Phys., **47**, 927–938 (1915).

—, "Die Grundgleichungen für die elektromagnetischen Vorgänge in bewegten Körpern," Nachr. Ges. Wiss. Göttingen, 53–111 (1908).

—, "Raum und Zeit," lecture delivered at the Congress of Scientists, Cologne, 21 Sept. 1908 translated in *P.R.C.*, pp. 75–91.

MØLLER, C., *The Theory of Relativity* (London: Oxford University Press 1952).

NEUMANN, G., "Die träge Masse schnell bewegter Elektronen," Ann. d. Phys., **45**, 529–579 (1914).

PANOFSKY, W. K. H., & M. PHILLIPS, *Classical Electricity and Magnetism* (Reading: Addison-Wesley 1961).

PAULI, W., *Theory of Relativity*, translated by G. FIELD (New York: Pergamon, 1958). This is a translation of PAULI's article "Relativitätstheorie" in Encykl. math. Wiss. Vol. V **19** (Leipzig: Teubner 1921).

PENDER, H., "On the Magnetic Effect of Electrical Conduction," Phil. Mag. **2**, 179–208 (1901).

PLANCK, M., "Das Prinzip der Relativität und die Grundgleichungen der Mechanik," Verh. d. p. Ges., **6**, 136–141 (1906).

—, "Zur Dynamik bewegter Systeme," Sitzungsber. d. k. preuss. Akad. Wiss., **13**, 542–570 (1907). This paper was also published in Ann. d. Phys. **26**, 1–34 (1908).

POINCARÉ, H., *Science and Hypothesis* (New York: Dover, 1952). This is a translation of Poincaré's *La Science et l'Hypothèse* (Paris: Ernest Flammarion 1902). This book is referred to as *S.H.*

—, *The Value of Science*, translated by GEORGE BRUCE HALSTED (New York: Dover 1958). This is a translation of POINCARÉ's *La Valeur de la Science* (Paris: Ernest Flammarion 1905). This book is referred to as *V.S.*

—, *Science and Method*, translated by FRANCIS MAITLAND (New York: Dover n.d.). This is a translation of POINCARÉ's *Science et Méthode* (Paris: Ernest Flammarion 1908). This book is referred to as *S.M.*

—, *Mathematics and Science: Last Essays*, translated by JOHN W. BOLDUC (New York: Dover 1963). This is a translation of POINCARÉ's *Dernières Pensées* (Paris, Ernest Flammarion 1913).

—, "Les Géométries non-euclidiennes," Revue générale des Sciences pures et appliquées **2**, 769–774 (1891).

—, "À propos de la théorie de M. Larmor," L'Éclairage électrique, **3**, 5–13, 289–295 (1895); *ibid.*, **5**, 5–14, 385–392 (1895). (Reprinted in *Œuvres de Henri Poincaré* (11 vols.; Paris: Gauthier-Villars 1934–1954), Vol. **9**, pp. 369–426; this volume is referred to as *Œuvres.*)

Poincaré, H., "La mesure du temps," Rev. Mét. Mor., **6**, 371–384 (1898); translated in *V.S.*, pp. 26–36.

—, "Des fondements de la Géométrie: À propos d'un Livre de M. Russell," Rev. Mét. Mor. **7**, 251–279 (1899). This is a book review of Bertrand Russell's *An Essay on the Foundations of Geometry* (Cambridge: Cambridge U. Press 1897; New York: Dover 1952). Pages 265–269 of Poincaré's review constitute pages 75–79 of Chapter V, "Experiment and Geometry", of Poincaré's *S.H.*

—, "Sur les rapports de la Physique expérimentale et de la Physique mathématique," *Rapports presentés au Congrès international de Physique réuni à Paris en 1900* (4 vols.; Paris: Gauthier-Villars 1900), Vol. **1**, pp. 1–29; translated as Chapters IX and X, pp. 140–182 of *S.H.*

—, "Sur les principes de la Mécanique," *Bibliothèque du Congrès international de Philosophie tenu à Paris du 1er au 5 août 1900* (Paris: Colin 1901), pp. 457–494; an expanded version of this paper is presented on pp. 123–139 of *S.H.*

—, "La théorie de Lorentz et le principe de réaction," *Recueil de travaux offerts par les auteurs à H. A. Lorentz* (The Hague: Nijhoff 1900), pp. 252–278. (Reprinted in *Œuvres*, pp. 464–488.)

—, "Sur la valeur objective de la Science," Rev. Mét. Mor., **10**, 263–293 (1902); an expanded version of this paper is presented in Part III, pp. 112–142 of *V.S.*

—, "L'état actuel et l'avenir de la Physique mathématique" delivered on 24 September 1904 at the International Congress of Arts and Science at Saint Louis, Missouri and published in Bull. Sci. Mat., **28**, 302–324; translated on pp. 91–111 of *V.S.*

—, "Sur la dynamique de l'électron," Comptes rendus de l'Académie des Sciences, 140, 1504–1508 (1905). (Reprinted in *Œuvres*, pp. 489–493.)

—, "Sur la dynamique de l'électron, Rend. del Circ. Mat. di Palermo, 21, 129–175 (1906). (Reprinted in *Œuvres*, pp. 494–550.)

—, "La dynamique de l'électron," Revue générale des Sciences pures et appliquées, **19**, 386–402 (1908). (Reprinted in *Œuvres*, pp. 551–586.) Excerpts from this paper have been translated in Book III, pp. 199–250 of *S.M.*

—, "La Mécanique nouvelle", the last of Poincaré's six Wolfskehl lectures given at Göttingen in 1909 and published in *Sechs Vorträge über ausgewählte Gegenstände aus der reinen Mathematik und mathematischen Physik* (Leipzig: Teubner 1910).

—, "Les rapports de la matière et de l'éther" given on 11 April, 1912 before the French Society of Physics and published in Journal de Physique théorique et appliquée, **2**, 347 (1912); translated in Chapter VII, pp. 89–101 of *Last Essays* ...

—, "L'Espace et le temps," given on 4 May, 1912 at the University of London; translated in Chapter II, pp. 15–24 of *Last Essays* ...

—, "Sur la théorie des Quanta," Comptes rendus de l'Académie des Sciences, **153**, 1103–1108). (Reprinted in *Œuvres*, pp. 620–625.)

—, "Sur la théorie des Quanta," Journal de Physique théorique et appliquée, **2**, 5–34 (1912). (Reprinted in *Œuvres*, 625–653.)

—, *La dynamique de l'électron* (Paris: Dumas 1913).

—, La Mécanique nouvelle (Paris: Gauthier-Villars 1924).

Reid, C., *Hilbert* (New York: Springer 1970).

Richardson, O. W., *The Electron Theory of Matter* (Cambridge: Cambridge University Press 1916).

Rohrlich, F., "Self-Energy and Stability of the Classical Electron," Am. J. Phys., **28**, 639–643 (1960).

—, *Classical Charged Particles* (Reading: Addison-Wesley 1965).

Russell, B., *An Essay on the Foundations of Geometry* (Cambridge: Cambridge U. Press 1897; New York: Dover Press 1952).

—, Mind, **14**, 412–418 (1905). This is a review of Poincaré's book *Science and Hypothesis*.

Schaffner, K., "The Lorentz Theory of Relativity," Am. J. Phys., **37**, 498–513 (1969).

—, *Nineteenth-Century Aether Theories* (New York: Pergamon 1972).

Schwartz, H., *Introduction to Special Relativity* (New York: McGraw-Hill 1968).

Schwartz, H., "Poincaré's Rendiconti Paper on Relativity, Part I," Am. J. Phys. **39**, 1287–1294 (1971); "II", *ibid.*, **40**, 862–872 (1972).

Schwarzschild, K., "Zur Elektrodynamik; I," Nachr. Ges. Wiss. Göttingen, 125–131 (1903); "II," *ibid.*, 132–141 (1903); "III," *ibid.*, 245–278 (1903).

Schweber, S., *An Introduction to Relativistic Quantum Field Theory* (New York: Row, Peterson and Company 1961).

Scribner, C., "Henri Poincaré and the Principle of Relativity," Am. J. Phys. **32**, 672–678 (1964).

Searle, F. C., "On the electromagnetic effects due to the motion of electrification through a dielectric," Phil. Mag., **27**, 324–339 (1889).

Serrin, J., "Mathematical Principles of Classical Fluid Dynamics," *Handbuch der Physik: Volume VIII, Part I* (Berlin-Göttingen-Heidelberg: Springer 1959).

Thomson, J. J., "On the electric and magnetic effects produced by motion of electrified bodies," Phil. Mag., **11**, 229–249 (1881).

—, *Notes on Recent Researches in Electricity and Magnetism* (Oxford; Clarendon 1893).

Volterra, V., Hadamard, J., Langevin, P., & Boutroux, P., Henri Poincaré (Paris: Librairie Félix Alcan 1914).

Weyl, H., "Gravitation und Elektrizität," Sitzungsber. d. k. preuss. Akad. Wiss., 465–480 (1918); translated in *P.R.C.*, 201–216.

—, *Space-Time-Matter*, translated by Henry L. Brose (New York; Dover 1952).

Whittaker, E. T., *A History of the Theories of Aether and Electricity* (2 vols.; London: Nelson, Vol. 1, 1951, Vol. 2, 1953). A version of volume 1 with a similar title had appeared in 1910 (Dublin: Dublin University Press 1910).

Wien, W., "Über die Möglichkeit einer elektromagnetischen Begründung der Mechanik," *Recueil de travaux offerts par les auteurs à H. A. Lorentz* (The Hague; Nijhoff 1900), pp. 96–107.

Wilson, W., "The Mass of a Convected Field and Einstein's Mass-Energy Law," Proc. Roy. Soc. (London), 734–740 (1936).

Zahn, C. T., & A. A. Spees, "A critical analysis of the classical experiments on the variation of electron mass," Phys. Rev., **53**, 511–521 (1938).

Department of Physics
Harvard University

(Received September 10, 1972)

Part II

A Technological Interlude

Essay 3

Unipolar Induction:
a Case Study of the Interaction between
Science and Technology

Summary

Unipolar induction, discovered in 1832 by Michael Faraday, is the case of electromagnetic induction in which a conductor and magnet are in relative rotatory motion. Attempts by scientists and engineers in the nineteenth and twentieth centuries to understand unipolar induction by using magnetic lines of force displayed striking national differences that influenced where the first large-scale unipolar dynamo was built. This episode is described, as well as the effect of unipolar induction on Albert Einstein's thinking toward the special theory of relativity, in sections 1–6. The analysis of electromagnetic induction in cases where the source of the magnetic field is in motion relative to the conductor is provided in sections 7–9.

Contents

1. Introduction 155
2. The problem of whether magnetic lines of force rotate: 1831–1901 157
3. National differences in the treatment of unipolar induction 164
4. How these national differences influenced where the first large-scale
   unipolar dynamo was built 173
5. The current status of unipolar dynamos 176
6. Some twentieth-century perspectives on unipolar induction 176
7. A conducting loop and magnet in relative inertial motion 184
8. The case of magnet and conductor in relative rotatory motion 184
9. Linear unipolar induction 188

1. Introduction

Since the phenomenon of unipolar induction may not be familiar to everyone, let me begin by describing it. Unipolar induction is the generation of current in a conductor for the case in which the conductor and/or the source of a uniform magnetic field are in relative rotatory motion. A typical case of unipolar induction is in Figure 1, where a conducting loop AC makes sliding contacts with one pole and the equator of a cylindrical permanent magnet.

Faraday's researches on electromagnetic induction in rotating systems led to intense debates concerning the location of the seat of the electromotive force (EMF); or, phrased equivalently, do the magnetic lines of force rotate with the magnet? For example, when in Figure 1 only the loop rotates, then the seat of the EMF is in the

ANNALS OF SCIENCE, 38 (1981), 155–189

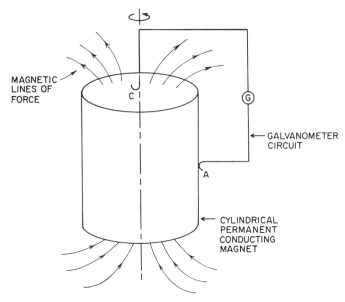

Figure 1.
An apparatus used by Faraday to investigate unipolar induction (see Faraday (footnote 1), vol. I, p. 65).

Figure 2.
The 87 ton unipolar dynamo constructed in 1912 at Westinghouse in Pittsburgh. (From Lamme (footnote 73 below), 1811.)

loop, and the current can be explained as caused by the loop cutting magnetic lines of force. When only the magnet rotates, then there are two possible seats of EMF, and two possible causes of the current measured in the loop. If the lines of force rotate with the magnet, then the seat of the EMF is once again in the loop because it is cut by lines of force. If the lines remain at rest, then the seat of the EMF is always in that part of the system which is rotating; in this case the seat of the EMF is in the magnet.[1]

The steps leading from Faraday's table top laboratory apparatus of 1832 in Figure 1 to the construction in 1912 of the 87-ton unipolar dynamo in Figure 2 is a fascinating case study in science, technology and society, demonstrating their interrelationship. The goal of this essay is to outline this interplay.

2. The problem of whether magnetic lines of force rotate: 1831–1901

Faraday's law of magneto-electric induction of 1831 is rooted in his experiments on the currents generated in conductors moving through magnetic fields—for example, the current generated in a copper disc rotating between the pole faces of a permanent magnet.[2] This law related the direction of the induction current to the direction of the conductor's motion as it cut magnetic lines of force which, Faraday wrote, could be 'depicted by iron filings'.[3]

But in order to clarify further and sharpen the law of induction in 1832, Faraday investigated cases in which the source of the magnetic fields could also move. In particular, he sought to determine whether for the production of current in a wire 'it was essential or not'[4] for a moving wire to cut lines of varying intensity 'or whether always intersecting curves of equal magnetic intensity, the mere motion was sufficient.'[5] The latter case Faraday had shown to be 'true'[6] when a copper plate, a copper sphere or a loop of wire rotated in the earth's magnetic field; but Faraday wished to 'prove the point with an ordinary magnet',[7] that is, under more controlled conditions.

Toward this goal Faraday's first experiment concerned a copper disc, a cylindrical permanent conducting magnet, and a galvanometer circuit that made sliding contacts with the disc's periphery and axis (Figure 3). I have found it useful to represent the results of electromagnetic induction experiments such as Faraday's where the disc and magnet were in relative motion with tables analogous to the one in Figure 3.

---

[1] We are taught that electromagnetic induction is a relative phenomenon, and yet most textbooks discuss it only from one point of view; namely, the generation of current in a conductor in inertial motion relative to a magnetic field. For interested readers I have outlined in sections 7, 8 and 9 the details for cases in which the source of the magnetic field is in motion relative to the conductor. In particular, the results for the case of relative rotatory motion may at first sight seem as puzzling today, as they did to many scientists and engineers of the 19th and early 20th centuries. There are two particularly informative modern physics texts that treat electromagnetic induction for the case in which the magnet is in motion: E. G. Cullwick, *Electromagnetism and relativity* (New York, 1959); and W. G. V. Rosser, *Classical electromagnetism via relativity* (New York, 1968), esp. pp. 190–203. Among the other interesting 20th-century analyses of electromagnetic induction see Swann (footnote 123 below) and Tate (footnote 123). Elsewhere I shall present an historical analysis of electromagnetic induction, a phenomenon which even at the beginning of the 20th century was not yet entirely understood by many physicists and engineers.

[2] Michael Faraday, *Experimental researches in electricity*, (3 vols., New York, 1965) vol. 1, 24 *ff*. A useful biography of Faraday is L. Pearce Williams, *Michael Faraday: a biography* (New York, 1971).

[3] Faraday *ibid.*, vol. 1, 32.

[4] *Ibid.*, vol. 1, 63, § 217.

[5] *Ibid.*

[6] *Ibid.*

[7] *Ibid.*

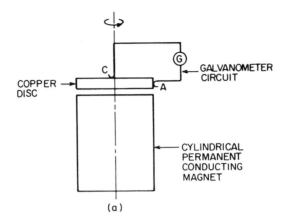

(a)

Experiment	Disc	Magnet	Current
1	R	R	Yes
2	R	X	Yes
3	X	R⁻¹ (or R)	No

(b)

Figure 3.

Faraday's apparatus (a) and results (b) for unipolar induction with a copper disc, a permanent magnet and a galvanometer circuit that makes sliding contacts with the disc at A and C (see Faraday (footnote 2), vol. I, 32 and 63). In the table (b), X = No Rotation, R = Rotation, R^{-1} = Rotation Inverse to R.

In experiment 1, the disc and magnet were cemented together: in experiments 2 and 3, they were separate.[8] Faraday concluded: 'Hence, rotating the magnet causes no difference in the results, for a rotatory and a stationary magnet produce the same effect upon the moving copper[9] [there is] a *singular independence* of the magnetism and the bar in which it resides.'[10] Thus, in his opinion, the lines of force did not rotate with the magnet. This conclusion violated Ampère's theory of magnetism, in which the atomic current whirls were primitive, and so the lines of force should have rotated with the magnet. For Faraday, on the other hand, the lines of force were primitive. It is important to note Faraday's emphasis that experiments 2 and 3 were not inverses of one another.[11] Faraday recognized that there was involved in all three experiments a three-part system—magnet, disc and galvanometer circuit. Therefore, the experiment inverse to 2 was a rotation of the magnet and galvanometer circuit, and then a current would have been induced.

---

[8] Faraday describes experiment 3 in *Faraday's diary* (6 vols., London, 1932), vol. 1, p. 402, § 257 (26 December 1831).

[9] Faraday (footnote 2), vol. 1, 63, § 218.

[10] *Ibid.*, 64, § 220; italics in original.

[11] Faraday (footnote 8), vol. 1, 402, § 256.

Another apparatus used by Faraday is in Figure 1.[12] He found in 1832 that rotating only the magnet about its axis caused a current in the galvanometer circuit. Then, by varying the distance between the ends A and C of the galvanometer's wires, he discovered that the maximum current resulted when the wires touched the magnet's equator and one of the magnet's poles. Consistent with his hypothesis of fixed lines of force, he concluded that the current was caused by the charges of unlike sign that accumulated on the magnet's pole and equator as a result of the magnet having rotated through its own lines of force.[13] With these experiments he considered that he had succeeded in making the law of induction more precise because it now depended upon only the intersection of lines of force and not, he wrote, 'on that peculiar condition, which I ventured to call the electro-tonic state'.[14]

Wilhelm Weber, in 1841, referred to the apparatus in Figure 1 as producing 'unipolar induction', since only one pole of the magnet was touched by a wire, and so only one magnetic fluid was in operation.[15] Being a proponent of Ampère's theory of magnetism, Weber considered the lines of force to rotate with the magnet.

By 1851 Faraday's investigations into atmospheric magnetism and diamagnetism, among other things, had convinced him of the physical reality of lines of force. So in 1851 Faraday once again investigated magnetic lines of force, focusing his researches upon their 'definite character; and of their distribution within a magnet and through space.'[16] In his *Diary* he posed the following question: 'Do the lines of force revolve with the magnet or do they not?'[17] His published reply is in the *Experimental researches*: 'Since current is induced when a magnet is pushed through a motionless wire loop, then lines of force participate in the magnet's translation.'[18] But, he continued, 'no mere rotation of a bar magnet on its axis, produces any induction effect on circuits exterior to it'.[19] His reason was that no lines of force were cut by the closed circuit because the lines of force did not rotate with the magnet 'any more than the rays of light which emanate from the sun are supposed to revolve with the sun'.[20] Thus, Faraday distinguished between the characteristics of a rotating and a translating magnet.

The crux of Faraday's experimental results were contained in his experiments with the unipolar apparatus of 1832 (see Figure 4). He interpreted the experiments in Figure 4 as follows:[21] the induced current arose in experiment 1 as a result of the unlike charges that accumulated on the rotating magnet's equator and periphery as a result of the magnet having rotated through its own lines of force. In experiment 2 the current arose as a result of the galvanometer circuit having cut lines of force.

---

[12] Faraday (footnote 2), vol. 1, 64–65.

[13] Faraday hypothesized that atmospheric effects such as the Aurora Borealis and Australis could also be explained in this manner by considering the earth to rotate through its own magnetic lines of force (footnote 2), vol. 1, 52–57). See section 8 for details concerning the volume and surface charge distributions as they arise on a rotating spherical magnet.

[14] Faraday (footnote 2), vol. 1, 67, § 231.

[15] W. Weber, 'Unipolare Induction', *Annalen der Physik*, **52** (1841), 353–386.

[16] Faraday (footnote 2), vol. 3, 327–349 (p. 328).

[17] Faraday (footnote 8) vol. 5, 398 (14 July 1851).

[18] Faraday (footnote 2), 336–337, §§ 3088–3090. See also section 7 below.

[19] *Ibid.*, 336, § 3090. See sub-section 8.1, where it is shown that the external electric field of a magnet rotating about an axis of symmetry parallel to the magnet's own magnetic field is totally electrostatic in origin; therefore, 'no mere rotation' of a magnet can generate a current in a stationary closed circuit.

[20] *Ibid.*, 336–337, § 3090.

[21] *Ibid.*, 337–338, §§ 3091–3122.

Experiment	Magnet	Circuit	Current
1	R	X	Yes
2	X	R^{-1}	Yes
3	R	R	No

Figure 4.
Faraday's results using the apparatus in Figure 1.

Faraday reasoned that if lines of force existed in the magnet and were continuous with those outside, then experiment 3 could be explained by the current in the exterior wire having been neutralized by a current in the magnet's interior that was equal in magnitude but was oppositely directed to the current in the exterior wire. The current in the magnet's interior arose as a result of the magnet having rotated through its own lines of force. He referred to this generation of a current within the magnet as the 'law of neutrality'.[22] In addition, he claimed that the same results for a rotating system could be obtained if the solenoid were substituted for the permanent magnet; however, he did not perform these experiments.[23]

Faraday's law of electromagnetic induction found its mathematical expression in the various electromagnetic theories of the 19th century.[24] In each of them the EMF in a closed circuit in inertial motion through a uniform magnetic field was the familiar result

$$\varepsilon = \frac{1}{c} \oint \mathbf{v} \times \mathbf{B} \cdot \mathbf{dl}, \tag{1}$$

where ε is the EMF, $\mathbf{v}$ is the relative velocity between the magnetic field $\mathbf{B}$ and the closed circuit, $\mathbf{dl}$ is an infinitesimal element of the closed circuit, c is the velocity of light in vacuum, and I am using Gaussian cgs units.

Most British electrodynamicists such as Oliver Heaviside,[25] Joseph Larmor,[26] Oliver Lodge,[27] Clerk Maxwell,[28] John Henry Poynting[29] and Joseph John

[22] Ibid., 342, § 3102.

[23] Ibid., 348, § 3120.

[24] See, for example, Sir E. Whittaker, A history of the theories of aether and electricity (2 vols., New York, 1951; repr. 1973), vol. 1. For a detailed calculation using Weber's theory of electromagnetism of the characteristics of a conducting sphere rotating about a uniform magnetic field directed parallel to the sphere's axis of rotation, see E. Jochmann, 'Über die durch einen Magnet in einem rotirenden Stromleiter inducirten elektrischen Ströme', Journal für die reine und angewandte Mathematik, 63, (1864), 158–178. An English translation is in Philosophical magazine, (4) 27 (1864), 506–528.

[25] For example, Oliver Heaviside, Electromagnetic theory (3 vols., London, 1893), vol. 1, 46–48.

[26] For example, James Larmor, 'Electromagnetic induction in conducting sheets and solid bodies', Philosophical magazine, (5) 17 (1884), 1–23; 'A dynamical theory of the electric and luminiferous medium—Part II', Philosphical transactions of Royal Society of London, (A) 186 (1895), 695–742, esp. pp. 727–731.

[27] For example, Oliver Lodge, Modern views of electricity (London, 1907), esp. pp. 166–171 which are unchanged from the first edition of 1889.

[28] For example, James Clerk Maxwell, A treatise of electricity and magnetism (2 vols., 3rd ed. Oxford, 1892; reprinted, New York, 1954) vol. 2, esp. pp. 179–189, 241–243.

[29] For example, J. H. Poynting, 'On the connexion between electric current and the electric and magnetic inductions in the surrounding field', Philosophical transactions of Royal Society of London, (A) 176 (1885), 277–396.

Thomson[30] considered lines of force, both electric and magnetic, as an interpretation of mathematical symbols useful for purposes of illustrating certain physical processes. However, to the best of my knowledge, whether magnetic lines of force rotated played no role in their calculations.[31] Yet when Lodge[32] or Thomson,[33] for example, referred to the physical characteristics of a rotating magnet, they considered it natural for the lines of force to participate in the magnet's rotation. Among their reasons was that even though Maxwell's theory was predicated upon an ether, electromagnetic induction must depend upon only the relative motion of magnet and conductor.

This relativist position concerning electromagnetic induction was held also by the other British physicists I mentioned,[34] and was eloquently put forth by their countryman S. Tolver Preston in two widely-read papers of 1885.[35] Like most everyone else at this time Preston could see lines of force. Preston considered, for example, the motion of iron filings whenever their source was either translated or rotated as support for his principal criticisms of Faraday's fixed-line hypothesis: (i) rotation was just a particular case of translation; (ii) if the lines of force rotated, then one was concerned only with relative and not at all with absolute motions; and (iii) Faraday's viewpoint violated Ampère's theory. Preston did not disagree with Faraday's data, but he emphasized that they were open to a double interpretation.[36] For example, Preston asserted that whereas in Faraday's viewpoint the current in the unipolar induction experiments 1 and 2 in Figure 4 was the effect of two different causes, in the moving line viewpoint the current arose under the same cause—the conductor was cut by or cut the lines of force, respectively. Therefore, the seat of the EMF was always in the wire loop AC. Rather than using Faraday's law of neutrality, according to the rotating line hypotheses the null result in experiment 3 was simply due to there having been no relative motion between magnet and conductor. (See section 8 for the details of an easily soluble case of unipolar induction.) Furthermore, Preston continued, the rotating line viewpoint was consistent with the complete equivalence between the effects of a rotating solenoid and a rotating permanent magnet on charges in their vicinity.[37]

For further support of his viewpoint Preston concluded a sequel paper of 1891 with quotations from letters of Lord Rayleigh and Weber.[38] Preston added, however, that Heinrich Hertz, with whom he had been in communication, was not optimistic over Preston's proposal to use an electrometer for determining whether a rotating magnet became statically charged. Hertz's reason was that existing

---

[30] For example, J. J. Thomson, *Recent researches in electricity and magnetism* (Oxford, 1893), esp. pp. 534–557.

[31] For example, see Larmor's paper of 1884 (footnote 26).

[32] Lodge (footnote 27) esp. pp. 166–171.

[33] Thomson (footnote 30), esp. pp. 22–23.

[34] Maxwell himself emphasizes this point several times in his *Treatise* (footnote 28)—for example, vol. 2, 179 and 241–243, See also pp. 11–13 of Larmor's paper of 1884 (footnote 26).

[35] S. Tolver Preston, 'On some electromagnetic experiments of Faraday and Plücker', *Philosophical magazine*, (5)19 (1885), 131–140; 'On some electromagnetic experiments, continued,—No. II. Diverse views of Faraday, Ampère, and Weber', *Ibid.*, 215–218.

[36] This had been previously pointed out by Beer, 'Über das Verhältniss des Laplace-Biot'schen Gesetzes zu Ampère's Theorie des Magnetismus; Vergleich der von Neumann und Plücker aufgestellen Theoren der magneto-elektrischen Induction', *Annalen der Physik*, **94** (1855), 177–192, esp. pp. 191–192.

[37] Preston (footnote 35), 133.

[38] S. Tolver Preston, 'The problem of the behaviour of the magnetic field about a revolving magnet *Philosophical Magazine*, (5) **31** (1891), 100–102.

electrometers lacked the accuracy sufficient for this purpose. Anyway Hertz, for whom Maxwell's theory was 'Maxwell's system of equations',[39] could not have been particularly interested in this controversy. Hertz believed that his recent Galilean-covariant theory of electrodynamics could account for the EMF in unipolar induction from the expected form of Faraday's law for conductors moving in uniform magnetic fields.[40] For example, according to Hertz the EMF in the wire loop AC of Figure 1 is

$$\varepsilon = \frac{1}{c} \int_A^C \mathbf{v} \times \mathbf{B} \cdot \mathbf{dl}. \tag{2}$$

On the other hand, many electrodynamicists did not believe that Faraday's induction law in Maxwell's theory was sufficiently general to discuss the current in open circuits such as those in unipolar induction.[41] In Hertz's view, whereas it was sometimes helpful to illustrate induction in a moving circuit with lines of force, it made no sense to attribute any physical reality to these lines.[42]

In a literature search through leading British physics and engineering journals for the last third of the 19th century, I found that with the exception of Preston's papers no others were concerned exclusively with whether magnetic lines of force rotated. On the other hand, the philosophical tradition in German-speaking countries, dating at least back to Immanuel Kant, of abstracting in the mind's eye to a higher degree of reality concepts from the world of perceptions, made lines of force with their representation using iron filings, and their wide explanatory powers, a most attractive notion. This process of abstraction was referred to often in the German scientific literature by the almost untranslatable term *Anschauung*.[43] In fact, Gustav Wiedemann's comprehensive review of unipolar induction in the fourth volume of his massive *Lehre von der Elektricität* of 1885 listed upwards of thirty papers on unipolar induction for the period 1860–85; most of them had appeared in the *Annalen*.[44] Wiedemann compared at length the *Anschauung* of Faraday and Weber.[45]

The eminent Austrian electrodynamicist Ernst Lecher published in the *Annalen* of 1895 what was considered to have been the most comprehensive review of unipolar induction since Wiedemann's.[46] Lecher listed in the period 1885–95 eight papers published on this problem, and seven of them had appeared in the *Annalen*. By 1895 all of Faraday's experiments on rotating magnetic systems had been repeated with

[39] Heinrich Hertz, *Electric waves* (trans. by D. E. Jones: London, 1893; reprinted, New York, 1962), 21. All references are to the reprint edition.

[40] H. Hertz, 'On the fundamental equations of electromagnetics for bodies in motion', *Annalen der Physik*, **41** (1890), 360; translated in (footnote 39), 241–268. All references will be to the English translation. See, in particular, pp. 250, 254–255.

[41] See, for example, Thomson (footnote 30); R. Clausius, 'On the theory of dynamo-electrical machines', lecture delivered at the Annual Meeting of the Schweitzerischein Naturforschende Gesellschaft at Zurich. 8 August 1883, in *Philosophical magazine*, (5) **17** (1884, 46–50, esp. p. 49. For a criticism of Thomson see August Föppl's widely-read text *Einführung in die Maxwell'sche Theorie der Elektricität* (Leipzig, 1894), esp. p. 325.

[42] Hertz (footnote 40), 255.

[43] I have previously discussed the importance of the notion of *Anschauung* in 'Visualization lost and regained: the genesis of the quantum theory in the period 1913–27', in J. Wechsler (ed.), *On aesthetics in science*, (1978, Cambridge, Mass.).

[44] Gustav Wiedemann, *Die Lehre von der Elektricität* (4 vols., Braunschweig, 1885), esp. vol. 4.

[45] *Ibid.*, vol. 4, 1118.

[46] Ernst Lecher, 'Eine Studie über unipolare Induction', *Annalen der Physik*, **54** (1895), 276–304.

apparatus more accurate than was available in Faraday's time. But Lecher demonstrated easily that they could all be interpreted using either Faraday's or Preston's *Anschauung*, and electrometric measurements lacked sufficient accuracy to decide between them. However, Lecher believed that Faraday's *Anschauung* had greater explanatory powers, and he demonstrated this with an experiment whose results had a consistent visual interpretation only with Faraday's *Anschauung*.

Lecher's apparatus (Figure 5 (*a*)) was a clever variation on Faraday's unipolar experiment.[47] The two electromagnets I and II could turn either together or separately. The wires a, c and e were fixed contacts and the wires b and d were in sliding contacts with the surfaces of magnets I and II, respectively. The contacts a, b, c, d and e could be connected in pairs with a galvanometer. Lecher represented the magnets with their lines of force in Figure 5 (*b*). Connecting cd to the galvanometer, and spinning magnet II with magnet I at rest, the galvanometer deflection was 38 units. The same result was obtained if de were connected instead. These results were independent of whether magnet I spun with II. In the next experimental run the current was generated by connecting bd to the galvanometer. If only I or II rotated in the same direction, then a deflection of 38 units was observed again on the galvanometer. If I and II rotated with equal velocities in the same direction, then no galvanometer deflection resulted. Lecher's interpretation of these results was: According to Faraday's *Anschauung* the site of the EMF was in the sections xy and $x'y'$ (see Figure 5 (*b*)), and the data was easily explained.

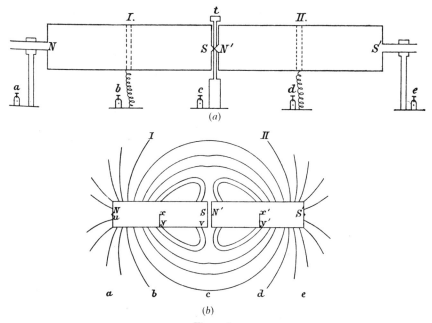

Figure 5.
(*a*) Lecher's apparatus for performing his unipolar induction experiment. (*b*) Lecher's drawing of the superposition of the magnetic lines of force generated by the two magnets in (*a*). (From Lecher (footnote 46), 295 and 296.)

[47] *Ibid.*, 295–302.

According to Preston's *Anschauung*, in the first experiment, even though the lines of force of both magnets 'flow into one another',[48] only the lines fixed to the rotating magnet could turn, while the other magnet's lines did not influence the process of induction. But this was not the case, Lecher continued, because removing I and then rotating II with either *cd* or *de* connected to the galvanometer resulted in a deflection less than the one obtained for the case when magnet I was present. Consequently in this experiment magnet I did reinforce the effect of magnet II whether or not they rotated together. 'In order to remain completely objective'[49] Lecher offered one final experiment. He coloured the lines of force of magnets I and II red and blue respectively. The lines were all symmetric about their generating magnets. According to Preston the lines were fixed to their magnets while each set of lines penetrated the other set, and yet were not influenced by the others' motion. Lecher wrote: 'Does it not appear easier to accept with Faraday that the lines of force are independent of any rotation of their generating magnet?'.[50]

Yet physicists and engineers in German-speaking countries continued to search intensely for the seat of the EMF in unipolar induction. It is perhaps in the introductory part of a paper by Otto Grotrian, in the *Annalen* of 1901,[51] that we can discern a principal reason for this quest. Grotrian wrote that 'exact knowledge of unipolar induction is of importance in electrical technology for the building of unipolar machines, that is, machines without commutators'.[52]

3. National differences in the treatment of unipolar induction

How can we account for the interest in German-speaking countries of physicists in problems of engineering, and in general of the interaction there between physicists and engineers? John Theodore Merz wrote of the 'completeness and thoroughness of research' in these countries, meaning that physicists and engineers in Germany, Austria and northern Switzerland were trained to consider problems from every possible viewpoint.[53] Joseph Ben-David has written that the rapid developments in German industry beginning in 1870 precipitated rapid developments in applied science at German universities.[54] The case of unipolar induction provides some fine structure to these replies. To see how this comes about I next survey the state of the electrical industry in England, America and Germany in the 19th century.

To the best of my knowledge, with few exceptions, there was in England little interaction between physicists and engineers. Some exceptions were Oliver Heaviside, Lord Kelvin and Lord Rayleigh, who focused their efforts primarily upon problems concerning telegraphy and electrical measurements, rather than the actual design of large-scale machines. In fact, in 1900 the British physicist Silvanus P. Thompson recalled that a small but 'pregnant' paper by Maxwell in 1867 on a

---

[48] *Ibid.*, 297.

[49] *Ibid.*, 301.

[50] *Ibid.*, 302.

[51] O. Grotrian, 'Elektrometrische Untersuchungen über unipolare Induction', *Annalen der Physik*, **6** (1901), 794–817.

[52] *Ibid.*, 794.

[53] J. T. Merz, *A history of European scientific thought in the nineteenth century* (2 vols., 1st ed. London, 1904–1912; reprinted New York, 1954), vol. 1, 213.

[54] J. Ben-David, *The scientist's role in society, a comparative study* (Englewood Cliffs, N.J., 1971), esp. pp. 126–127.

theoretical investigation of Werner Siemens's groundbreaking dynamo-electric principle 'was almost totally forgotten'.[55]

Well known were Thomas Edison's poor relations with the American academic community, undoubtedly reinforcing his belief that theoreticians could be of use to him only for performing complex calculations.[56] M. B. Snyder of the Central High School of Philadelphia aptly summed up the world-wide opinion of the state of science and technology in America when he wrote, in 1887, of 'America's worship of everything practical'; of her bold faith in haphazard progress; and of the lack of relations between those involved in pure research and in engineering as existed, for example, in Germany.[57] Indeed, Edison's steamroller method of research was considered, wrote Georg Siemens, the biographer of the House of Siemens, to be 'truly American fashion'.[58]

Werner Siemens, founder of the house of Siemens, considered himself to be more than merely an inventor. In even stronger words Georg Siemens has written that Werner Siemens was 'more than merely an inventor and a man of business, e.g., on the pattern of an Edison.'[59] In his autobiography, *Personal recollections*, Werner Siemens never mentioned Edison, despite their meetings in 1889, and the partnership in 1885–1888 of Siemens & Halske with the Deutsche Edison Company.[60] In contrast with Edison, Siemens energetically cultivated the friendship and colleagueship of physicists. Siemens often recalled his gratitude for having been accepted into the circle of the great men of science like Emil du Bois-Reymond, Hermann von Helmholtz, Rudolf Clausius, Wiedemann and Gustav Kirchhoff,[61] despite his own relatively poor education. Siemens often engaged the services of these men. He was one of the founders of the Berlin Physical Society and often lectured there, in addition to publishing frequently in the *Annalen*.

One of Siemens's greatest inventions is the 'dynamo-electric principle' of 1866 for self-exciting 'dynamo-electric machines', or 'dynamos' for short.[62] This principle opened up new vistas for d.c. machines which up to that time were useful chiefly for telegraphy. But until the late 1880s the design of d.c. dynamos was an art form. It should come as no surprise that one of the first theories of the dynamo was proposed in 1880 by the physicist O. Fröhlich, who was employed by Siemens & Halske.[63] In 1884 another physicist and friend of Siemens, Clausius, formulated a more complete theory of the dynamo.[64]

[55] Silvanus P. Thompson, *Dynamo-electric machinery* (New York, 1900), 20. Maxwell's paper is 'On the theory of the maintenance of electric currents by mechanical work without the use of permanent magnets', *Proceedings of the Royal Society of London*, **91** (1867); reprinted in *The scientific papers of James Clerk Maxwell* (ed. W. D. Niven; 2 vols. Cambridge, 1890; reprinted, New York, 1954), vol. 2, 79–85.

[56] For example, M. Josephson, *Edison: a biography* (New York, 1959).

[57] M. B. Snyder, 'The Electrical Exhibition and pure research', *The electrician and electrical engineer*, 6 (February 1887), 44–49.

[58] Georg Siemens, *History of the house of Siemens* (trans. by A. F. Rodger: 2 vols., (Munich, 1957), vol. 1, 87.

[59] *Ibid.*, 111.

[60] Werner von Siemens, *Personal recollections* (trans. W. C. Coupland: New York, 1893). See Josephson (footnote 56) and Siemens (footnote 58) for discussions of the interactions between Edison and Siemens.

[61] Siemens (footnote 60), 44.

[62] *Ibid.*, 329–333. See also Siemens (footnote 58), 78–82. For a detailed discussion of the dynamo-electric principle replete with the original papers, see Heinrich Schellen, *Magneto electric and dynamo-electric machines: their construction and practical application to electric lighting and their transmission of power* (trans. by N. S. Keith and P. Neymann and 'with very large additions and notes relating to American machines' by N. S. Keith: New York, 1884). The relevant portion is Part IV.

[63] Schellen (footnote 62), 458.

[64] Clausius (footnote 41).

I conclude this biographical sketch of Werner Siemens with some of his other accomplishments for the cause of science and engineering in German speaking countries. In 1882 he donated funds to establish chairs of electrical engineering in Germany. He was also responsible for urging the establishment of a class of nobility for those who have distinguished themselves in industry, science and literature. Thus, it was that in 1873 he became Werner von Siemens. In 1883 Siemens donated land for the Physical-Technical Imperial Institute to be directed by, as he wrote: 'the first physicist of our time, Privy Chancellor von Helmholtz'.[65] In 1886 Siemens was instrumental in establishing a similar institute at the ETH in Zurich, to be directed by another good friend, Heinrich Friedrich Weber. From 1896 to 1900 Albert Einstein attended classes there, taught by Weber among others, and observed at first hand the interaction between engineers and physical scientists.[66] The stamp of Siemens on engineering in German-speaking countries is clear—a strong interaction between scientists and engineers.

On the basis of his early experiences in electrolytic research, Werner Siemens was asked in 1877 by the chief engineer of the Royal Ironworks at Oker in the Harz mountains to design a dynamo for refining copper. In electrolytic processes it is important that the current be high and that the voltage be as low as possible. By late 1878 there were three Siemens & Halske dynamos operating in the mines at Oker

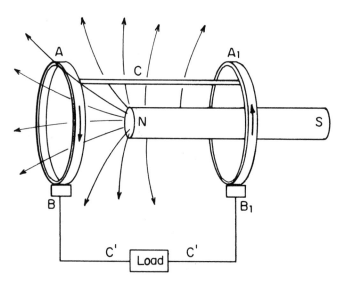

Figure 6.
An elementary unipolar dynamo where, C is the armature conductor; A, A_1 are collector rings; B, B_1 are brushes; C is a frame conductor; and the arrows on A, A_1 denote the direction of current flow.

[65] Siemens (footnote 60), 376. One of Siemens's sons, Arnold, married von Helmholtz's daughter.
[66] 'Obituary of Prof. Dr. Heinrich Friedr. Weber', *Schweizerischen Naturforschenden Gesellschaft Verh.*, **95** (1912), 43–53, esp. p. 47. Einstein's education at the ETH is discussed by R. McCormmach in the 'editor's foreward', *Historical studies in the physical sciences*, **7** (1976).

producing 994A at 3·5V.[67] Siemens knew, however, that a unipolar dynamo was ideal for this work since it was intrinsically a high current but low voltage dynamo. In 1881 he described the 1887 design of a commercial unipolar dynamo, with which Siemens wrote, he was assisted by 'my friend Kirchhoff'.[68] It was a self-exciting dynamo composed of two hollow copper cylinders that rotated about the side pieces of a horseshoe magnet. Kirchhoff suggested slicing the copper tubes lengthwise into strips and then using nonconducting material to join them together into a tubular structure, thereby reducing losses due to Foucault or eddy currents. Each strip was attached to a pair of collecting rings that were concentric with the tube. Current was taken off the rings with brushes that slid over their surfaces (see Figure 6). The current was divided at A to flow around the conducting ring to the brush B. This became essentially the method for current collection in all unipolars. The frame conductors could be connected in series or in parallel. The EMF of a unipolar dynamo was calculated directly from Faraday's law. Four defects appeared immediately which could not be understood without a theory of the dynamo: saturation of the iron in the field magnets, and three others which by the 1890s were known as ring reaction, armature reaction and poor regulation. An improved version of the 1878 machine (Figure 7) was exhibited in Paris in 1881, and a larger one was placed in operation in 1884. Although the machine of 1884 never appeared on the price list of Siemens & Halske,[69] S. P. Thompson reported in 1900 that a unipolar had been used successfully for many years at Oker.[70]

Siemen's machine was referred to as a unipolar dynamo because its armature produces d.c. current by rotating through a uniform magnetic field. It was equivalent to the apparatus in Faraday's rotating disc experiment (Figure 3), except that here the disc was transformed into a sheath over the magnet (see Figure 8). However, the term 'unipolar' was objectionable to many engineers, and so in the 1890s the names 'homopolar' and 'acyclic' came into use.[71]

In the period 1885–1890, a number of prominent engineers had recognized the importance of designing unipolar dynamos, especially for electrolytic processes; among them were George F. Forbes in England,[72] Elihu Thomson in the United States,[73] and C. E. L. Brown of the Oerlikon Works in Zurich.[74] The reason was that the unipolar dynamo did not require a commutator. The commutator was the most expensive part of a dynamo, frequently requiring repairs due to rubbing of the

[67] These machines are described in Siemens and Halske, 'Grosse dynamoelektrische Maschine für Rein-Metall-Gewinnung im hüttenmännischen Betriebe', *Elektrotechnische Zeitschrift*, **2** (February 1881), 54–55. Siemens's intense involvement in electrochemistry is discussed throughout W. Siemens (footnote 60) and in G. Siemens (footnote 58), esp. chapter XIII.

[68] W. Siemens, 'Die dynamoelektrische Maschine', lecture delivered to the Akademie der Wissenschaften of Berlin on 18 November 1880, in *Elektrotechnische Zeitschrift*, **2** (March 1881), 89–95. esp. pp. 94–95 (p. 94).

[69] These machines are described in F. Uppenborn, 'Über Unipolarmaschinen'. *Centralblatt für Elektrotechnik*, **7** (1885), 324–329.

[70] Thompson (footnote 55), 476.

[71] To the best of my knowledge Uppenborn (footnote 69) was one of the first to criticize the term unipolar induction as a 'lucus a non lucendo' (p. 325).

[72] 'Electrical engineering at the Inventions Convention'. *The engineer*. **60** (July 1885). 47–48.

[73] Elihu Thomson, 'Discussion at Boston' to the paper of B. G. Lamme, 'Development of a successful direct current 2000-kW unipolar generator', lecture deliverd to the 29th Annual convention of the American Institute of Electrical Engineers, Boston, 28 June 1912, in *American Institute of Electrical Engineers*, **30** (1912), 1811–1835. The 'discussion' is on pp. 1836–1840, and Thomson's comments are on pp. 1839–1840.

[74] Thomson describes Brown's designs in *ibid*.

Figure 7.
The Siemens & Halske unipolar dynamo exhibited in Paris in 1881. (From Schellen (footnote 62), 456.)

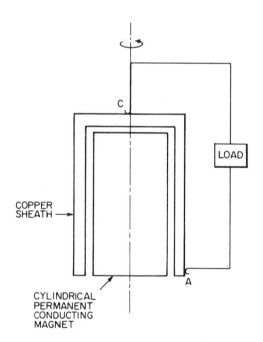

Figure 8.
The copper disc in Figure 3 is here transformed into a sheath over the cylindrical permanent conducting magnet.

copper leaf brushes over its uneven copper and mica surfaces. The major obstacle toward constructing large-scale unipolars was considered to have been the collection of current, because increasing the current required a higher rate of armature rotation, for example, 500–1,000 rpm. At these speeds the standard copper leaf brushes on the small prototypes tended to wear down and/or heat. Forbes, after having tried unsuccessfully to use mercury as a liquid collector, claimed in 1885 to have solved the collection problem with carbon brushes. He suggested that carbon brushes be used also on commutator machines. But by 1886 Elihu Thomson found that carbon brushes were not completely satisfactory for unipolars because their inability to follow vibrations of the armature required high pressures to hold them in contact, resulting in high i^2R losses. In 1885–1886 Thomson had built two small unipolars for lighting purposes. However, their regulation was poor, due, as he realized some years later, to insufficient knowledge of compounding.

Unaware of Thomson's results, Brown's prototypes of 1889 using carbon brushes also failed. Ironically, as of 1890 a principal result of research on unipolars was the improvement of commutator machines.

In an address to the American Institute of Electrical Engineers (AIEE) in 1894 the eminent American engineer F. B. Crocker, with C. H. Parmly, reported a drop of interest in the design of unipolars, and that thus far only Siemens's unipolar dynamo had been successful commercially.[75] But with the introduction of the steam turbine they believed the time was ripe to investigate the possibility of using unipolars for lighting purposes. Besides their simplicity of construction, Crocker and Parmly emphasized that the ring shaped form of the unipolar's magnetic circuit reduced the reluctance to a minimum. An efficient design for a tube dynamo is in Figure 9. Crocker and Parmly demonstrated that the relationship among the design parameters D (the armature's diameter), N (rev/min), ε, B, v, were

$$D \sim \frac{\varepsilon}{vB} \tag{3}$$

and

$$D \sim \sqrt{\frac{\varepsilon}{NB}}. \tag{4}$$

Around 1894 the maximum B was of the order of 8,000 gauss, and D, N or v were considered the quantities to be varied. For example, to produce 130V for lighting at 200 rpm required that $D = 18$ ft; but two 65V unipolars in series rotating at 800 rpm permitted a reduction of D. Despite S. P. Thomson's researches, Crocker and Parmly had confidence that Forbes had indeed solved the collection problem. They also predicted that owing to the uniformity of the primary magnetic field, losses due to hysteresis, eddy currents and armature reaction could be minimized. But they could not substantiate these claims, owing to the lack of data from large-scale unipolars.

---

[75] F. B. Crocker and C. H. Parmly, 'Unipolar dynamos for electric light and power', lecture delivered to the Eleventh General Meeting of the American Institute of Electrical Engineers, Philadelphia, 16 May 1894, in *American Institute of Electrical Engineers*, **11** (1894), 406–429.

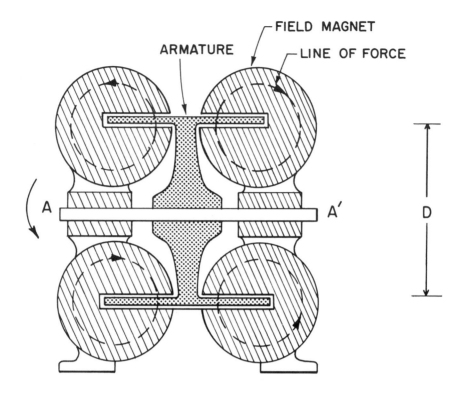

Figure 9.
The reluctance of a magnetic circuit (that is, the field magnet's resistance to lines of force) is defined analogously to the electric resistance in a wire. Consequently, the reluctance is directly proportional to the magnet's length and inversely proportional to the magnet's cross-sectional area. Since the unipolar's d.c. current arises from the rotating armature's continuous cutting of lines of force from a single pole, then the unipolar's magnetic circuit can be designed to have a reluctance that is much less than that of a multipolar dynamo. For example, in this schematic of a unipolar dynamo the magnetic circuit of each ring-shaped electromagnet (whose cross-section is indicated by the slanted lines) is designated by the dashes which represent lines of force. The electromagnets are slit in order to enable the rotation of the armature of diameter D about the line AA'.

Crocker and Parmly nowhere discussed the question of whether lines of force rotated. Their only comment was that current was induced in the armature through its cutting lines of force. But this question nevertheless lingered in the background whenever unipolar dynamos were discussed. For hysteresis and eddy current losses in the iron of the field magnet were due to the effects of the magnetic field whose origin was the copper conducting strips of the rotating armature. Consequently, the problem arose of whether this reaction occurred because: (1) the secondary magnetic field swept past the field magnet; or (2) the secondary field fluctuated in space. Parmly and Crocker chose not to concern themselves with these statements of the problems of hysteresis and eddy current losses in the field magnet.

In the discussion session their optimism over unipolars was endorsed strongly by another well known American electrical engineer Arthur E. Kennelly, who had been Edison's theoretician from 1887 to 1893. One of the forward-looking young men of a.c. engineering, Charles Proteus Steinmetz, also agreed with Crocker and Parmly. Steinmetz commented as well upon the notion of lines of force. He asserted that lines of force were 'nothing else than a physical hypothesis,[76] and it made no sense to discuss whether they rotated or not. For example, the magnetic field in the vicinity of a bar magnet remained constant in magnitude and direction regardless of the magnet's state of motion. So, Steinmetz continued, certain recent attempts to eliminate brush problems by having only the field magnet rotate were doomed to failure, 'you can very easily get a machine which can never do anything.'[77] Evidently Faraday's warning in the experiment concerning the disc and magnet had gone unheeded by many engineers; for (referring to Figure 8) if only the magnet rotated, then no current was generated in the external *closed* circuit. Another of Steinmetz's favourite illustrations of the labyrinthine webs that one could fall into in the moving line controversy was that the magnetic field in the vicinity of a bar magnet was uniform both to observers at rest in the laboratory and rotating with the magnet. Therefore, one could conclude that lines of force did and did not rotate.[78]

In Germany the problem of whether lines of force rotated or not was considered, as the engineer C. L. Weber wrote in the *Elektrotechnische Zeitschrift* of 1895, to concern 'a fundamental *Anschauung* of immediate importance and technical meaning, and not an academic moot point'.[79] The reason was that designers principally in Germany, but also in England and Switzerland, were studying the feasibility of building large-scale a.c. unipolar dynamos.[80] In these machines the field magnet was a rotating toothed wheel, which was wound so that only, for example, north poles were opposite the stationary armature (see Figure 10). The problem here was whether the reaction of the field magnet on the armature could be attributed to the lines sweeping past the armature or fluctuating in space. In other words, as Weber and others (among them the well-known professor of electrical engineering at Karlsruhe, E. Arnold) wrote, was one to use the *Anschauung* of Lecher and Faraday or of Preston? As far as I know, a.c. unipolar dynamos were never produced in Germany or in Switzerland for commercial purposes. However, this sort of dynamo was operating at the City of London lighting station in 1895; furthermore, as was to be expected, the British engineer W. M. Mordey's description of the machine did not mention any problems concerning rotating lines of force.[81]

---

[76] *Ibid.*, 427.

[77] *Ibid.*

[78] See C. P. Steinmetz, 'Discussion on "An imperfection in the usual statements of the fundamental law of electromagnetic induction"', *American Institute of Electrical Engineers*, **27** (1908), 1352–1354. For a discussion of unipolar machines by Steinmetz see his *Steinmetz Electrical Engineering Library*, vol. 6 (New York, 1917), chapter 22.

[79] C. L. Weber, 'Über unipolare Induktion', *Electrotechnische Zeitschrift*, **16** (August 1895), 513–514 (p. 514).

[80] For further details see E. Arnold, 'Über die unipolare Induktion und Wechselstrommaschinen mit ruhenden Wickelungen', *Electrotechnische Zeitschrift*, **16** (7 March 1895), 136–140. Arnold supported the rotating line viewpoint. Weber (footnote 79) criticized Arnold's supporting statements in the light of Lecher's recently published paper (footnote 46).

[81] *The engineer*, **80** (1895), 56–58.

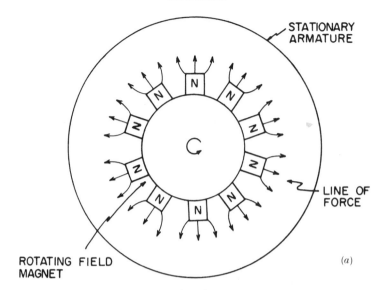

STATIONARY
ARMATURE

N

LINE OF
FORCE

ROTATING FIELD
MAGNET

(a)

(b)

(c)

4. How these national differences influenced where the first large-scale unipolar dynamo was built

At the close of the 19th century, among engineers it was believed that the collection problem could somehow be minimized in a unipolar dynamo whose armature was of a sufficiently large diameter. The editors of the British journal *Electrical world and engineer* wrote that such 'a large machine at a single step forms somewhat of a commercial venture'.[82] The British, who were always complaining over the rapid growth of the German electrical industry compared with their own, did not have the capital for a large venture which could be unfruitful commercially.[83]

Siemens and Halske had the capital. However, it is reasonable to make the following conjecture as to why it was that neither this firm nor any other in German-speaking countries produced the first large-scale unipolar machines. In these countries physicists and engineers worked in the close collaboration supported by Werner Siemens. This collaboration was almost certainly one of the foundations of the characteristic of research in Germany that Merz referred to as 'completeness and thoroughness of research';[53] furthermore, Merz emphasized that 'the German man of science was a philosopher'.[84] As a matter of course, therefore, physicists and engineers in German-speaking countries considered problems from every possible viewpoint, and so it was not surprising that, as I have discussed above, they were enmeshed in the controversy over the various *Anschauung* concerning lines of force. In fact, so enmeshed were they in this controversy that the first large-scale commercial unipolar machine was not built in Germany, Austria or Switzerland. It was designed and built in the United States where, in the first decade of the twentieth century, the electrical industry underwent a period of rapid growth. And in America the Kantian notion of *Anschauung* was not seriously considered in an interaction between scientists and engineers that also lacked the intensity of the one in German speaking countries.

F. B. Crocker, who had served as President of the AIEE from 1897 to 1898, wrote, in 1901, in a vein so modern to our eyes: 'Electricity is, to my mind, the only mechanical pursuit that has "soul". The successful electrician is born'.[85] In particular, Crocker continued, the American engineer was a special breed being interested in 'things that are new *because they are new* [on the other hand] a foreigner considers newness itself an objection'.[86] Crocker emphasized another factor that he considered as important in the education of an American engineer; namely, after graduation a period of apprentice courses in industry, often at General Electric or Westinghouse. But we must add one more ingredient, for American engineering had

---

[82] *Electrical world and engineer*, **45** (1905), 228–229. (p. 228).

[83] E.g., *The electrical review*, **43** (2 December 1898), 833; *ibid.*, **47** (14 December 1900), 965–966.

[84] Merz (footnote 53), 215.

[85] Quoted in *The story of electricity* (ed. T. C. Martin and S. L. Coles: New York, 1919), 92, from F. B. Crocker's article in *Saturday evening post*, 22 June 1901.

[86] Crocker (footnote 85); italics in original.

Figure 10.

(*a*) The rotating toothed field magnet wound so that only north poles face the stationary armature. (*b*) and (*c*) are prototypes of the rotating toothed magnet and stationary armature, respectively. These prototypes were designed by E. Arnold at the Oerlikon Works in Zurich. (From Arnold (footnote 80), 138.)

the benefit of one of Edison's greatest inventions—Menlo Park, which Edison created solely for producing inventions and then constructing prototypes. The industrial research laboratories of G.E. at Schenectady and Westinghouse at Pittsburg were descendents of Menlo Park. By the turn of the century, G.E. employed Steinmetz and Westinghouse employed Nikola Tesla. Although the research methods of such men as Steinmetz and Tesla differed from those of Edison, nevertheless much of the spirit of Edison's typically American approach to industrial research pervaded the atmosphere at G.E. and Westinghouse.[87] So it was that the first large-scale unipolar dynamo was designed and built at G.E. in Schenectady in 1905; and for no purpose other than to show that it could be done. Furthermore, the question of whether lines of force rotated was ignored. The engineer in charge was Joseph E. Noeggerath, an American who had been educated since the age of ten in Germany. Noeggerath's Doctoral Dissertation at the Technische Hochschule, Hannover, concerned a method for collecting large currents from the fast rotating armature of a unipolar dynamo. It is significant that in order to continue his work on unipolar dynamos he had to leave Germany to work for G. E., where his efforts were supported by Steinmetz.[88] At G. E. Noeggerath designed and built a tube dynamo producing 300 kW at 500V in a primary magnetic field of 10,000 gauss, with a steam turbine as prime mover. The armature was 30 feet in length and 5 feet in diameter. Fulfilling everyone's hopes the losses in the brush contacts were of the same order of magnitude as those due to hysteresis, eddy currents, ring and armature reaction, and regulation. It turned out that the wind from the rotating armature tended to lift the brushes from the rings just enough to reduce friction losses. Noeggerath's assessment of a unipolar dynamo was similar to that of Crocker and Parmly; namely, the machine's simplicity, in particular its lack of a commutator, 'should speak favorably for the new type' of dynamo'.[89] Noeggerath's description of the new machine was presented at the 1905 meeting of the AIEE and was received enthusiastically in England and Germany. By 1905 problems with commutators in the large d.c. dynamos linked with steam turbines had become serious. Usually inefficient means had to be used to obtain high d.c. current and voltage—for example, an a.c. dynamo with a converter, or small d.c. dynamos in series. The unipolar dynamo was seen as a panacea.[90]

[87] See also Josephson (footnote 56) and E. E. Morison, *From know-how to nowhere* (New York, 1974), esp. Chapter VII.

[88] Steinmetz's support of Noeggerath was recalled by a co-worker E. F. W. Anderson in Anderson's 'Inventors I Have Known', in P. Alger. *The human side of engineering* (Mohawk Duplicating Service) (see p. 133). Noeggerath's Doctoral Dissertation is entitled. 'Über die Stromabnahme mit besonderer Berücksichtigung höher Geschwindigkeiten' (Oldenbourg).

[89] J. E. Noeggerath, 'Acyclic (homopolar) dynamos', lecture deliverd to the 193rd meeting of the American Institute of Electrical Engineer's New York, 27 January 1905, in *American Institute of Electrical Engineers*, **24** (1905), 1–18, with discussions at New York and Boston on pp. 19–27; the quotation is from p. 5. See also E. W. Moss and J. Mould, 'Homopolar generators', *Journal of the Institution of Electrical Engineers*, **49** (24 April 1912), 804–816; and (footnote 82). Unfortunately the picture of Noeggerath's unipolar dynamo is too dark for reproduction.

[90] 'The Barbour homopolar dynamo', *Engineering*, **92** (9 September 1911), 318–319, esp. p. 318. For an interesting description of attempts to construct in Russia prototypes of unipolars with mercury contacts by the German educated engineer Boris Umgrimoff, see 'The unipolar dynamo and mercury contacts', *Engineering*, **92** (25 August 1911), 265. Materials for the dynamo were obtained from Germany.

As to whether the lines of force rotated or not, Noeggerath in a matter of fact manner stated that 'the secondary fluxes cut the frame as the armature revolves causing hysteresis';[91] he said nothing more. At the discussion session Kennelly commented favourably upon unipolar dynamos. He referred to the question of whether lines of force rotated as merely a 'theoretical contest',[92] and was of the opinion that lines of force had 'in a certain sense' physical existence.[93] Furthermore, Kennelly considered that the problem of the seat of the EMF in unipolar induction had not yet been settled decisively, but his remarks revealed that he was not overly preoccupied with this point. H. E. Heath, a coworker of Noeggerath, replied that since induction is 'a purely relative matter',[94] it made little difference whether the magnet or armature rotated because the result was the same; namely, a current was produced in the armature. As we have seen in discussing Figure 8, this was a risky statement.

Despite the success of Noeggerath's dynamo, the first large-scale unipolar dynamo was built for commercial purposes in 1912 at that bastion of a.c. technology, Westinghouse at Pittsburgh. It was designed by their chief engineer B. G. Lamme,[95] who had designed the a.c. dynamos at Niagara Falls. In a fascinating presentation to the 1912 meeting of the AIEE Lamme described the design and onsite debugging of this 2,000 kW, 264V, 7700A unipolar dynamo, powered by a 1,200 rpm steam turbine (see Figure 2). The machine was built for a cement factory in the town of Easton, Pennsylvania.

Straightaway Lamme disavowed any interest in the question of whether lines of force rotated. In words almost identical to those of Kennelly, Heath and Noeggerath, Lamme wrote: 'whether the magnetic flux rotates or travels with respect to the rotor or the stator ... makes no difference ... the result is the same on either assumption'.[96] Lamme discovered that when properly lubricated the old style copper mesh brushes served best.

Almost as if by its very success the unipolar dynamo became in 1912 a dinosaur. By 1912 there were relatively lighter and more economical high speed a.c. generators, driven by steam turbines, with converters for d.c. current; and more efficient systems of reduction gears for the direct connection of commutator machines to steam turbines. But more than anything else it was simply a matter of cost accounting that effectively ended further research and development on unipolar dynamos. In his assessment of Lamme's machine, W. A. Dick wrote that station operators were already thoroughly familiar with the commutator machines, and 'a new type always involves additional training and experience'.[97] From time to time Westinghouse manufactured a unipolar dynamo—for example, in 1934 one was ordered for pipe welding by an American steel mill. The machine produced 1,125 kW, 7.5V, 150 kA at 514 rpm.[98]

[91] Noeggerath (footnote 88), 5.

[92] *Ibid.*, 24.

[93] *Ibid.*

[94] *Ibid.*, 26.

[95] Lamme (footnote 73).

[96] *Ibid.*, 1811.

[97] W. A. Dick, 'Note on a large unipolar generator', *Electrical journal*, **9** (September 1912), 732–738 (p. 738).

[98] See *Standard handbook for electrical engineers* (ed. A. E. Knowlton: New York, 1957), 833.

5. The current status of unipolar dynamos

In the past 25 years there has been a resurgence of interest in unipolar dynamos for use in such fields as elementary-particle physics and ship propulsion. In 1950, an Australian physicist, M. L. Oliphant, built a unipolar dynamo to provide a huge current for discharge at the proper moment into the coils of a synchrotron, in order to produce an enormous magnetic field for the purpose of accelerating nucleons up to a laboratory kinetic energy of 10 GeV.[99] Oliphant's reason for using a unipolar dynamo was that increasing the strength of the magnetic field meant decreasing the synchrotron radius, thereby permitting Australia to participate in the frontiers of high energy physics at a cost she could afford (see Figure 11). The dynamo's armatures were four steel discs, each of them were 139 inches in diameter, 10·5 inches thick and 20 tons in weight. They rotated in opposite directions at 900 rpm in the field of a large electromagnet. After ten minutes a peak current of 1.7×10^6 amperes was collected by jets of liquid sodium and could be conducted to the synchrotron coils, where a magnetic field of 62,000 gauss was expected. However, as far as I know Oliphant's unipolar dynamo was never used for research into elementary particle physics.

Today the problem of collecting current is considered as solved completely by the use of the eutectic alloy NaK. Furthermore, superconducting magnets with fields of the order of 50,000 to 100,000 gauss open new vistas in the miniaturization of unipolar dynamos. Unipolar dynamos using superconducting magnets with NaK as a collector and with an efficiency of 98% are available for use commercially from G.E. (see Figure 12.)[100] The Naval Ship Research and Development Center at Annapolis, in conjunction with G.E. and Westinghouse, is actively considering the use of unipolar dynamos with superconducting magnets for ship propulsion. In the near future a 50-foot ship will be launched powered by a 3,000 hp unipolar engine coupled to a unipolar dynamo producing 150V at 20,000A.[101]

6. Some twentieth-century perspectives on unipolar induction

Having brought the story of the unipolar dynamo to a happy conclusion, let us return to the rotating line controversy in the late 19th century. In a widely-read electromagnetic theory text of 1894, the German engineer and physicist August Föppl was led by his relativism to adopt the moving line *Anschauung* for the description of unipolar induction.[102] These portions were totally rewritten by the

[99] M. L. Oliphant, 'The cyclosynchrotron: acceleration of heavy particles to energies above 1,000 MeV.. and the homopolar generator as a source of very large current pulses', *Nature*, **165** (1950), 466–468; 'The acceleration of protons to energies above 10 GeV', *Proceedings of the Royal Society of London*, **(A) 234** (1956), 441–456. The suggestion of homopolar generators as sources of large magnetic fields for particle physics research is due to R. L. Garwin, thesis No. T167 (Case Institute of Technology, 1947); for subsequent elaborations upon Garwin's work, see R. I. Strough and E. F. Shrader, 'Pulsed air core series disk generator for production of high magnetic fields', *Physical review*, **22** (1959), 578–582. I am grateful to Professor Edward Purcell for bringing Oliphant's work to my attention.

[100] L. M. Harvey and R. D. Fulmer, 'A new concept of electric ship propulsion', lecture presented to the New England Section of The Society of Naval Architects and Marine Engineers; C. F. Jones, 'The potential of superconductors for shipboard power applications', *Naval engineers journal*, (1967), 791–797; E. F. McCann and C. J. Mole, 'Superconducting propulsion systems for merchant and naval ship concepts', lecture presented at the Annual Meeting, New York, 15–17 November 1973, of the Society of Naval Architects and Marine Engineers. I thank Dr. L. J. Chmura of the Electric Boat Company for alerting me to this research.

[101] Personal communication from Dr. E. Quandt of the Naval Ship Research and Development Center, Annapolis, Maryland.

[102] Föppl (footnote 41), 327–329.

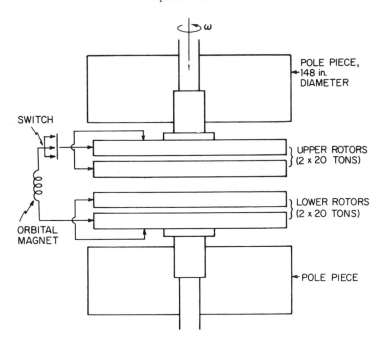

Figure 11.
Oliphant's unipolar dynamo. Rotation is about the vertical axis.

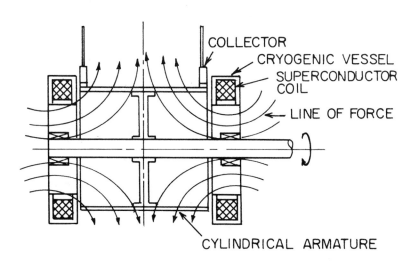

Figure 12.
The unipolar dynamo in 1979.

Göttingen physicist Max Abraham's popular 1904 revision.[103] Abraham's viewpoint was totally Hertzian. For Abraham the measurable quantity, that is, the induced current, was the same whether the lines rotated or not, and the EMF in unipolar induction was given by (2). Earlier, in 1900, Henri Poincaré had also dismissed the problem of whether lines of force rotated as meaningless.[104]

The great Dutch physicist H. A. Lorentz, in a monographic essay of 1903, considered lines of force to be useful aids for a simpler discussion of electromagnetic phenomena.[105] His preference for an ether-based dynamical theory of electromagnetic phenomena, in which velocities relative to the ether play a role, for example, to explain the Michelson-Morley experiment,[106] led him to prefer Faraday's viewpoint—although Lorentz emphasized that either viewpoint is valid. But Lorentz neither used lines of force in his calculations, nor did he refer to them with the term *Anschauung*.[107]

We know that by mid 1905 Einstein had read the book of Föppl[108] and almost certainly Abraham's revised edition of 1904;[109] he had attended classes at Heinrich Weber's Institute at the ETH; he was an avid reader of the *Annalen*; a favourite uncle was an electrical engineer, and his father was in the electrical business; and at the patent office he frequently assessed patents concerning electrical machinery.[110] If one were interested around 1905 especially in processes concerning electromagnetic induction—most major physicists like Lorentz and Poincaré were not[111]—one could analyse them using Lorentz's electromagnetic theory. Depending upon the particular process of electromagnetic induction, the researcher was confronted by at least one of the following problems: (1) whether the external electric field of a moving magnet was generated by some combination of the three possible types of electrons in Lorentz's theory—conduction electrons, polarization electrons,

[103] M. Abraham and A. Föppl, *Einführung in die Maxwellsche Theorie der Elektrizität* (Leipzig, 1904), esp. pp. 398–409.

[104] Henri Poincaré, 'Sur l'induction unipolare', *L'éclairage électrique*, **23** (April 1900), 41–53.

[105] H. A. Lorentz, 'Maxwell's elektromagnetische Theorie', in *Encyklopädie der mathematischen Wissenschaften*, vol. 5 (Leipzig, 1904–1922), 63–144, esp. pp. 99–100; Lorentz's monograph was received June 1903. Lorentz had written previously on unipolar induction in 'Remarques au sujet d'induction unipolare', *Arch. néerl*, **9** (1904), 380; reprinted in *Collected papers* (9 vols., The Hague, 1935–1939), vol. 3, 177–179.

[106] See A. I. Miller, 'On Lorentz's methodology', *The British journal for the philosophy of science*, **25** (1974), 29–45.

[107] See Lorentz's paper of 1903 (footnote 105), esp. pp. 118–122.

[108] See G. Holton, 'Influences on Einstein's early work', in his *Thematic origins of scientific thought: Kepler to Einstein* (Cambridge, 1973), pp. 197–217.

[109] I have begun to investigate the effect of Abraham on Einstein in A. I. Miller, 'On Einstein, light quanta, radiation and relativity in 1905', *American journal of physics*, **44** (1976), 912–923.

[110] For example, M. Flückiger, *Albert Einstein in Bern* (Bern, 1974), esp. p. 63.

[111] They were interested particularly in formulating an electromagnetic world-picture. See A. I. Miller, 'A Study of Henri Poincaré's "Sur la dynamique de l'électron"', *Archive for history of exact sciences*, **10** (1973), 207–328; A. I. Miller, 'The physics of Einstein's relativity paper of 1905 and the electromagnetic world-picture of 1905', *American journal of physics*. **45** (1977), 1040–1048; my essay in (footnote 106) and in detail in my book, *Albert Einstein's special theory of relativity: emergence (1905) and early interpretation (1905–1911)* (Reading, Mass., 1981).

or magnetic electrons;[112] (2) whether Faraday's law was valid for unipolar induction; and (3) whether the *Anschauung* of Preston and Faraday was correct. We can now better appreciate Einstein's recollection in 1919 that 'the phenomena of electromagnetic induction forced me to postulate the [special] relativity principle'.[113] It was his genial idea in 1905 to be concerned, as a first approximation, with determining a relationship between the fields and forces that observers in relative inertial motion on a wire loop and magnet, respectively, used to calculate a quantity common to both observers—the induced current.[114] Having presented this problem from a new perspective in the introduction to his relativity paper of 1905, Einstein solved it in a later section; and then boldly dismissed as 'meaningless' questions concerning the seat of the EMF in unipolar induction.[115] Thus, set in their historical context, the problems of unipolar induction serve to further illuminate Einstein's thinking toward the special relativity theory as a process tempered delicately by the interplay among science, technology and society.

In the period 1912–1922, the locus of the search for a crucial experiment to decide whether lines of force rotated shifted to the United States. Samuel J. Barnett,[116] Edwin H. Kennard[117] and George B. Pegram[118] performed the first truly open circuit experiments. The set of experiments utilized cylindrical condensers coaxial with the sources of a magnetic field, which were either a cylindrical magnet (Figure 13) or a solenoid (Figure 14). The systems were arranged so that their parts could rotate separately. During a run the condenser plates were connected, the rotation was stopped, and the plates were disconnected and then tested for a charge. The

[112] H. A. Lorentz, 'Weiterbildung der Maxwellschen Theorie. Elektronentheorie,' in *Encyklopädie der Mathematischen Wissenschaften* (Leipzig, 1904–1922), vol. 14, pp. 145–280; this monograph is a sequel to the one in (footnote 105 above) and was received December 1903. The relevant pages are 206–208. For further discussion of these three types of electrons, see H. A. Lorentz, 'Alte und neue Fragen der Physik,' *Physikalische Zeitschrift*, **11** (1910), 1234–1257; reprinted in *Collected papers* (footnote 105), vol. 7, 205–257, esp. pp. 225–228. See also H. A. Lorentz, *Lectures on theoretical physics* (trans. by L. Silberstein and A. P. M. Trivelli: 3 vols. London 1931), vol. 3, esp. Chapter VIII, which is from lectures delivered by Lorentz at Leiden in the period 1910–1912; and see the section 9 to this essay.

See my book (footnote 111) for further analysis of problems concerning electromagnetic induction. Suffice it to say here that the symmetry most striking to Lorentz and Poincaré was the one between the electric and magnetic phenomena produced by moving magnets and dielectrics, respectively, that is, the 'duality between electrical and magnetic phenomena' (see p. 99 of Lorentz's *Encyklopädie* paper of 1903 in (footnote 105). This symmetry was first emphasized by Heaviside and Hertz (see O. Heaviside. *Electrical papers*. vol. 1 (London. 1894) 451–455; H. Hertz. 'On the relations between Maxwell's fundamental electromagnetic equations and the fundamental equations of the opposing electromagnetics', *Annalen der Physik*, **23** (1884) 84–103, in H. Hertz, *Miscellaneous papers* (trans. by D. E. Jones and G. A. Schott: London, 1896), 273–290; and Föppl (footnote 41), 121–122).

[113] Quoted from an unpublished manuscript of Einstain in G. Holton, 'On trying to understand scientific genius', in his (footnote 108), 353–380, esp. p. 364.

[114] See Albert Einstein, 'Zur Elektrodynamik bewegter Körper', *Annalen der Physik*, **17** (1905), 891–921. A new translation is in my (footnote 111), 391–415; see p. 392. All references will be to my book which analyzes Einstein's special relativity paper in its historic context. The version of Einstein's paper in the often-used Dover reprint volume, *The principle of relativity: a collection of original memoirs on the special and general theory of relativity by H. A. Lorentz, A. Einstein, H. Minkowski and H. Weyl* (trans. by W. Perrett and G. B. Jeffery: New York, n.d.), 37–65, contains mistranslations and misprints.

[115] *Ibid.*, 408.

[116] S. J. Barnett, 'On electromagnetic induction and relative motion', *Physical review*, **35** (1912), 323–336; 'On electromagnetic induction and relative motion. II', *Physical review*, **12** (1918), 95–114.

[117] E. H. Kennard, (a) 'Unipolar induction', *Philosophical magazine*, (6) **23** (1912), 937–941; (b) 'The effect of dielectrics on unipolar induction', *Physical review*, **1** (1913), 355–359; (c) 'On unipolar induction: another experiment and its significance as evidence for the existence of the aether', *Philosophical magazine*, (6) **33** (1917), 179–190.

[118] G. B. Pegram, 'Unipolar induction and electron theory,' *Physical review*, **10** (1917), 591–600.

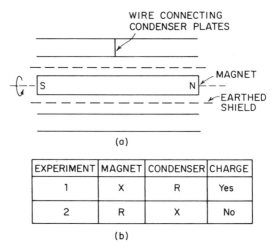

(a)

EXPERIMENT	MAGNET	CONDENSER	CHARGE
1	X	R	Yes
2	R	X	No

(b)

Figure 13.
The apparatus (a) and data (b) for the system of a permanent magnet and a condenser in relative rotatory motion about a common axis.

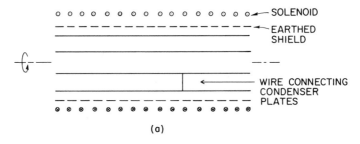

(a)

EXPERIMENT	SOLENOID	CONDENSER	CHARGE
1	X	R	Yes
2	R	X	No
3	R	R	Yes

(b)

Figure 14.
The apparatus (a) and data (b) for the system of a current carrying solenoid and a condenser in relative rotatory motion about a common axis.

results were interpreted as follows: if the lines of force rotated, then a charge would have been induced on the condenser plates; if the lines did not rotate, then an electrostatic charge would have appeared on the magnet's surface. In order to prevent this charge from affecting the condenser, and also to eliminate any spurious effects, an earthed shield was placed between the magnetic system and the condenser. The results of the experimental runs are in the tables in Figures 13 and 14.

Everyone expected identical results for the first two experiments in Figures 13 and 14. Kennard wrote that there was no reason to expect a 'radical difference between induced and permanent magnetism.'[119] But in the opinion of Kennard, the lack of a charge on the condensers in the experiments 2 of Figures 13 and 14 contradicted the moving line theory, and supported Lorentz's theory with its all-pervasive resting ether. Although Barnett recognized correctly the importance of electromagnetic induction to Einstein's thinking toward the relativity theory of 1905, he incorrectly interpreted Einstein's special relativity paper of 1905. Barnett had concluded that, according to the theory of relativity, all phenomena were measurably relative. Yet experiments 1 and 2 in Figures 13 and 14 contradicted what he considered as Einstein's principle of relativity. In order to maintain the rotating line viewpoint, which was consistent with some sort of principle of relative motion, Barnett invoked an ether with the appropriate dielectric properties to neutralize the charge induced due to the rotating lines of force in experiments 1 and 2 of the tables in Figures 13 and 14. Thereby he considered that he had also explained the lack of symmetry between these experiments. But he went further in claiming that his viewpoint precluded even open circuit experiments for disproving the moving line hypothesis.

For Kennard, on the other hand, the decisive experiment was the third one in Figure 14 which he performed in 1917. He considered that the experiment 3 of Figure 14 was 'a stumbling block in the way of those ultra-relativists who would abandon the conception of an aether altogether',[120] that is, Lorentz's ether which has no dielectric properties. Kennard's reason was that electromagnetic induction occurred for a case in which there was no relative motion between the ponderable parts of the system; therefore, electromagnetic induction depended at least in part upon rotation through the ether.

In 1917 Pegram reminded Barnett and Kennard as well as 'many well-trained physicists and engineers' that as long ago as 1895 Joseph Larmor had shown that the magnetic field of a current carrying solenoid is unaltered when the solenoid rotated about its axis; and therefore a rotating or resting solenoid could only affect a moving charge.[121] Furthermore, Pegram continued, this result was valid in any electron theory. Pegram concluded, in his paper of 1917, that Table 2 of Figure 14 was consistent with the 'Lorentz–Einstein relativity theory.'[122]

In 1920 William Francis Gray Swann reminded everyone that in any electron theory the external field of a rotating conducting magnet is electrostatic in origin.

[119] Kennard (footnote 117 (c)), 180.

[120] *Ibid.*, 190.

[121] Pegram (footnote 118), 597. The 1895 paper of Larmor is in footnote 26 above. The argument is: a steady current can not affect a resting charge, and the uniform rotation of a current carrying solenoid about its axis only superposes another uniform motion on its current.

[122] Pegram (footnote 118), 596. This is indicative of how many American physicists understood special relativity theory around 1917. See also footnote 123 below.

EXPERIMENT	MAGNET	CONDENSER	CHARGE
1	X	R	Yes
2	R	X	Yes

(NO EARTHED SHIELD)

EXPERIMENT	SOLENOID	CONDENSER	CHARGE
1	X	R	Yes
2	R	X	No

(EARTHED SHIELD IS IRRELEVANT HERE)

Figure 15.

The results of the elecrromagnetic induction experiments using the apparatus in Figure 13 and 14, *without* the earthed shields.

Therefore, no result could have been expected in experiment 2 of Figure 13 because of the earthed shield around the magnet.[123]

Concluding this sketch of the work of Barnett, Kennard, Pegram and Swann, I present for comparison the results of the open circuit experiments for the magnet and solenoid without the earth shield (see Figure 15). Clearly caution must be exercised in asserting without further explanation that electromagnetic induction is a matter simply of relative motion between ponderable bodies. But the asymmetry between the results for the magnet and solenoid does not violate relativity theory. In the years to come physicists and engineers would have to be reminded repeatedly of the fundamental difference between unipolar induction effects arising from a rotating magnet and a rotating solenoid. This is a good illustration of the incorrect expectations resulting from relying only upon lines of force and not taking into account their source. In fact, the case of unipolar induction serves as a caveat that in scientific research a visual representation, and more generally an *Anschauung*, need not always be a fruitful method of approach. With regard to the production of large-scale unipolar generators, a preference for these modes of thinking led in German-speaking countries to an unproductive interaction between scientists and engineers.

[123] W. F. G. Swann, 'Unipolar induction', *Physical review*, **15** (1920), 365–398. In fact, another goal of Swann's paper is to demonstrate that, contrary to the belief of many physicists, Faraday's law in conjunction with the other equations of electromagnetic theory can account exactly for the current induced in an open circuit moving through a uniform magnetic field, for example, unipolar induction. Swann accomplishes this by using Lorentz's electron theory supplemented with a principle of relativity and assumptions concerning the constitution of a permanent magnet—for example, that it is composed of either Ampèrian current whirls or magnetic dipoles. He prefers to deduce the electromagnetic field quantitites of a moving magnet from the Lorentz electron theory rather than the more straightforward methods employing the transformation formulae of Einstein's special relativity theory. Swann's reason is that although derivations dependent upon assumptions concerning the constitution of matter are more laborious, nevertheless they are more rigorous and hence more convincing. This attitude toward Einstein's special relativity theory was shared by other American physicists and will be treated elsewhere. For other examples of this attitude see Pegram (footnote 122); and J. T. Tate, 'Unipolar induction', *Bulletin of the National Research Council.* **4** (1922), 75–95, esp. p. 78.

By 1936 the controversy concerning whether lines of force rotated had degenerated into the proponents of either the moving line or fixed line hypotheses enumerating the names of well-known physicists who, in their opinion, did not consider as meaningless the question of whether lines of force have physical characteristics.[124]

Today we know that electromagnetic theory places stringent restrictions upon assumptions concerning the state of motion of lines of force (see sections 8 and 9). Minkowski's electrodynamics can be used in the laboratory system to calculate the electromagnetic quantities of moving magnets. In this theory one may use lines of force to discuss effects of the magnet's external electric field on conductors, but the internal state of a rotating magnet can be described by rotating lines of force only in certain limiting cases, for example, a very long cylindrical magnet.[125] Yet using Minkowski's equations in a rotating reference system leads to an external electric field in violation of Gauss's law. The reason is that Minkowski's equations are not applicable in rotating reference systems; rather, general relativity theory with its cosmological implications is required in noninertial reference systems.[126] So in 1979 the problem of unipolar induction still awaits a complete solution. In the course of 147 years it has moved across many different disciplines including physics and engineering, in addition to affecting society; and today into cosmology as well. I think that Faraday would have been pleased with these developments.

---

[124] W. Cramp and E. H. Norgrove, 'Some investigations on the axial spin of a magnet and on the laws of electromagnetic induction', *Journal of the Institution of Electrical Engineers*, **78** (1936), 481–491; note also the correspondence in *ibid.*, **79** (1936), 344–348. Cramp and Norgrove believed in the fixed-line hypothesis. In support of their viewpoint they offer the following experiment: a bar magnet, free to rotate, is hung by a string above another bar magnet whose rotation can be measured. The axes of both magnets are collinear. Rotating the upper magnet produces no effects on the lower magnet, other than the expected longitudinal force between the opposing poles. Thus, they conclude that lines of force remain fixed because if not, then the lines belonging to the upper magnet would have grabbed hold of those generated by the lower one, thereby causing the latter magnet to turn (p. 487).

[125] Briefly, the reason is that with a very long cylindrical rotating magnet the entire internal electric field is due to the apparent volume and surface charge distributions. Therefore, no true charge is displaced; in addition, the external electric field is zero. For details see Cullwick (footnote 1), esp. pp. 144–149.

[126] For example, from (10) the external electric field $\mathbf{E}'$ in a reference system Σ' rotating with the magnetized sphere is

$$\mathbf{E}' = -\frac{\mathbf{v}}{c} \times \mathbf{B} + \mathbf{E}.$$

Since in the Σ' system there is no electric dipole moment, the electric displacement $\mathbf{D}' = \mathbf{E}'$. Since the conducting sphere is isolated, an observer in Σ' should measure no net charge, i.e., $\nabla' \cdot \mathbf{D}' = 0$, where the divergence is calculated in Σ'. But instead we obtain

$$\nabla' \cdot \mathbf{D}' = \nabla' \cdot \mathbf{E}' = 2\frac{\omega}{c} \cdot \mathbf{B},$$

and charge conservation is violated. For further discussions of electro-dynamics in rotating reference systems, see Cullwick and Rosser (footnote 1); Arnold Sommerfeld, *Electrodynamics* (trans. E. G. Ramberg: New York, 1964), esp. pp. 287–289, 333–334, 359–363; T. Schlomka and G. Schenkel, 'Relativitätstheorie und Unipolarinduktion', *Annalen der Physik*, **5** (1949), 57–62; and M. G. Trocheris, 'Electrodynamics in a rotating frame of reference', *Philosophical magazine*, (7) **40** (1949), 1143–1154, esp. pp. 1152–1154. For examples of electrodynamical calculations using general relativity, see. L. I. Schiff, 'A question in general relativity', *National Academy of Sciences proceedings*, **25** (1939), 391–395; A. Yildiz and C. H. Tang, 'Electromagnetic cavity resonances in accelerated systems', *Physical review*, **146** (1966), 947–954; and A. Yildiz, 'II', *Nuovo cimento*, **61** (1969), 1–11.

7. A conducting loop and magnet in relative inertial motion

The EMF (ε) in a conducting loop in inertial motion relative to a uniform magnetic field **B** is calculated from (1) in section 2. The EMF can be interpreted as due to the conductor cutting the magnet's lines of force; in fact, (1) can be rewritten as

$$\varepsilon = -\frac{1}{c}\frac{d\Phi}{dt} \tag{5}$$

where Φ is the magnetic flux.

In the inverse case the EMF should be the same because the direction and magnitude of the induced current depends upon only the relative velocity of the magnet and conductor. Thus, replacing **v** with $-$**v** in (1) yields for the case of the magnet in inertial motion with velocity $-$**v** relative to the stationary conducting loop the result

$$\varepsilon = \frac{1}{c}\oint \mathbf{B} \times \mathbf{v}\cdot\mathbf{dl}. \tag{6}$$

But what is the physical meaning of the term $\mathbf{B}\times\mathbf{v}/c$? From considerations based upon either the relativity transformations for the electromagnetic field quantities, or by assuming that a permanent conducting magnet consists of either Ampèrian current whirls or of magnetic dipoles, it can be shown that a permanent conducting magnet in inertial motion is the source of an electric field $\mathbf{E}=\mathbf{B}\times\mathbf{v}/c$. (6) follows because the EMF in the conducting loop can also be written as $\varepsilon=\oint\mathbf{E}\cdot\mathbf{dl}$, where **E** is the electric field in the loop's vicinity. With Faraday this inverse case can also be given a line cutting interpretation since (6) can be expressed in the form of (5).[127]

8. The case of magnet and conductor in relative rotatory motion

In order to assist the reader in becoming accustomed to electromagnetic phenomena arising from relative rotatory motion between magnet and conductor, I shall here outline the particular case of an isolated spherical conducting magnet rotating about a diameter along the direction of its magnetic field. I shall take the liberty of discussing this case in detail as Lorentz would have, for his theory is the pre-relativistic electromagnetic theory that is most familiar to us, Then I shall sketch the solutions in the theories of Minkowski and of Maxwell. The isolated rotating permanent conducting magnetic sphere in Figure 16 is of radius R. All calculations are done in the laboratory reference system whose origin is coincident with the sphere's center. The sphere's magnetic field **B** is parallel to the line NS.[128]

8.1. *Lorentz's theory.* The equilibrium condition for the conduction electrons inside the rotating sphere is the vanishing of the net Lorentz force **F** acting upon each of them, that is,

$$\mathbf{F}=q\mathbf{E}+q\frac{\mathbf{v}}{c}\times\mathbf{B}, \tag{7}$$

where $\mathbf{v}=\omega\times\mathbf{r}$, ω is the sphere's angular velocity, and q is the magnitude of the

---

[127] For further discussion of the inverse case see Cullwick (footnote 1), esp. pp. 47–48.
[128] See footnote 126 for the calculation carried out in the sphere's rest system.

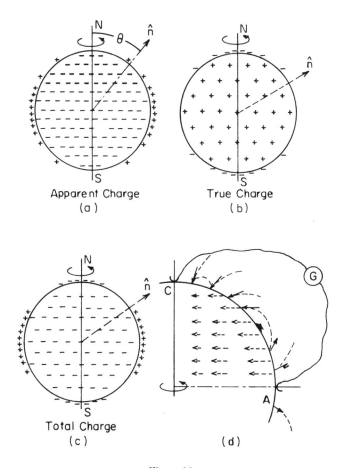

Figure 16.

An isolated permanent conducting spherical magnet of radius R rotates about the line NS. In (a) the apparent volume and surface charge distributions are due to the apparent electric polarization of the rotating magnet. In (b) the true volume and surface charge distributions are due to the displacement of conduction electrons. The charge distributions in (a) and (b) are calculated from Minkowski's theory. In (d) the total volume and surface charge distributions are the sum of those calculated in (a) and (b), or calculated directly from Lorentz's electromagnetic theory. In (d) the external electric field is denoted by solid arrows, and the internal electric field by dotted arrows whose lengths are proportional to the electric field strengths.

electron's charge. Consequently, there arises within the sphere an electric field **E**,

$$\mathbf{E} = \mathbf{B} \times \frac{\mathbf{v}}{c}, \tag{8}$$

which annuls the term $\mathbf{v}/c \times \mathbf{B}$ due to the electrons being acted upon by the sphere's magnetic field. The resultant internal electric field is shown in Figure 16(d); its sources are real volume and surface charge distributions ρ_{total} and σ_{total}, respectively. These quantities are calculated in the standard fashion. The volume charge distribution is

$$\rho_{\text{total}} = \frac{1}{4\pi} \nabla \cdot \left(\mathbf{B} \times \frac{\mathbf{v}}{c} \right) = -\frac{\omega \mathbf{B}}{2\pi c}. \tag{9}$$

In order to determine σ_{total} the external electric field must be calculated from Poisson's equation; due to the spherical boundary this calculation is straightforward. The result for the exterior electric field arising from the rotating magnetic sphere is

$$\mathbf{E} = -\nabla\phi = \frac{\omega B R^5}{2cr^4} [-\hat{r}(3\cos^2\theta - 1) - \hat{\theta}\sin 2\theta], \tag{10}$$

where the sphere's potential ϕ is

$$\phi = -\frac{\omega B R^5}{6cr^3}(3\cos^2\theta - 1).^{129} \tag{11}$$

Equations (8) and (10) yield

$$\sigma_{\text{total}} = \frac{3\omega B R}{8\pi c} \left(1 - \frac{5}{3}\cos^2\theta \right). \tag{12}$$

The charge distributions ρ_{total} and σ_{total} are shown in Figure 16(c). Therefore, points at different latitudes on the sphere's surface differ in potential by an amount calculable from (12). For example, the potential difference (P.D.) between the pole N ($\theta = 0°$) and a latitude at an angle θ is

$$\theta_\theta - \phi_{\theta=0°} = \frac{\omega B R^2}{2c}\sin^2\theta. \tag{13}$$

If $\theta = 90°$, then there is a potential difference of

$$\frac{\omega B R^2}{2c}. \tag{14}$$

If a conducting wire of arbitrary shape, and at rest in the laboratory, is brought into sliding contacts with the rotating sphere's pole N and equator, then from (14) there will arise an EMF in the wire of amount

$$\varepsilon = \frac{\omega B R^2}{2c}, \tag{15}$$

and a current will flow in the stationary wire from the equator to the pole. The current causes an accumulation of true positive (negative) charge at the pole (equator) in an attempt to equalize the potentials at these two points. The result is an imbalance between $q\mathbf{E}$ and $q\mathbf{v}/c \times \mathbf{B}$, and current will flow within the sphere from C to A to try to restore the equilibrium charge distributions in Figure 16(c).

---

[129] See Rosser (footnote 1), 201.

Since $\omega = 2\pi v$, where v is the number of revolutions per second, (15) can be rewritten as

$$\varepsilon = \frac{v}{c} B(\pi R^2) \qquad (16)$$

or

$$\varepsilon = \frac{v}{c} \Phi, \qquad (17)$$

where Φ is the magnetic flux cutting the surface of revolution containing the point A. If $\theta = 45°$, for example, then (17) is obtained again, only with $\Phi = B(\pi d^2)$ with $d = R/\sqrt{2}$.

Equivalently, (19) can be written as

$$\varepsilon = \frac{1}{c} \int_{R \sin 90°}^{0} \mathbf{B} \times \frac{\mathbf{v}}{c} \cdot \mathbf{dl}. \qquad (18)$$

In the inverse case where the sphere is stationary and the wire loop rotates in the opposite direction—(18) becomes

$$\varepsilon = \frac{1}{c} \int_{R \sin 90°}^{0} \frac{\mathbf{v}}{c} \times \mathbf{B} \cdot \mathbf{dl}. \qquad (19)$$

Therefore, in either case the EMF can be interpreted as arising from flux cutting. But the magnet's internal state—that is, its true volume and surface charge distributions—cannot be interpreted according to the fixed line or rotating line hypotheses of Faraday or Preston, respectively.

Furthermore, the integrals in (18) and (19) are not only independent of which part of the system is rotating, but path independent as well. These two results were emphasized, in particular, by Abraham in 1904.[130]

8.2. *Minkowski's theory.* The total charge distributions, external electric field and EMF in the stationary wire loop are the same here as in Lorentz's theory. However, in Minkowski's theory the total volume and surface charge distributions are composed of two parts: true volume and surface charge distributions that are the source of the dielectric displacement $\mathbf{D}$ (see Figure 16 (*b*)): and apparent volume and surface charge distributions (see Figure 16 (*a*)) which result from the relativity of simultaneity where according to the special relativity theory magnetic matter in motion appears in the laboratory system to have an electric polarization $\mathbf{P}$. Just as is the case for Lorentz's theory, the current in the wire loop can be interpreted from the viewpoint of flux cutting; but even more striking in Minkowski's theory is that the magnet's internal state does not lend itself at all to an interpretation using lines of force.

8.3. *Pre-Lorentz results.* The external and internal electric fields and total charge distributions for a spherical conductor rotating about a diameter parallel to an externally imposed uniform magnetic field are the same as for the spherical conducting magnet treated above. This problem had been solved in 1863 by Jochmann using Weber's theory and by Larmor using Maxwell's theory.[131] It

[130] See footnote 103.
[131] See Weber (footnote 24); and Larmor (footnote 26), first reference.

provided a model for investigating atmospheric effects such as the Aurora Borealis and Australis. The atmosphere serves as the wire connecting the equator and pole.[132]

9. Linear unipolar induction

There is at least one case of unipolar induction that cannot be treated satisfactorily using the pre-1905 formulation of Lorentz's electromagnetic theory; namely, to determine the source of the external electric field of a conducting bar magnet whose dimensions are much larger in the xz plane than in the y direction, and which is in inertial motion in the direction of the positive x-axis (see Figure 17). There is an external electric field because current flows in the wire loop AC that is stationary in the laboratory and makes sliding contacts with the magnet. But there is no displacement of real charges in the magnet because every conduction electron experiences the same force. In 1907, Lorentz wrote of the impossibility in this case 'to distinguish sharply between polarization- and magnetization-electrons'.[133] According to Minkowski's theory the magnet's external electric field arises from the apparent charge, which is the source of an electric polarization (see subsection 8.2). The first fully relativistic treatment that I have found of this problem is by Richard Becker.[134] Lorentz's treatment in 'Alte und neue Fragen' is an interesting mixture of Einstein's special relativity theory and Lorentz's own theory of electrons.[135]

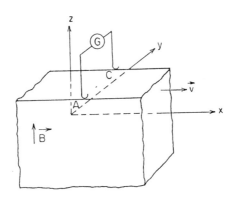

Figure 17.
A conducting permanent magnet moving in the direction of the positive x-axis. The galvanometer circuit is stationary in the laboratory and makes sliding contacts with the moving magnet at A and C. B is the magnet's magnetic field as measured in the laboratory system.

---

[132] See also Hertz's detailed calculations on these problems in his Inaugural Dissertation, Berlin, 15 March 1880, 'On induction in rotating spheres', in his *Miscellaneous papers* (footnote 112), 35–126. Hertz's results were not widely circulated (see Larmor (footnote 26), first reference, 21).

[133] Lorentz (footnote 112), 225.

[134] R. Becker. 'Unipolar-Induktion als Folge des relativischen Zeitbegriffs'. *Die Naturwissenschaften.* **51** (1932), 917–919. See also Rosser (footnote 1). 190–195; and W. K. H. Panofsky and M. Phillips. *Classical electricity and magnetism* (Reading, Mass., 1962). 165–167, 337–339.

[135] See Lorentz (footnote 105).

Acknowledgements

For useful conversations it is a pleasure to thank S. Belostock, L. J. Chmura, G. Holton, E. M. Purcell, E. Quandt and A. Yildiz. I am grateful to Professor Gerald Holton for his encouragement and comments: for the hospitality offered to me by the Department of Physics, Harvard University, of which I am an Associate; and for a Fellowship for 1979–1980 from the John Simon Guggenheim Memorial Foundation. This research was supported in part by the National Science Foundation's Program for the History and Philosophy of Science. The paper is based on a lecture entitled 'On unipolar dynamos, the electrical industry and relativity theory', presented on 27 April 1977 to the 'Technology Studies Seminar series' at the Massachusetts Institute of Technology.

Part III

Special Relativity

Essay 4

<u>On Einstein's Invention of Special Relativity</u>

Albert Einstein distinguished sharply between the context of discovery and the context of invention. For example, in a letter of 6 January 1948 to Michele Besso, Einstein wrote:

> Mach's weakness, as I see it, lies in the fact that he believed more
> or less strongly, that science consists merely of putting experimen-
> tal results in order; that is, he did not recognize the free con-
> structive element in the creation of a concept. He thought that
> somehow theories arise by means of *discovery* [durch Entdeckung] and
> not by means of *invention* [nicht durch Erfindung]. (1972, p. 391,
> italics in original).

By "invention" Einstein meant the mind's ability to leap across what he
took to be the essential abyss between perceptions and data on the one
side in order to create concepts and axioms on the other. Although
sometimes Einstein interchanged the terms discovery and invention, he
deemed invention to be the route of creative scientific thinking.
Today I shall develop that to a good approximation this motif describes
Einstein's invention of special relativity in contrast to the efforts of
Max Abraham, H.A. Lorentz and Henri Poincaré, among others, who sought
to discover an electromagnetic basis for physical theory.[1]

One of the problems in dealing with the emergence of special relativity is the dearth of extant archival documentation from the period
1901-1905. The scenario that I shall present here is woven from conjectures that receive support over the widest possible number of archival, primary and secondary sources.[2] It is apropos to quote from
Einstein's own view of what constitutes worthwhile history of science —
namely, that history of science should "convey the thought processes
that led to discoveries" and thus focus on how physicists "struggle
with their problems, their trying to find a solution which came at
last often by very indirect means, is the correct picture." The historian of science, continued Einstein, just like the scientist, can

construct a scenario that is at best "only very likely correct."[3]

The three main paths to special relativity are philosophy, physics and technology. Before 1905 many German scientists and engineers considered these paths to intersect, but Einstein combined them in a unique way. First I shall review these paths as they confronted Einstein in 1905 through works with which he was acquainted. This will enable us to glimpse the dynamics of thinking that led Einstein to realize that the contemporaneous notion of time had to be re-examined, which led him to invent the concept of the relativity of simultaneity, and then to propose the axioms of special relativity. This sequence squares with Einstein's own words to the effect that his creative thinking occurred in a predominantly nonformal manner, usually with the aid of visual thinking, and axiomatization followed.[4] From analysis of his published papers during 1901–1907, and his correspondence, there emerges a portrait of someone who was aware of developments in philosophy, technology and physics. Biographies that depict him as virtually cut off from the world of science do an injustice to his achievements in 1905. In fact, his awareness of contemporary research renders all the more dazzling how he opted for a course of action so different from that of other physicists in 1905.

1. Philosophy

By the end of the 19th century Mach's criticisms in his 1883 *Science of Mechanics* of Newtonian absolute space and time as idle metaphysical conceptions had been heeded by most major physicists (e.g., Mach 1883, pp. 273, 279–284). On the other hand, Mach's empiricistic emphasis did not deter philosopher-scientists such as Heinrich Hertz and Henri Poincaré from exploring the Kantian notion of a priori organizing principles. Compared with Mach these men placed a strong premium on the primacy of the imagination and the deep meaning of mathematics. Before 1905 Einstein had read at least the introduction to Hertz's 1894 *Principles of Mechanics* (Sauter 1965). Although Einstein, as did others, disagreed with Hertz's implementation of his program, by 1905 Einstein may well have been impressed with the power of an approach to physical theory that emphasized axioms and mental pictures over empirical data.

During 1902–1904 Einstein had read Poincaré's 1902 book *Science and Hypothesis* (Seelig 1954, p. 69). In this display of intellectual virtuosity Poincaré developed arithmetic, geometry, classical mechanics and electromagnetic theory within a neo-Kantian framework that emphasized the close connection between the construction of pre-scientific and scientific knowledge.[5] Although the connection in the hands of Poincaré turned out to be too close for Einstein, variations on this theme would pervade his scientific work. In fact, it is reasonable to conjecture that Poincaré's writings impressed Einstein no less than did Mach's "incorruptable skepticism" (Einstein 1946, p. 21). (See Miller 1981b and in press for evidence to support this conjecture.)

I turn next to survey those experimental data that we can suggest with some certainty Einstein was aware of in 1905, and how they were explained by current physics.

2. Data

In the 1905 special relativity paper entitled, "On the Electrodynamics of Moving Bodies," Einstein referred to "unsuccessful attempts to discover any motion of the earth relatively to the 'light medium'." (Einstein 1905, p. 392).[6] He gave no explicit citations to these ether-drift experiments; in fact, there are no literature citations anywhere in the paper. Then he specialized to the class of experiments accurate to the "first order of small quantities" — that is, accurate to order (v/c) where v is the velocity of the earth relative to the ether and c is the velocity of light in the free ether that is measured by an observer at rest in the ether. (Einstein 1905, p. 392). For example, Martinus Hoek's (1868) experiment is an ether-drift experiment of first-order accuracy. Among the wider class of "unsuccessful attempts" was the second-order experiment of Albert A. Michelson and Edward W. Morley (1887). No serious historian has argued that Einstein was unaware of this experiment before 1905 (Holton 1969). Then there were the second-order null experiments of Lord Rayleigh (1902) and Dewitt B. Brace (1904) to detect double refraction in isotropic crystals. Rayleigh and Brace's experiments were analyzed in Max Abraham's widely-cited paper of 1904 in the *Annalen der Physik* entitled, "On the Theory of Radiation and of Radiation Pressure." Since it is inconceivable that someone would publish in a journal to which he had no access, we can safely conjecture that Einstein had at least perused Abraham's 1904 paper. Another group of first-order experiments are positive experiments that did not attempt to detect the earth's motion through the ether — for example, observations of stellar aberration and experiments to verify Fresnel's dragging coefficient that were performed by Hippolyte Fizeau (1851), and Michelson and Morley (1886). In these experiments v was the velocity of ponderable matter relative to the laboratory.

Einstein's emphasis in the special relativity paper on the first-order experiments supports his later comments that stellar aberration and Fizeau's experiment had been the most influential of the often-cited empirical data to his thinking toward the special relativity theory. In a number of places Einstein later recalled of these data, "They were enough."[7] Their explanation, as we shall discuss in a moment, turned about a quantity that also explained systematically the failure of first-order ether-drift experiments — namely, Lorentz's local time coordinate. In historical context, Einstein's predilection for first-order data is not surprising. Others also emphasized these data over the 1887 Michelson-Morley data. For example, Lorentz especially framed his electromagnetic theory of 1892 to systematically include Fresnel's dragging coefficient. Second-order data are nowhere mentioned in his seminal opus, *Maxwell's Electromagnetic Theory and Its Application to Moving Bodies* (1892a). In a short sequel publication (1892b) Lorentz proposed the ad hoc hypothesis of contraction to explain the irksome 1887 Michelson-Morley experiment. Another example is Max Abraham's widely-read book *Theory of Electricity* (1904c), where he considered Fizeau's 1851 experiment to be critical for deciding between Lorentz's and Hertz's theories of the electrodynamics of moving bodies (pp. 435-436). Almost certainly Einstein had read Abraham's 1904 book, and we know from a letter of 19 February 1955 to his biographer Carl Seelig (Born 1969, p. 104) that before 1905 Einstein had read Lorentz's 1892 presentation of his new version of Maxwell's theory as well as Lorentz's monograph entitled, *Treatise on Electrical and Magnetic Phenomena in Moving Bodies* (1895). In this treatise Lorentz reviewed available first-order experiments and explained them in a more systematic manner than he had in 1892.

3. H.A. Lorentz's Electromagnetic Theory

Since the results of Einstein's analysis of Lorentz's electromagnetic theory played a key role in his thinking toward the special relativity theory, it is useful to analyze how Lorentz treated data on the optics of moving bodies. Lorentz postulated that the sources of the electric and magnetic fields are submicroscopic particles that move about in an all-pervasive absolutely resting ether. The five fundamental equations of Lorentz's theory are:

$$\vec{\nabla} \times \vec{E} = -\frac{1}{c}\frac{\partial \vec{B}}{\partial t}$$

$$\vec{\nabla} \times \vec{B} = \frac{1}{c}\frac{\partial \vec{E}}{\partial t} + \frac{4\pi}{c}\rho\vec{v}$$

$$\vec{\nabla} \cdot \vec{E} = 4\pi\rho$$

$$\vec{\nabla} \cdot \vec{B} = 0$$

Maxwell-Lorentz Equations

$$\vec{F} = \rho\vec{E} + \rho(\vec{v}/c) \times \vec{B}$$

Lorentz Force Equation

where $\vec{E}$ and $\vec{B}$ are the electric and magnetic fields, respectively, and ρ is the electron's volume density of charge. Since Lorentz's fundamental equations are written relative to a reference system at rest in the ether, which we shall call S, then c is the velocity of light measured in S, and $\vec{v}$ is the electron's velocity relative to S. Lorentz's electromagnetic field equations possess the property expected of a wave theory of light; namely, that relative to a reference system that is fixed in the ether, the velocity of light is independent of the source's motion and is a determined constant c. But this may not necessarily be the result of measuring the velocity of light in an inertial reference system. Therefore, the reference systems in the ether are preferred reference systems. But experiments had not revealed any effect of the earth's motion through the ether on optical or electromagnetic phenomena.

The situation concerning the velocity of light was as follows: Newtonian mechanics postulates that the velocity of light emitted from a moving source should differ from the velocity of the light emitted from a source at rest by the amount of the source's velocity; consequently, the velocity of light c' from a source moving with velocity v is given by Newton's law for the addition of velocities, $\vec{c}' = \vec{c} + \vec{v}$. On the other hand, according to the wave theory of light, the quantity c' measured by an observer at rest in the ether is c' = c, and Lorentz's equations agreed with this requirement. But the effect of the ether on measurements of the velocity of light done on the moving earth was expected to yield a result in agreement with Newtonian mechanics, i.e., observers on the moving earth should measure the velocity c' = c + v, where now c' is the velocity of light relative to the earth and v is the ether's velocity relative to the earth. However, experiments accurate to second-order in (v/c) led to the result c' = c. To this order of accuracy, optical and electromagnetic phenomena occurred on the moving earth as if the earth were at rest in the ether. Therefore, to second-order accuracy in (v/c), Newtonian mechanics and electromagnetism are inconsistent with optical phenomena occurring in inertial reference systems.

In the 1895 treatise Lorentz responded systematically to the failure of the first-order ether-drift experiments to detect any effects of the earth's motion on optical phenomena as follows. For regions of the ether that were free of matter, or within neutral matter that is neither magnetic nor dielectric, the Lorentz equations for the electric field $\vec{E}$ and magnetic field $\vec{B}$ in the ether-fixed reference system is the set of Equations (S) in Figure 1.

<u>Lorentz's Theorem of Corresponding States</u>

S (Ether-fixed) S_r (Inertial System)

Lorentz's Modified
Space and Time
Transformations

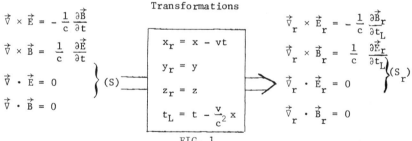

FIG. 1

Lorentz's modified space and time transformations contain the "local time coordinate" t_L and Lorentz referred to the electromagnetic field quantities $\vec{E}_r = \vec{E} + \vec{v}/c \times \vec{B}$ and $\vec{B}_r = \vec{B} - \vec{v}/c \times \vec{E}$ as "new" vectors.

Applying the modified space and time transformations to the Eqs. (S), Lorentz obtained their analogues in the inertial reference system S_r. We can appreciate Lorentz's achievement at a glance because to first order in the quantity (v/c), the Lorentz equations have the same form in the inertial system S_r as in the ether-fixed system S, and thus the same physical laws pertain to both these reference systems; in other words, to this order of accuracy optical experiments could not reveal the motion of the system S_r. Lorentz called this stunning and desirable result the "theorem of corresponding states" (p. 84); it rested on the hypothesis of the mathematical "local time coordinate" t_L. The real or physical time was still the one from the transformations of classical mechanics that leave Newton's second law covariant and to which I shall refer as the Galilean transformations. Thus, although the physical time of electromagnetic theory was not absolute in the meaning of Newton (1687, p. 6), i.e., of flowing without reference to anything external, there was an absoluteness associated with the time in Lorentz's theory because there was no reason to believe that the time in different reference systems should differ, in agreement with our perceptions.

In summary, to order (v/c) the velocity of light in S_r was the same as in S, i.e., c' = c, and so to this order of accuracy Lorentz's theorem of corresponding states removed the inconsistency between Newton's prediction and that of electromagnetic theory, in favor of electromagnetic theory.

In the final chapter of the 1895 treatise Lorentz presented the ad hoc hypothesis of contraction. This blemish on Lorentz's theory was emphasized in several of Poincaré's philosophic-scientific criticisms of the state of physical theory, e.g., in *Science and Hypothesis* (1902, p. 172). Nevertheless, Poincaré was impressed with the theorem of corresponding states because absolute motion had no place in his philosophy or physics, a point that he addressed with great vigor in *Science and Hypothesis*.

4. The Frontier of the Physics of 1905: The Electromagnetic World-Picture

Owing to the wide range of impressive successes of Lorentz's theory in 1900 Wilhelm Wien suggested the "possibility of an electromagnetic foundation for mechanics" (1900, p. 97), i.e., pursuance of an electromagnetic world-picture based on Lorentz's electromagnetic theory, instead of the relatively unsuccessful inverse research effort called the mechanical world-picture. A far-reaching implication of the electromagnetic world-picture was that the electron's mass originated in its own electromagnetic field, and should therefore be velocity dependent. During 1901-1902 this implication was verified by Walter Kaufmann (1901, 1902a, 1902b). Theoretical developments of Kaufmann's data were rapid. During 1902-1903 Kaufmann's colleague at Göttingen, Max Abraham, formulated a theory of a rigid sphere electron that agreed with Kaufmann's latest data (1902a, 1902b, 1903). However, Abraham's theory offered no explanation for the Michelson-Morley experiment and disagreed with the data of Rayleigh and Brace.

Prompted by new second-order data and by Kaufmann's measurements, as well as Poincaré's criticisms, Lorentz (1904b) proposed his own theory of a deformable electron that also agreed with Kaufmann's data. In this theory the contraction hypothesis was deemed no longer to be ad hoc because it became one of several hypotheses that could explain more than one experiment accurate to second order in v/c — that is, in addition to Michelson and Morley the experiments of Rayleigh, Brace and of F.T. Trouton and H.R. Noble (1903). Poincaré agreed (1904). But in 1904 Abraham immediately countered with a severe criticism of Lorentz's theory (1904b): Lorentz's deformable electron was unstable because it could explode under the enormous repulsive forces between its constituent parts.

Einstein's predictions in the special relativity paper for the transverse and longitudinal masses for the electron are ample evidence that he was aware of Kaufmann's data and Abraham's elegant attempts in 1902-1903 to explain them (1905, p. 414). In fact, Abraham's most complete exposition of his theory of the electron was published in the *Annalen der Physik* in 1903. But before 1905 Einstein had neither seen Lorentz's 1904 paper on his new theory of the electron nor was he aware of Poincaré's elegant (1905) version of Lorentz's theory in which Poincaré ensured the stability of Lorentz's electron.[8]

To summarize: By 1905 physicists believed that fundamental physical theory was proceeding in the correct direction. The most successful theory was Lorentz's. It was what Einstein would refer to as a "constructive theory" (1919) because it explained such effects as the directly unobservable contraction of length, the observed variation of mass with velocity, and the fact that the measured velocity of light always turned out to be the same as if the earth were at rest in the ether, all to be caused by the interaction of constituent electrons with the ether.

The stage was set for a great new era in science to emerge from what everyone considered to be the cutting edge of scientific research — namely, high-velocity phenomena. But as we shall see this turned out not to be the case. We move next to the technological path, which is an area where basic problems were deemed unimportant for progress on the frontier of physics — the area of electromagnetic induction. In German-speaking countries problems in this area interfaced technology and basic research; they received a particularly interesting treatment because, as the intellectual historian J.T. Merz has written, the "German man of science was a philosopher." (Merz, Vol. 1, p. 215).

5. Electromagnetic Induction, Kant, Electrons, and Relativity

Michael Faraday's law of electromagnetic induction states that the rate at which the lines of force are cut determines the strength of the current induced; furthermore, the direction and magnitude of the induced current depend on only the relative velocity between the circuit and magnet. But Faraday's interpretation of electromagnetic induction differed when circuit and magnet were rotating relative to each other.

384

An apparatus of the sort in Figure 2 was important to Faraday because by 1851 he had convinced himself that lines of force participated in the magnet's linear motion, but not in its rotation.[9]

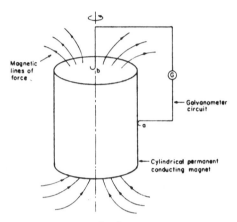

Magnetic lines of force

Galvanometer circuit

Cylindrical permanent conducting magnet

FIG. 2
An example of a unipolar dynamo.

In Figure 2 the wire loop makes sliding contact with the rotating magnet's periphery and touches one of the magnet's poles; hence, the sort of electromagnetic dynamo in Figure 2 became known as a unipolar dynamo. Faraday's interpretation of unipolar induction is: if the loop rotates counterclockwise, a current appears in it owing to its cutting the lines of force. If the loop remains at rest, and the magnet rotates clockwise, a current of the same magnitude and direction appears in the loop; the current was thought to originate in the magnet owing to the magnet's rotating through its own lines of force. When magnet and loop turn together in the same direction, there is no net current in the loop because the loop's current is cancelled by the magnet's internal current. Thus, two different explanations were required for the current induced in the loop, depending on whether the loop or magnet rotated. Clearly, the experimental data are more easily understood in terms of relative motion between the wire loop and the magnet with its co-moving lines of force.

Representing the magnetic field by lines of force found fertile ground in the German-speaking countries because the Kantian philosophical background of German scientists and engineers led them to consider lines of force to be a fundamental *Anschauung* (Miller 1981a). In this context *Anschauung* refers to the intuition through pictures formed in the mind's eye from previous visualizations of physical processes in the world of perceptions; *Anschauung* is superior to viewing merely with senses. In short, lines of force were seen everywhere. There ensued a controversy in the German-speaking scientific-engineering community on the merits of the *Anschauungen* of Faraday versus the rotating-line view; experiments were offered to distinguish between them. One engineer emphasized that this controversy was "not an academic moot point" for two chief reasons (Weber 1895, p. 514): (1) There was intense research and development toward unipolar direct-current and alternating-current generators owing to the relative simplicity of these machines vis-à-vis multipolar machines. (2) Calculating the effect of a rotating magnet on stationary armature coils required knowing whether the magnetic field lines remained stationary; this problem appeared also in multipolar machines with stationary field magnets. So seriously did this controversy rage that despite available technology in Germany the first large-scale unipolar dynamo was built elsewhere.

With confidence I can conjecture that Einstein encountered this controversy during his independent and classroom studies at the ETH, Zurich, and then while a Patent Clerk. Among the evidence to support this conjecture is that while at the ETH Einstein had read August Föppl's 1894 book, *Introduction to Maxwell's Theory of Electricity* (Holton (1967-68). There Föppl discussed unipolar induction and frankly confessed that fundamental conceptual, and perhaps also quantitative problems, remained in electromagnetic induction (Chapter 5). But the relativity paper itself contains irrefutable evidence that Einstein had thought about unipolar induction, because in the paper's Section 6 he audaciously dismissed as "meaningless" problems concerning the seat of emf in "unipolar machines." (Einstein 1905, p. 406). Einstein had come to realize that the crux of the conceptual problems in unipolar induction could be resolved through studying the simple case of a permanent magnet and conducting loop in relative inertial motion. But we are getting ahead of ourselves.

On the theoretical side of the problem of electromagnetic induction, Föppl's 1894 book and the 1904 edition rewritten by Abraham (1904c) emphasized that the laws of mechanics and electromagnetism are used to explain electromagnetic induction as follows (pp. 398-409): (1) The laws of mechanics are used to discuss the motion of circuit and magnet; (2) the current in the loop depends on only the relative velocity between loop and magnet, and this agrees with the principle of relative motion from mechanics. This principle asserts that physical phenomena occur as a result of the relative motion of material bodies, and that the laws of mechanics are the same for all inertial systems. Abraham stressed that experiments supported the principle of relative motion for electrodynamics only to first order in (v/c) (p. 404).

In a widely-cited review paper entitled *Extensions of Maxwell's Theory. Theory of Electrons* (1904a), Lorentz moved toward extending his electromagnetic theory to take account of the bulk magnetic and dielectric properties of moving bodies. Lorentz's extension required his postulating three different sorts of electrons: conduction electrons responsible for electric current; polarization electrons that produce dielectric properties; and magnetization electrons to explain magnetic properties of matter. Once again the local time coordinate played a key role. For example, using the local time coordinate Lorentz calculated the electric field due to the moving magnet, and its effect on a stationary current loop — that is, a constructive explanation for electromagnetic induction (pp. 235-239). Lorentz himself, however, found his calculation puzzling because it meant that a permanent conducting magnet, supposedly comprised only of conduction and magnetization electrons, developed dielectric properties (see also Lorentz 1910, p. 225).

Besides appending hypotheses of additional electrons to an already overburdened superstructure, Lorentz by the end of 1904 was still unable to explain the stability of his electron (1905, p. 100). Einstein recalled in 1946 that he had been aware of this shortcoming which he took to be a "fundamental crisis" (p. 37). He probably learned of this defect in Lorentz's theory from Abraham's 1904 paper in the *Physikalische Zeitschrift* where Abraham first proposed it (1904b). Then there was Max Planck's 1900 explanation for the radiation from hot bodies (Planck 1900, 1901), which brings us to the final ingredient necessary for describing Einstein's unique view of the physics of 1905 — problems concerning radiation.

6. Radiation

Planck's radiation law shocked Einstein. For not only did it violate classical electromagnetism and mechanics, but Einstein could not even adapt classical physics to it, as he recalled (1946, p. 53). To Einstein alone Planck's law was a "second fundamental crisis" in addition to the instability of Lorentz's electron (1946, p. 37). He chose to accept Planck's law and to see what general conclusions could be inferred concerning the electromagnetic and mechanical world-pictures. It is reasonable to conjecture that Planck's law piqued Einstein's interest in problems concerning radiation. Evidence in support of this conjecture and of Einstein's 1946 recollections can be found in his 1907 paper "On the Consequences of the Principle of Relativity for the Inertia of Energy" (1907a).[10] To the best of my knowledge this paper contains Einstein's earliest published views on the physics of 1905. Einstein wrote that his investigations of the structure of light revealed that classical electromagnetism failed in volumes of the order of the electron's. Thus, the electromagnetic world-picture could not succeed.[11] His Brownian motion investigations had yielded a similar result for mechanics; hence, exit any possibility for a mechanical world-picture. From these startling results Einstein concluded that the laws of mechanics, thermodynamics and electromagnetism can be employed in space and time regions large enough so that fluctuations can be neglected. This is precisely what he did in the special relativity paper. But prior to this point, wrote Einstein in 1946, he "despaired of the possibility of discovering the true laws by means of constructive efforts based on known facts." (1946, p. 53).

In summary, by the end of 1904 Einstein was aware of most of the key "known facts", e.g., the ether-drift experiments and Kaufmann's data, of philosophical currents (of which I shall have more to say in a moment), of fundamental problems in electromagnetic induction, and had concluded, as he said later, that the physicists of 1905 were theorizing "out of [their] depth." (1923, p. 484).

7. Mental Pictures and Concepts

At this juncture the three paths to special relativity merged in Einstein's predilection for visual thinking. It is safe to conjecture that the value of this sort of thinking for a fundamental analysis of science had been underscored for Einstein chiefly by Ludwig Boltzmann and Hermann von Helmholtz.[12] Two of Einstein's more reliable biographers, Philipp Frank and Carl Seelig, wrote of Einstein's having read these master philosopher-scientists on his own at the ETH (Frank 1947, p. 38; Seelig 1954, p. 30). Einstein read others of Helmholtz's "papers and lectures" at the informal study group called the "Olympia Academy" of which he was a charter member along with Conrad Habicht and Maurice Solovine (Einstein 1956, p. viii).

In his 1897 *Lectures on the Principles of Mechanics*, Boltzmann suggested that "unclarities in the principles of mechanics [derive from] not starting at once with hypothetical mental pictures but trying to link up with experience from the outset." (Boltzmann 1897a, p. 225). Elsewhere in 1897 Boltzmann delved into an analysis of perceptions in order to set the notion of useful mental pictures as best as he could. He defined useful mental pictures to be common to various perceptual complexes or groups of phenomena, thereby providing an understanding of the complexes. Boltzmann referred to useful mental pictures as "concepts." (Boltzmann 1897b, p. 59).

In 1894 Helmholtz had given a similar definition of a concept in an essay where he discussed the important role of mental pictures and non-verbal thinking in the creative process: "...the memory images of pure sense impressions can also be used as elements in combinations of ideas, where it is not necessary or even possible to describe those impressions in words and thus to grasp them conceptually...the idea of the stereometric form of a material object plays the role of a concept formed on the basis of the combination of an extended series of sensuous intuition images." (Helmholtz 1894, p. 507).

In the special relativity paper the measuring rods and inertial reference systems of the thought experimenter play the role of concepts in the meaning of Boltzmann and Helmholtz. The relativistic notion of simultaneity and the two principles of relativity are concepts that function also as a priori organizing notions in a sense analogous to that of Hertz and Poincaré.[13] Three decades later Einstein would define a concept in much the same way as did Boltzmann and Helmholtz — namely, as a "memory-picture" that serves as an ordering element for perceptions (Einstein 1946, p. 7). Needless to say, I am not claiming that Einstein consciously took his notions of "concept" and of the dynamics of thinking from these philosophers. Rather the relationship of Einstein to Boltzmann and Helmholtz, among the other philosophers and scientists that he had read before 1905, was best described by Frank, who wrote that from their works Einstein "learned how one builds up the mathematical framework and then with its help constructs the edifice of physics." (1947, p. 38). It is to how Einstein built up the edifice of physics that I turn next.

8. The Trend Toward Axiomatization

As Wien put it in 1900, Hertz's program was diametrically opposite to the goal of an electromagnetic world-picture. Yet Wien hoped that in due course the new research effort would measure up to the logical structure of Hertz's mechanics. Abraham wrote similarly of Hertz in the *Annalen* paper of 1903 in which he presented his electron theory. Wien's suggestion for an electromagnetic world-picture appeared in the 1900 Lorentz *Festschrift* and also in the *Annalen* of 1901. We can conjecture that before 1905 Einstein had read the *Annalen* version of Wien's paper. The trend toward axiomatics was in the air in 1905. It was to be achieved by degrees, after all extant data were analyzed. Yet owing to his research on fluctuation phenomena, by 1905 Einstein knew that the electromagnetic world-picture could not succeed.

That notions of axiomatization were mentioned at key places in the scientific literature permits us to begin to fill in Einstein's 1946 recollection of a turning point in his thinking toward special relativity. Rather tersely in his "Autobiographical Notes" he wrote that in the midst of the "despair" of 1904 he decided to try his hand at a purely axiomatic theory that could offer a "universal formal principle ...the example I saw before me was thermodynamics." (1946, p. 53). A problem ripe for historical analysis is how Einstein realized that the principles of thermodynamics were exemplars for the double-edged sword that he would use in his Gordian resolution of problems confronting the physics of 1905 — that is, his realization that the laws of thermodynamics could be wielded as restrictive principles that *also* make no assumptions on the constitution of matter. In the "Autobiographical Notes" he mentioned only their usefulness as restrictive principles. A possible reply to this problem is: Does not everyone know these two properties of the laws of thermodynamics? Although this reply may be the case, it omits assessing the problem in historical context. For implicitly knowing the double-edged power of thermodynamics is quite different from applying it to the physics of 1905. I add that the straight-forward reply is also uninteresting to the historian who seeks to place Einstein in the matrix of science in 1905.

A conjecture that may be closer to the mark is that in Abraham's 1904 *Annalen* paper "On the Theory of Radiation and of Radiation Pressure," Einstein saw the laws of thermodynamics applied specifically because they make no assumptions on the constitution of matter. Abraham resorted to this strategy because he refused to permit the data of Rayleigh and Brace to falsify his theory of the electron. After acknowledging this disagreement between data and theory Abraham went on to emphasize the need for further research on the optics of moving bodies. His route was to investigate the thermodynamics of radiation in order to analyze a quantity indigenous to every theory of the electron — namely, the Poynting vector, all the while working within a theory that made no assumptions on the constitution of matter.

Seventy-odd pages after Abraham's 1904 paper in the *Annalen* is Einstein's paper in which he calculated that black-body radiation could exhibit observable fluctuations (1904). Then, as I conjectured earlier (with support, in part, from Einstein's own testimony in 1907), Einstein went on in unpublished calculations to conclude that classical electromagnetism failed in spatial regions of the size of the electron. Thus, Abraham's elegant applications of the thermodynamics of radiation may well have struck a responsive chord in someone who "despaired of discovering the true laws [through] constructive efforts based on known facts."

9. Einstein's Method for Axiomatization

Einstein's next step was to decide what was to be axiomatized. He recalled that he found the clue in a thought experiment over which he had been pondering since 1895, an experiment that permitted him to supplement the "known facts." (1946, p. 53).[14]

The essence of this experiment is: (1) Current physics asserts that an observer who is moving alongside a light wave whose source is in the ether should be able to discern the effects of his motion by, for example, measuring the velocity of light — that is, this observer can perform all possible ether-drift experiments. (2) But to Einstein it was "intuitively clear" that the laws of optics could not depend on the state of the observer's motion (1946, p. 53). Statements (1) and (2) are mutually contradictory, and to Einstein this thought experiment contained a paradox.

In their own way, Lorentz and Poincaré were also attempting to resolve this paradox, one degree of accuracy in (v/c) at a time that is, by proposing hypotheses such as the local time and the ad hoc contraction of moving bodies. Einstein, on the other hand, would realize that the key to the paradox lay, as he said later, in the "axiom of the absolute character of time, viz., of simultaneity [which] unrecognizedly was anchored in the unconscious." (Einstein 1946, p. 53).

The scientific-philosophic-technological route toward this realization we can conjecture was as follows. Of all the ether-drift experiments that had been performed, those accurate to first-order in (v/c) were explainable by Lorentz's theorem of corresponding states. Lorentz's theorem was based on a set of modified space and time transformations that included the mathematical local-time coordinate (see Figure 3). Although Lorentz's equations remained unchanged under the modified space and time transformations, this was not the case for the equations of mechanics. Thus, according to Lorentz's modified transformations, the equations of mechanics were not the same in every inertial reference system. But this result violated Newton's exact principle of relative motion (Corollary V from the *Principia*, 1687, p. 20), which states that no mechanical experiment could reveal an inertial system's motion. The mathematical statement of Newton's principle of relative motion lies in

Galilean Transformations	Lorentz's Modified Space and Time Transformations
$x_r = x - vt$	$x_r = x - vt$
$y_r = y$	$y_r = y$
$z_r = z$	$z_r = z$
$t_r = t$	$t_L = t - (vx/c^2)$
(a)	(b)

FIG. 3

(a) The coordinates (x_r, y_r, z_r, t_r) and (x, y, z, t) refer to the two *inertial* reference systems S_r and S.

(b) The coordinates $(x_r, y_r, z_r, t_r = t)$ and (x, y, z, t) refer to the inertial reference system S_r and to the ether-fixed reference system S; and t_L is the mathematical "local time coordinate."

*   *   *

the Galilean transformations of Figure 3(a), where for mechanics the reference systems S_r and S are both inertial reference systems. Consequently, the transformation rules for the laws of mechanics and of electromagnetism depended on two different notions of time — one physical and the other mathematical — contrary to Lorentz's goal of reduction. Thus whereas most scientists in 1905 considered the inexactness in electromagnetism of the principle of relative motion from mechanics as basic to the tension between these disciplines,[15] Einstein delved deeper and found that current physics rendered mechanics and electromagnetism *incompatible*.

The texts of Föppl and Abraham emphasized to Einstein the intimate connection between mechanics and electromagnetism for interpreting electromagnetic induction. Faraday's law was basic also to Lorentz's electromagnetic theory where the local time made it possible to calculate the moving magnet's effect on either a resting conductor or the open circuit in unipolar induction. But, as we recall, this calculation rested on certain special assumptions on the constitution of matter. Moreover, Lorentz's electromagnetic theory explained electromagnetic induction in two different ways, depending on whether the conductor or the magnet was moving, even though the physically measurable effect was a function of only their relative velocity. For Einstein this was, as he wrote in 1905, an asymmetry that was "not inherent in the phenomena." (1905, p. 392). The new ingredient in Einstein's notion of symmetry, compared to those found in the contemporaneous scientific literature, was that Einstein's was essentially aesthetic and concerned physical processes and not mathematics.[16] Furthermore, symmetry considerations enabled Einstein to further supplement the "known facts." In 1919 Einstein recalled that he found the asymmetry in the interpretation of electromagnetic induction to be so "unbearable" that it forced him to focus on the necessity for an equivalence of viewpoints between observers on the wire loop and on the magnet, rather than on the source of the moving magnet's electric field as had Lorentz (Holton 1971-1972, p. 364).

10. Einstein's Invention of the Special Theory of Relativity

All of the ingredients were now present for Einstein to invent special relativity. As we saw, the Galilean transformations could not explain the velocity of light measurements. So Einstein assumed that the modified transformations of Lorentz with their apparently ubiquitous local time coordinate would have to play a role in relating phenomena between reference systems. From Lorentz's modified transformations in Figure 3(b), Einstein could have deduced a new result for the addition of velocities:[17] $w_r = (w - v)/(1 - vw/c^2)$, where w_r is the velocity of a moving point relative to S_r and w is its velocity relative to S. For the case of $w = c$, then $w_r = c' = c$, instead of Newton's addition law of velocities, $c' = c - v$. Thus Einstein could have realized that, to first order in (v/c), the addition law for velocities from Lorentz's modified transformations produced a result that agreed with the intuition of his thought experimenter. Since Lorentz's modified transformations differed from the Galilean transformations only in the local time coordinate, Einstein recalled in a review paper of 1907 asking himself whether the local time might be *the* time (1907b, p. 413)? But this step required asserting that the times in inertial reference systems differed because the local time coordinate depends on their relative velocity. Yet the absoluteness of time had always been accepted. Furthermore, the thought experimenter's intuition demanded an examination of the mathematical relation between Newton's principle of relative motion and Lorentz's theorem of corresponding states: after all, the spatial coordinates of Lorentz's modified transformation equations of 1895 were mathematically the same as the ones in the Galilean transformations, and the local time coordinate had been invented for use in electromagnetic theory. Einstein's imposing a Newtonian unity on Lorentz's modified transformation equations meant also his asserting the equivalence of the reference systems S and S_r. This was a big step, for it meant rejecting Lorentz's ether, and with it the dynamical interpretations of an enormously successful and, for the most part, satisfying theory. To his surprise, Einstein had found that the notion of time was both the central point and the Achilles heel of the electrodynamics of moving bodies.

For aid in analyzing the nature of time, Einstein recalled that he benefited from the "critical reasoning [in] David Hume's and Ernst Mach's philosophical writings." (Einstein 1946, p. 53). In the 1948 letter to Besso from which I quoted earlier, Einstein wrote that "Hume had a greater effect on me" than Mach. In his 1944 paper entitled, "Remarks on Bertrand Russell's Theory of Knowledge," Einstein provided a discussion of Hume's insights that enable me to offer the following conjecture. [We know that Einstein had read Hume's *Treatise of Human Nature* during 1902-1904 at the informal study group called the "Olympia Academy" (Einstein 1956, p. viii; Seelig 1954, p. 69).] My conjecture is that Hume's analysis of sense-perceptions offered strong evidence that exact laws of nature could not be induced from empirical data.

And Hume's analyses of the limits imposed by sense-perceptions on notions of causality and of time enabled Einstein to realize that the
high value of the velocity of light, compared with the other velocities we encounter daily, had prevented our appreciating that "the absolute character of time, viz., of simultaneity, unrecognizedly was anchored in the unconscious." (Einstein 1946, p. 53). Poincaré's pregnant statement in *Science and Hypothesis* to the effect that we have no
direct intuition of the simultaneity of two distant events may also
have been helpful here (p. 90).

But Einstein's fundamental analysis of physical theory went far beyond science as it is normally conceived: from an analysis of physics
and technology per se as regards, for example, electromagnetic induction, into an analysis of sensations, and then much as Boltzmann had
suggested into the "mode of our own thinking." (Boltzmann 1899, p.
104). Einstein concluded that the customary sensation-based notions
of time and simultaneity resulted in a physics burdened with asymmetries, hypotheses concerning the constitution of matter, unobservable
quantities, and ad hoc hypotheses. Thus prevented from lapsing into a
dogmatic slumber, Einstein found suggestive a combination of axiomatics
with a neo-Kantian view that was predicated on the usefulness of organizing principles such as the second law of thermodynamics. From the
eclectic philosophical view that had been tempered by over four decades
of scientific research, in 1944 Einstein wrote that in his opinion
Kant's reply to Hume's crushing message was that in thinking one must
use "concepts to which there is no access from the materials of sensory
experience, if the situation is viewed from the logical point of view."
(1944, p. 287).

And to a good approximation this is what Einstein had done in 1905,
based in part on a nascent form of his mature philosophical position.
For example, he enlarged Newton's principle of relativity to include
Lorentz's theory even though existing data did not indicate this to be
the case. Einstein's masterstroke was to accomplish this extension of
Newton's principle of relativity by linking the low-velocity experiment
of magnet and conductor in relative inertial motion, that he analyzed
in the first paragraph of the paper, to the unnamed ether drift experiments. He reasoned that electromagnetic induction depends on the laws
of mechanics and electromagnetism, which covers optics too; then he
"conjectured" that Newton's principle of relativity covers these three
disciplines to order (v/c). (1905, p. 392). He boldly continued by
raising this widened principle of relativity to a postulate or axiom
to be applied in its fullest thermodynamic sense. Then he went on to
propose another postulate that is basic to every wave theory of light:
in Lorentz's theory this postulate asserts that in the ether-fixed
system S the velocity of light is independent of the source's motion
and is always c. In special relativity which does not contain Lorentz's
ether, this is axiomatic in every inertial reference system although
empirical data gave support to only $(v/c)^2$. [18]

On these two axioms Einstein built a theory of pristine beauty that
accounts for but does not explain phenomena, and it makes no assump-

tions on the constitution of matter. Like thermodynamics it was a theory of principle (Einstein 1919). In the special relativity paper he deduced from the two axioms the relativity of time and of simultaneity. Thus, the order of development of these concepts in the published paper is the reverse of their invention.

11. Concluding Remarks

We can depict Einstein's approach to axiomatization in the special relativity theory as a hybrid version of the views of Boltzmann who emphasized mental pictures, but who was not daring enough in raising concepts to axiomatic status owing to his anti-Kantian stance; of Hertz's brilliant use of axioms as organizing principles, but within a scheme that was inapplicable to real mechanical phenomena; of Poincaré's far-reaching neo-Kantian organizing principles, but which placed too much emphasis on perception and empirical data; of Mach, who, with Poincaré, presented to Einstein paths not to follow without care — that is, not to emphasize perceptions; of Wien's suggestion of axiomatics as a goal; and of Abraham's 1904 paper that suggested an approach both to fundamentals and to not permitting data to decide the issue, a lesson soon to be useful to Einstein.

All of this enabled Einstein to cull away nonessentials such as those arising from excessive preoccupation with ether-drift experiments and to move boldly counter to the prevailing currents of theoretical physics by resolving problems in a Gordian manner — by inventing a view of physics in which certain problems do not occur, a view in which the 1895 paradox became mere fiction. He accomplished this feat by realizing the necessity for a demarcation between data and the mental constructs that are concepts or axioms, in order to pluck out of the air the version of space-time that was out there beyond our perceptions.

Today I have tried to demonstrate how complex is the mosaic from which special relativity emerges. Further work remains to be done to better understand how this scenario was played out against the backdrop of one man's view of philosophy, physics and technology in 1905.

Notes

[1] See Miller (1981b) for details concerning the theoretical and experimental physics of the 19th and early 20th centuries that were germane to the emergence of the special theory of relativity. The book's bibliography contains an extensive listing of secondary sources.

[2] The problem of source materials for the emergence of special relativity is analyzed in my (1981b, esp. Chapter 1). See also Holton (1960).

[3] Report of an interview with Einstein on 4 February 1950 (Shankland 1963, p. 48).

[4] Curiously enough, in their contribution to this session, John Earman, Clark Glymour and Robert Rynaciewicz completely deemphasize the philosophical component in Einstein's thinking toward special relativity. Instead these three philosophers of science depict Einstein as having discovered special relativity as a result of manipulating primes and exchanging v with −v in certain transformation rules for the electric and magnetic fields from Lorentz's electromagnetic theory. The result of their speculations is a caricature of Einstein that is antithetical to Einstein's own description of his creative thinking and to his own emphasis on the importance of philosophical considerations toward inventing special relativity.

[5] By neo-Kantian I mean that Poincaré based his philosophical view on two synthetic a priori organizing principles, namely, the principle of mathematical induction (1902, p. 13) and the notion of continuous groups of transformation (1902, p. 70). For further discussion of Poincaré's philosophy of science see Miller (1981b and in press).

[6] For the purpose of serious historical analysis I had to retranslate Einstein's relativity paper from the *Annalen* version for my book (1981b), where the retranslation appears in the Appendix (pages 391-415). The hitherto most frequently quoted English translation is by W. Perrett and G.B. Jeffery in the Dover reprint volume, The Principle of Relativity. But that contains some substantive mistranslations, infelicities, and outdated Britishisms. For example, Einstein's second principle of the relativity theory is mistranslated as: "Any ray of light moves in the 'stationary' system of co-ordinates with the determined velocity c, whether the ray be emitted by a stationary or by a moving body." The correct translation is: "Any ray of light moves in the 'resting' coordinate system with the definite velocity c, *which is independent* of whether the ray was emitted by a resting or by a moving body" (italics added to indicate a key phrase that was omitted in the Dover translation).

The Dover translation was made from a retypeset version of Einstein's relativity paper that had appeared in a Teubner reprint volume, thereby adding to the misprints in the original *Annalen* version. In addition, the Dover translation does not distinguish between Einstein's footnotes and those added to the Teubner edition by Arnold Sommerfeld. [The history of the Dover volume is in my 1981b, pages 391-392.] This state of affairs is an example of the pitfalls inherent in using a translation that was not made from the original paper and of the importance of going back to the original papers.

Hereinafter all citations to Einstein's relativity paper are to the translation in my book.

[7]Report of an interview with Einstein of 4 February 1950 (Shankland 1963, p. 48). Further evidence from Einstein's own writing (e.g., Einstein 1907b, p. 413) is discussed in Miller (1981b, Chapter 3).

[8]Needless to say, one can be aware of the results of a paper without having actually seen it. Thus, it is pertinent to mention that Wilhelm Wien (1904) summarized certain results in Lorentz's new theory of the electron and Emil Cohn (1904) discussed Lorentz's (1904b) space and time transformations. Although Einstein probably read Wien's 1904 *Annalen* paper, there is no extant documentation that he had seen the more detailed paper of Cohn that contained the exact Lorentz transformations. [See Miller (1981b) for discussion of the papers of Wien and Cohn.]

[9]Faraday (1852, pages 336-337, §§3088-3090). See Miller (1981a) for discussion of Faraday's experiments with rotating magnets and for analysis of problems concerning unipolar induction. Miller (1981b, esp. Chapter 3) focuses on the importance of unipolar induction to Einstein's thinking toward the special theory of relativity.

[10]I have analyzed this important paper in Miller (1981b).

[11]See Miller (1981b, Chapter 2) for historical evidence to support the conjecture that in 1904 Einstein had deduced the wave-particle duality for light, which he did not publish until 1909 (Einstein 1909a, 1909b).

[12]See my (1981b, 1983, in press) for a development of this conjecture for Boltzmann and Helmholtz. The importance of visual thinking for Einstein was first emphasized by Holton (1971-1972).

[13]See my (1981b and in press) for a comparative analysis of the philosophies of science of Poincaré and Einstein.

[14]Einstein's stunning use in thought experiments of images constructed from objects that had actually been perceived is an example of his utilizing the notion of *Anschauung* toward inventing the special theory of relativity. [See my (1983, in press) for further development.]

[15]This problem is connected with the inability of mechanics to explain the measured velocity of light and to whether the principle of action and reaction from mechanics is applicable to electromagnetic theory. [See Abraham (1904c).] For example, whereas Hertz's theory is compatible with both principles of mechanics it fails to explain Fizeau's 1851 experiment. By 1904 mechanical models of the ether were not seriously discussed because most physicists deemed them to be unfruitful.

[16]See my (1981b), especially Chapters 1 and 3, for a discussion of the notions of symmetry used by Abraham, Lorentz and Poincaré.

[17]This result turned out to be valid to all orders in (v/c). It is noteworthy that the new addition law for velocities is not necessarily a relativistic result. In fact, the new addition law is valid in every theory of a deformable electron that transforms according to the generalized Lorentz transformations:

$$x' = \ell\gamma(x - vt)$$
$$y' = \ell y$$
$$z' = \ell z$$
$$t' = \ell\gamma\left(t - \frac{v}{c^2}x\right)$$

where $\gamma = (1 - v^2/c^2)^{-\frac{1}{2}}$ and ℓ is a function of the magnitude of the relative velocity v between the two reference systems related by the transformations. For Lorentz's theory of the electron and for special relativity $\ell = 1$. For Paul Langevin's theory of a deformable electron $\ell = \gamma^{-\frac{1}{3}}$. Thus, Langevin's electron when in motion undergoes a contraction in the direction of its motion and an expansion in the transverse direction so that its volume remains unchanged. But the same velocity addition law holds because $\ell\gamma$ cancels when the quantity dx/dt is calculated. Poincaré (1906) investigated the class of electron theories that transform according to the generalized Lorentz transformations where $\ell \neq 1$ (see Miller 1973, 1981b esp. Chapter 1).

[18]Einstein went on to remove the asymmetries inherent in the contemporaneous treatment of electromagnetic induction in Section 6 of the relativity paper. Incidentally, contrary to Earman, Glymour and Rynaciewicz, Part II of the relativity paper (comprised of Sections 6, 7, 8, 9 and 10) is not a virtual potpourri of topics in dire need of an editor. Rather Part II is a quite "logical" application of the kinematics that Einstein developed in Part I: Sections 6, 7 and 8 discuss problems that concern electromagnetic fields and not their sources — for example, the relativity of the electric and magnetic fields, stellar aberration, Doppler's principle and radiation pressure; Section 9 introduces as a source a macroscopic charged body that is nondielectric, nonmagnetic and nonconducting; and Section 10 specializes the source of Section 9 to the submicroscopic electron with no assumptions on its structure. Nor, contrary to Earman, Glymour and Rynaciewicz, is there any extant archival evidence to the effect that Einstein wrote Part II before Part I, or that he realized the relativity of simultaneity as a result of deliberations on calculational details concerning topics in Part II.

References

Abraham, Max. (1902a). "Dynamik des Elektrons." Göttinger Nachrichten 20-41.

————————. (1902b). "Principien der Dynamik des Elektrons." Physikalische Zeitschrift 4: 57-63.

————————. (1903). "Principien der Dynamik des Elektrons." Annalen der Physik 10: 105-179.

————————. (1904a). "Zur Theorie der Strahlung und des Strahlungsdruckes." Annalen der Physik 14: 236-287.

————————. (1904b). "Die Grundhypothesen der Elektronentheorie." Physikalische Zeitschrift 5: 576-579.

————————. (1904c). Theorie der Elektrizität: Einführung in die Maxwellsche Theorie der Elektrizität. Leipzig: Teubner. (Revision of Föppl (1894).)

Boltzmann, Ludwig. (1897a). Vorlesungen über die Principe der Mechanik. I. Theil. Leipzig: Barth. (The Preface and §§1-12 are reprinted in Boltzmann (1974). Pages 223-254.)

————————. (1897b). "Über die Frage nach der objectiven Existenz der Vorgänge in der unbelebten Natur." Sitzungsberichte der mathematisch-naturwissenschaftlichen classe der kaiserlichen Akademie der Wissenschaften. Wien 106(Abt. 2a): 83-109. (As reprinted as "On the Question of the Objective Existence of Processes in Inanimate Nature." In Boltzmann (1974). Pages 57-76.)

————————. (1899). "Über die Grundprinzipien und Grundgleichungen der Mechanik." In Clark University 1889-1899, decennial celebration. Worcester, Mass.: Clark University. Pages 261-309. (As reprinted as "On the Fundamental Principles and Equations of Mechanics." In Boltzmann (1974). Pages 101-128.)

————————. (1974). Theoretical Physics and Philosophical Problems. (ed.) B. McGuiness, (trans.) P. Foulkes. Boston: Reidel. (This contains English versions of selected essays from Boltzmann's, Populäre Schriften. Leipzig: Barth, 1905 as well as others of Boltzmann's writings.)

Born, Max. (1969). Physics in my Generation. New York: Springer-Verlag.

Brace, Dewitt B. (1904). "On Double Refraction in Matter moving through the Aether." Philosophical Magazine 7: 317-329.

Cohn, Emil. (1904). "Zur Elektrodynamik bewegter Systeme." Berlin Berichte 40: 1294-1303.

Einstein, Albert. (1904). "Allgemeine molekulare Theorie der Wärme." Annalen der Physik 14: 354-362.

——————————. (1905). "Zur Elektrodynamik bewegter Körper." Annalen der Physik 17: 891-921. (Translated in Miller (1981b). Pages 392-415.)

——————————. (1907a). "Die vom Relativitätsprinzip geforderte Trägheit der Energie." Annalen der Physik 23: 371-384.

——————————. (1907b). "Relativitätsprinzip und die aus demselben gezogenen Folgerungen." Jahrbuch der Radioaktivität und Elektronik 4: 411-462.

——————————. (1909a). "Zum gegenwärtigen Stande des Strahlungsproblems." Physikalische Zeitschrift 10: 185-193.

——————————. (1909b). "Enwicklung unserer Anschauungen uber das Wesen und die Konstitution der Strahlung." Physikalische Zeitschrift 10: 817-825.

——————————. (1919). "What is the Theory of Relativity." The London Times November 28, 1919. (Versions appear in A. Einstein, Ideas and Opinions. New York: Bonanza Books, n.d. Pages 227-232; A. Einstein, Out of My Later Years. Totowa, New Jersey: Littlefield Adams and Co., 1967. Pages 54-57.)

——————————. (1923). "Grundgedanken und Probleme der Relativitatstheorie." Stockholm: Imprimerie royale (An address delivered before the Nordische Naturforscherversammlung, Göteborg, July 11, 1923.) (As reprinted as "Fundamental Ideas and Problems of the Theory of Relativity." In Nobel Lectures, Physics: 1901- 1921. New York: Elsevier, 1967. Pages 479-490.)

——————————. (1944). "Remarks on Bertrand Russell's Theory of Knowledge." In The Philosophy of Bertrand Russell. (The Library of Living Philosophers. Volume V.) Edited by P.A. Schilpp. Evanston: The Library of Living Philosophers. Pages 277-291.

——————————. (1946). "Autobiographical Notes." In Albert Einstein: Philosopher-Scientist. (The Library of Living Philosophers. Volume VII.) Edited by P.A. Schilpp. Evanston: The Library of Living Philosophers, 1949. Pages 2-94.

——————————. (1956). Lettres à Maurice Solovine. (ed.) M. Solovine. Paris: Gauthier-Villars. (Translated into French by M. Solovine who also wrote an introduction.)

——————————. (1972). Albert Einstein -- Michele Besso: Correspondance 1903-1955. Paris: Hermann. (Translated into French by P. Speziali who also supplied notes and an introduction.)

Faraday, Michael. (1852). "On Lines of Magnetic Force; their definite character; and their distribution within a magnet and through space." Philosophical Transactions of the Royal Society of London 142: 25-56. (As reprinted in Experimental Researches in Electricity, Volume 3. New York: Dover, 1965. Pages 328-370.)

Fizeau, Hippolyte. (1851). "Sur les hypothesès relatives à l'éther lumineaux, et sur une expérience quî parait démontrer que le mouvement des corps change la vitesse avec laquelle la lumière se propage dans leur intérieur." Comptes rendus hebdomadaires des séances de l'académie des sciences 33: 349-355.

Föppl, August. (1894). Einführung in die Maxwell'sche Theorie der Elektricität. Leipzig: Teubner.

Frank, Philipp. (1947). Einstein: His Life and Times. New York: Knopf.

Helmholtz, Hermann von. (1894). "Über den Ursprung der richtigen Deutung unserer Sinneseindrucke." Zeitschrift für Psychologie und Physiologie der Sinnesorgane VII: 81-96. (As reprinted as "The Origin and Correct Interpretation of our Sense Impressions." In Selected Writings of Hermann von Helmholtz. (ed.) (trans.) R. Kahl. Middletown, CT: Wesleyan University Press, 1971. Pages 501-572.)

Hertz, Heinrich. (1894). Die Prinzipien der Mechanik in neuen zusammenhange. (Gesammelte Werke, bd. III.) Edited by P. Lenard. Leipzig: J.A. Barth. (As reprinted as The Principles of Mechanics. (trans.) D.E. Jones and J.T. Walley. New York: Dover, 1956.)

Hoek, Martinus. (1868). "Détermination de la vitesse avec laquelle est entrainée une onde traversant un milieu en mouvement." Archives Néelandaises des Sciences Exactes et Naturelles 3: 180-185.

Holton, Gerald. (1960). "On the Origins of the Special Theory of Relativity." American Journal of Physics 28: 627-636. (As reprinted in Holton (1973). Pages 165-183.)

--------------. (1967-1968). "Influences on Einstein's Early Work." The American Scholar 37: 59-79. (As reprinted in Holton (1973). Pages 197-217.)

--------------. (1969). "Einstein, Michelson, and the 'Crucial' Experiment." Isis 60: 133-197. (As reprinted in Holton (1973). Pages 261-352.)

--------------. (1971-1972). "On Trying to Understand Scientific Genius." The American Scholar 41: 95-110. (As reprinted in Holton (1973). Pages 353-380.)

--------------. (1973). Thematic Origins of Scientific Thought: Kepler to Einstein. Cambridge, MA: Harvard University Press.

Kaufmann, Walter. (1901). "Die magnetische und electrische Ablenkbarkeit der Becquerelstrahlen und die scheinbare Masse der Elektronen." Göttinger Nachrichten 143-155.

--------------. (1902a). "Über die elektromagnetische Masse des Elektrons." Göttinger Nachrichten 291-303.

--------------. (1902b). "Die elektromagnetische Masse des Elektrons." Physikalische Zeitschrift 4: 54-57.

Lorentz, H.A. (1892a). "La théorie électromagnétique de Maxwell et son application aux corps mouvants." Archives Néelandaises des Sciences Exactes et Naturelles 25: 363-551. (As reprinted in Lorentz (1935-1939), Volume 2. Pages 164-343.)

------------. (1892b). "The relative Motion of the Earth and the Ether. Koninklijke Akademie van wetaschappen te Amsterdam 1: 74. (As reprinted in Lorentz (1935-1939), Volume 4. Pages 219-233.)

------------. (1895). Versuch einer Theorie der elektrischen und optischen Erscheinungen in bewegten Körpern. Leiden: Brill. (As reprinted in Lorentz (1935-1939), Volume 5. Pages 1-137.)

------------. (1904a). "Weiterbildung der Maxwellschen Theorie. Elektronentheorie." Encyklopädie der mathematischen Wissenschaften 14: 145-288.

------------. (1904b). "Electromagnetic Phenomena in a System Moving with any Velocity Less than that of Light." Proceedings of the Academy of Sciences of Amsterdam 6: 809-831. (As reprinted in Lorentz (1935-1939), Volume 5. Pages 172-197.)

------------. (1905). "Ergebnisse und Probleme der Elektronentheorie." Elektrotechnischen Verein zu Berlin. 1904. Berlin: Springer-Verlag. Pages 76-124. (As reprinted in Lorentz (1935-1939), Volume 8. Pages 79-124.)

------------. (1910). "Alte und neue Fragen der Physik." Physikalische Zeitschrift 11: 1234-1257. (As reprinted in Lorentz (1935-1939), Volume 7. Pages 205-257.)

------------; Einstein, A.; Minkowski, H.; and Weyl, H. (1923). The Principle of Relativity: A Collection of Original Memoirs on the Special and General Theories of Relativity. (trans.) W. Perrett and G.B. Jeffery. London: Metheuen. (As reprinted New York: Dover, 1952.)

------------. (1935-1939). Collected Papers. 9 vols. The Hague: Nijhoff.

Mach, Ernst. (1883). Die Mechanik in ihrer Entwicklung historisch-kritisch dargestellt. Leipzig: F.A. Brockhaus. (As reprinted from the 9th German edition as The Science of Mechanics: A Critical and Historical Account of Its Development. (trans.) T.J. McCormack. La Salle, IL: Open Court, 1960.)

Merz, John T. (1904-1912). A History of European Scientific Thought in the Nineteenth Century. 2 vols. Edinburgh: Blackwood. (As reprinted New York: Dover, 1965.)

Michelson, Albert A. and Morley, Edward. (1886). "Influence of Motion of the Medium on the Velocity of Light." American Journal of Science 31: 377-386.

--. (1887). "On the Relative Motion of the Earth and the Luminiferous Ether." American Journal of Science 34: 333-345.

Miller, Arthur I. (1973). "A Study of Henri Poincaré's 'Sur la Dynamique de l'Electron'." Archive for History of Exact Sciences 10: 207-328.

------------------. (1981a). "Unipolar Induction: A Case Study of the Interaction Between Science and Technology." Annals of Science 38: 155-189.

------------------. (1981b). Albert Einstein's Special Theory of Relativity: Emergence (1905) and Early Interpretation (1905-1911). Reading, MA: Addison-Wesley.

------------------. (1983). "On the Origins, Methods, and Legacy of Ludwig Boltzmann's Mechanics." Scheduled to appear in Proceedings of the International Conference on Ludwig Boltzmann. Braunschweig: Vieweg.

------------------. (in press). On the Nature of Scientific Discovery. Boston: Birkhäuser.

Newton, Isaac. (1687). Philosophiae Naturalis Principia Mathematica. London: Royal Society. (As reprinted as Sir Isaac Newton's Mathematical Principles of Natural Philosophy and his System of the World. 2 vols. (trans.) A. Motte, revised by F. Cajori. Berkeley: University of California Press, 1973.)

Planck, Max. (1900). "Zur Theorie des Gesetzes der Energieverteilung im Normalspektrum." Verhandlungen der Deutschen Physikalischen Gesellschaft 2: 237-245.

------------. (1901). "Über das Gesetz der Energieeverteilung in Normalspektrum." Annalen der Physik 4: 553-563.

Poincaré, Henri. (1902). La Science et l'Hypothese. Paris: Flammarion. (As reprinted as Science and Hypothesis. (trans.) W.J. Greenstreet. New York: Dover, 1952.)

——————————. (1904). "L'état actuel et l'avenir de la Physique mathématique." Bulletin des Sciences mathématiques et astronomiques 28: 302-324. (As reprinted in The Value of Science. (trans.) G.B. Halsted. New York: Dover, 1958. Pages 91-111. This is a translation of Poincaré's, La Valeur de la Science. Paris: Flammarion, 1905.)

——————————. (1905). "Sur la dynamique de l'électron." Comptes rendus hebdomadaires des seances de l'academie des sciences 140: 1504-1508. (As reprinted in Oeuvres de Henri Poincare. Paris: Gauthier-Villars, 1934-1953, Volume 9. Pages 489-493.)

——————————. (1906). "Sur la dynamique de l'electron." Rendiconti del Circolo matematico di Palermo 21: 129-175. (As reprinted in Oeuvres de Henri Poincare. Paris: Gauthier-Villars, 1934-1953, Volume 9. Pages 494-550.)

Lord Rayleigh. (1902). "Does Motion through the Aether case Double Refraction?" Philosophical Magazine 4: 678-683.

Sauter, Joseph. (1965). Erinnerungen an Albert Einstein. This phamphlet (unpaginated) was published in 1965 by the Patent Office in Bern, and contains documents pertaining to Einstein's years at that office as well as a note by Sauter.

Seelig, Carl. (1954). Albert Einstein: Eine dokumentarische Biographie. Zürich: Europa-Verlag.

Shankland, Robert S. (1963). "Conversations with Albert Einstein." American Journal of Physics 31: 47-57.

Trouton, F.T. and Nobel, H.R. (1903). "The Mechanical Forces Acting on a Charged Electric Condenser moving through Space." Philosophical Transactions of the Royal Society London 202: 165-181.

Weber, C.L. (1895). "Über unipolare Induktion." Elektrotechnische Zeitschrift 16: 513-514.

Wien, Wilhelm. (1900). "Über die Möglichkeit einer elektromagnetischen Begründung der Mechanik." In Recueil de travaux offerts par les auteurs à H.A. Lorentz. The Hague: Nijhoff. Pages 96-107. (As reprinted in Annalen der Physik (1901) 5: 501-513.)

——————————. (1904). "Erwiderung auf die Kritik des Hrn. Abraham." Annalen der Physik 14: 635-637.

Part IV

On Some Contemporary Approaches to Special Relativity

Essay 5

ON LORENTZ'S METHODOLOGY *

Elie Zahar, in his recent paper [1973], presents an attempt to use Lakatos's methodology of 'research programmes' to analyse the work of Einstein and Lorentz. An attempt to interpret or to explain episodes in the history of science within some larger context using an imaginative conceptual framework can result in exciting and provocative studies. However, a necessary prerequisite to this type of study is the presentation of relevant scientific literature within its proper historical context, and to take account of its internal structure. This Zahar has not done, with the result that his study becomes at best misleading, and his criticisms of the researches of others becomes empty. One reason may be that Zahar's application of the methodology of research programmes seems intended to follow the form proposed by Lakatos in his [1970]. However, Zahar takes account only of Lakatos's first rule of procedure for the use of historical case studies[1]—'(1) *one gives a rational reconstruction* [according to the methodology of research programmes]'—but not of the second one—'(2) *one tries to compare this rational reconstruction with actual history and to criticize one's rational reconstruction for lack of historicity.* . . .'

Zahar states that his goal in his [1973] is:[2]

> . . . I propose to refute *all* the charges of *ad hocness* which have been levelled at the *L.F.C.* [Lorentz-Fitzgerald contraction hypothesis] and show that Lorentz's programme progressed until after 1905.

Zahar describes Lorentz's 'ether programme' as a progressive research programme, the 'hard core' of which is 'Maxwell's equations for the electromagnetic field [and] Newton's laws of motion,' the Galilean transformation and the Lorentz force equation, and an heuristic which 'arises from the overall *metaphysical* principle that all physical phenomena are governed by actions transmitted by the ether.'[3]

Then, consonant with the methodology of research programmes, Zahar considers Lorentz's ether programme as consisting of 'three consecutive theories', T_1, T_2, T_3, a series which exhibits a progressive theoretical problem shift. According to Zahar, the theory T_1 is that proposed in Lorentz's [1892a]; theory T_2 is that sketched out in Lorentz's [1892b] and then presented in detail in Lorentz's [1895]; and theory T_3 is supposed to be the one presented in Lorentz's famous [1904]. Zahar gives the following description of T_1, T_2 and T_3:[4]

---

* This work was written while I was on sabbatical leave for 1972–73 as a National Endowment for the Humanities Fellow in the Department of Physics at Harvard University.

[1] Lakatos [1970], p. 138; italics in original.

[2] Zahar [1973], p. 99; italics in original.

[3] *Ibid.*, p. 100; italics in original. According to Zahar [1973], pp. 99–100 (and Lakatos):

> The hard core consists of assumptions which, by methodological decision, as it were, are kept unfalsified. Each theory in the programme is a conjunction of, on the one hand, the hard core and, on the other, of auxiliary hypotheses to which the modus tollens is directed whenever anomalies arise. A programme also has a heuristic which consists of a set of suggestions and hints which govern the construction or modification of the auxiliary hypotheses. The heuristic, which sets out a research policy, is less rigid than the hard core.

[4] Zahar [1973], p. 100; italics in original.

Reprinted from *The British Journal for the Philosophy of Science* **25** (1974), 29–45.

T_1 consists of the hard core as defined above together with the (tacit!) assumptions (*i*) that moving clocks are not retarded and (*ii*) that material rods are not shortened by their motion through the ether.

T_2 is obtained from T_1 by substituting the *L.F.C.* for assumption (*ii*). According to the *L.F.C.* a body moving through the ether with velocity $\vec{v}$ is shortened by the factor $\sqrt{1-v^2/c^2}$.

T_3 is the conjunction of the hard core, of the *L.F.C.* and of the assumption, that, contrary to (*i*), clocks moving with velocity $\vec{v}$ are retarded by the factor $\sqrt{1-v^2/c^2}$.

He then restates his goal as:[1]

I claim that both the shift from T_1 to T_2 and that from T_2 to T_3 were non *ad hoc*. This implies in particular that the introduction of the *L.F.C.* which took Lorentz from T_1 to T_2 was not an *ad hoc* manoeuvre.

This is essential for Zahar to prove, because if it were not the case, then Zahar would have to consider Michelson's result as 'an obstacle for Lorentz's programme';[2] thus, Zahar would have trouble arguing that the shift from T_1 to T_2 is a progressive problem shift, rather than the beginning of a new research programme. The reason is that the *L.F.C.*, stated as dimensions in the direction of motion shortened by the amount $\sqrt{1-v^2/c^2}$ (or to second order in v/c by the amount $1-v^2/(2c^2)$) is not a plausible auxiliary hypothesis because it is not in 'the spirit of the heuristic of the programme.'[3] According to Zahar, the *L.F.C.*, tagged in this way onto the hard core, would have to be considered as *ad hoc$_3$*.[4]

Zahar's argument for the non *ad hocness* of the *L.F.C.* turns on two points: one which concerns methodology; the other involves, as I shall demonstrate, a convenient distortion of historical facts which enables Zahar to fit Lorentz's work of 1892–1904 into the structure of the methodology of research programmes. The point of methodology is:[5]

Ad hocness in research programmes is defined not as a property of an isolated hypothesis but as a relation between two consecutive theories. A theory is said to be *ad hoc$_1$* if it has no novel consequences as compared with its predecessor. It is *ad hoc$_2$* if none of its novel predictions have been actually 'verified'; for one reason or another the experiment in question may not have been carried out, or—much worse—an experiment devised to test a novel prediction may have yielded a negative result. Finally the theory is said to be *ad hoc$_3$* if it is obtained from its predecessor through a modification of the auxiliary hypotheses which does not accord with the spirit of the heuristic of the programme.

He then defines what he means by a 'novel fact':[6]

A fact will be considered novel with respect to a given hypothesis if it did not belong to the problem-situation which governed the construction of the hypothesis.

It will become evident that these redefinitions of '*ad hoc*' and 'novelty' are contrived merely to support Zahar's contention that the *L.F.C.* is not *ad hoc*, and, therefore, that the shift from T_1 to T_2 fits into the mold defined by the method-

---

[1] *Loc. cit.;* italics in original. [2] *Ibid.*, p. 107. [3] *Ibid.*, p. 101.
[4] *Loc. cit.* [5] *Loc. cit.* [6] *Ibid.*, p. 103; italics in original.

ology of research programmes. In fact, according to Henri Poincaré's 'logical (*i.e.* 'rational')' judgement of what constitutes *ad hoc* statements (to be discussed below), Zahar's notion of *ad hoc* is itself *ad hoc*.

Zahar's strategy is to argue that Lorentz added to T_1 an hypothesis H_2,[1] which is not *ad hoc* relative to T_1, *i.e.* is a generalisation of a statement already contained in T_1, and, moreover, H_2 does not pertain to Michelson's result. The conjunction of H_2 with the hard core of T_1 is the theory T_2. Then, according to Zahar, Lorentz *deduced* the *L.F.C.* from T_2. The *L.F.C.* is then considered to be a novel fact relative to H_2. This leads us to the second point in Zahar's argument —the one which concerns a distortion of historical fact. Zahar states it in two parts:[2]

(Z_a): Lorentz deduced the *L.F.C.* from a deeper theory, namely from what I call the Molecular Forces Hypothesis (hereafter referred to as the *M.F.H.*) and which can be loosely formulated as follows: 'Molecular forces behave and transform like electromagnetic forces.' Moreover, in his deduction of the *L.F.C.*, Lorentz made use of his famous transformation, as is clearly indicated by the following passage from his [1895]:

'For, if we now understand by S_1 and S_2 not, as formerly, two systems of charged particles but two systems of molecules—the second at rest and the first moving with velocity v in the direction of the axis of x- between the dimensions of which the relation subsists as previously stated; and if we assume that in both systems the x-components of the forces are the same, while the y- and z-components differ from one another by the factor $\sqrt{1-v^2/c^2}$, then it is clear that the forces in S_1 are in equilibrium whenever they are so in S_2 ... The displacement would naturally bring about this disposition of the molecules of its own accord and thus effect a shortening in the direction of motion in the proportion of 1 to $\sqrt{1-v^2/c^2}$ *in accordance with the formulae given in the above-mentioned paragraph.*'

I further maintain that for anybody prepared to accept the assumption of an ether at rest, the *M.F.H.* is a plausible auxiliary hypothesis which introduces no alien elements into Lorentz's programme. Putting it more objectively, the theory T_2 proposed by Lorentz is non *ad hoc*₃, because the *M.F.H.* is structured in accordance with the heuristic of the ether programme, which requires that physical phenomena be explained in terms of actions propagated in the ether.

(Z_b): Moreover, *the M.F.H. arose out of considerations which had nothing to do with Michelson's experiment.* The *M.F.H.* arose out of mathematical considerations pertaining to the transformation properties of Maxwell's equations. *Hence Michelson's null result is a novel fact relative to the M.F.H.!* The *M.F.H.* is consequently non *ad hoc*₂; it constituted both theoretical and empirical progress.

[1] H_2 will turn out to be Lorentz's hypothesis that molecular forces transform like electromagnetic forces, *i.e.* the Lorentz force.

[2] *Ibid.*, pp. 106–7. The quotation of Lorentz is from the portion of the *Versuch* [1895] that pertains to Michelson's results and translated in *P.R.C.*, p. 7. The italics are Zahar's. The passage from Lorentz's [1895] in which the Michelson and Morley experiment is discussed is translated in *P.R.C.*, pp. 3–7. All page references will be to *P.R.C.*

Zahar then attempts to buttress the line of argument presented in (Z_a) and (Z_b) by drawing upon Lorentz's [1892a]:[1]

> (Z_c): In Lorentz's [1892a] there is no mention of the 'crucial' experiment first performed by Michelson in 1881, then repeated by Michelson and Morley in 1887. This should not surprise us once we have realised that the Lorentz transformation was originally used as a mathematical device of which Lorentz gave no physical interpretation, in much the same way as we might use the expression *ict* without attaching any physical meaning to the multiplication of the distance *ct* by $\sqrt{-1}$.

To comment systematically on Zahar's analysis, it is thus necessary to analyse (Z_c) and the statements which follow, and then return to (Z_a) and (Z_b). The majority of my critical comments will be directed toward Zahar's arguments for the non *ad hocness* of the *L.F.C.*

Zahar's characterisation of the foundations of Lorentz's theory is incorrect. Newton's three laws of motion were subsequently found not to be an integral part of the theory. Lorentz, in his [1895], deduced that Newton's third law was inconsistent with the fundamental equations of the electromagnetic theory.[2] Furthermore, it is quite ahistorical to consider Maxwell's equations but not the ether as part of the hard core. The reason is that for Lorentz the Maxwell equations had no meaning without the ether. Lorentz considered the ether as a real quantity, even though its physical effects might never be detected directly. Aside from the fact that the ether constituted an integral part of electromagnetic theory *c*. 1900, it is clearly evident from even a cursory study of Lorentz's papers on electromagnetic theory that his 'research policy' was guided by the Maxwell equations and the Lorentz force equation, which were inseparable from the ether —in fact, the Maxwell equations were supposed to be the means by which one could calculate the state of the ether.

In (Z_c) is the comment that Lorentz did not mention Michelson's 'crucial' experiment in his [1892a]. This is not an oversight on Lorentz's part; indeed, Lorentz was very concerned about not being able to account for Michelson's result by use of his electromagnetic theory. In a letter to Lord Rayleigh on 18 August 1892, shortly after completing his [1892a], Lorentz wrote:[3]

> Fresnel's hypothesis [of a stagnant aether] taken conjointly with his [partial dragging] coefficient $1-1/n^2$, would serve admirably to account for all the observed phenomena were it not for the interferential experiment of Mr Michelson, which has, as you know, been repeated after I published my remarks on its original form, and which seems decidedly to contradict Fresnel's views. *I am totally at a loss to clear away this contradiction, and yet I believe if we were to abandon Fresnel's theory, we should have no adequate theory at all*, the conditions which Mr Stokes has imposed on the movement of aether being irreconcilable to each other.
>
> *Can there be some point in the theory of Mr Michelson's experiment which has as yet been overlooked?*

---

[1] Zahar [1973], pp. 111–12.
[2] Lorentz [1895], pp. 24 ff. For further discussion of the incompatibility of Newton's third law with Lorentz's electromagnetic theory see Miller [1973].
[3] Quoted from K. Schaffner [1972], p. 103. Italics added.

This information contradicts Zahar's suggestion that Lorentz in 1887 might already have thought of the contraction hypothesis, 'but would have considered such a simplistic contraction hypothesis unacceptable'.[1] Simplistic indeed, and such a hypothesis, moreover, would have been considered as *ad hoc* according to anyone's interpretation of the term. Lorentz was too good a physicist to put forth such a hypothesis without at least being able to support it with a plausibility argument; it appears that none existed in 1887.

Zahar, on the other hand, implies that Lorentz waited until he could deduce the *L.F.C.* from another auxiliary hypothesis. However, in the light of Lorentz's letter to Lord Rayleigh, this discussion is quite beside the point.

To comment upon the next sentence in (Z_e) we must discuss briefly a certain mathematical transformation used by Lorentz in [1892a]. Zahar is correct in noting that a version of what later became known as the Lorentz transformation was introduced by Lorentz in [1892a] as a mathematical aid in solving the in-homogeneous second order partial differential equations for the electromagnetic field quantities in an inertial reference system.[2] The transformation is:[3]

$$x' = \gamma x_r \tag{1}$$

$$y' = y_r \tag{2}$$

$$z' = z_r \tag{3}$$

$$t' = t - \gamma^2 \frac{v}{c^2} x_r \tag{4}$$

where $x_r = x - vt$, $y_r = y$, $z_r = z$, $t_r = t$; $\gamma = 1/\sqrt{1 - v^2/c^2}$; the coordinates x, y, z, t are measured relative to a reference system S fixed in the ether; x', y', z', t' are measured relative to a reference system Q' (Zahar calls it S'), which has no physical meaning; and x_r, y_r, z_r, t_r refer to a real inertial system S_r. In fact, in Q' the speed of light is not c to all orders in v/c. This can easily be shown as follows. The transformations inverse to (1)–(4) that relate S and Q' are:

$$x = \gamma x' + vt' \tag{5}$$

$$y = y' \tag{6}$$

$$z = z' \tag{7}$$

$$t = t' + \gamma \frac{v}{c^2} x' \tag{8}$$

To facilitate the remainder of this proof, I resort to the following device: in S we have

$$x^2 + y^2 + z^2 = c^2 t^2 \tag{9}$$

which means that the speed of light is c in S. Transforming (9) to Q' by use of (5)–(8) yields

$$x'^2 + y'^2 + z'^2 = c^2(1 - v^2/c^2)t'^2 \tag{10}$$

thus, in Q' the speed of light is

$$c\sqrt{1 - v^2/c^2}, \tag{11}$$

[1] Zahar [1973], p. 115.
[2] *Ibid.*, p. 112.
[3] See Lorentz [1892a], p. 297. Zahar quotes these equations on his p. 112.

c

and not c. Lorentz was aware of this shortcoming of the system Q', although, of course, not by the method I outlined above. Lorentz, as a result of inspecting the functional form of the fields obtained by use of transforming the field equations from S_r to Q', could not have failed to see that the electromagnetic field quantities propagate with the velocity $\sqrt{c^2 - v^2}$, and, therefore, may have no physical interpretation. However, to first order in v/c these fields do propagate with the velocity c. Lorentz gave two reasons for making this approximation: it facilitates subsequent calculations; and, more important, it permitted him to prove '*un théorème général*'[1]—namely, that to first order in v/c the electromagnetic field quantities for an harmonically bound microscopic charged particle (as he referred to the electron in his [1892a]) have the same form in S_r as they do in S. To this order of accuracy the spatial coordinates in Q' become identical to those in S_r, and the time coordinate becomes what Lorentz was to define in his [1895] as the 'local time'.[2] The '*théorème général*' can be considered as an earlier version of the 'theorem of corresponding states'[3] from Lorentz's [1895].

In summary: Q' is a fictitious coordinate system which can be considered as at rest in the ether and in which the speed of light is c only to order v/c. Q' is obtained from S_r by extension of the coordinates along the x_r-axis, *i.e.* by a change of units.

In the second sentence of (Z_c) Zahar asserts that Lorentz accounted for Michelson's result by using Q'. Zahar then amplifies this statement:[4]

(Z_d): However, soon after writing his [1892a], Lorentz realised that the transformation equations lent themselves to an interpretation which he immediately set out in his [1892b] and which he then expounded in greater detail in his [1895].

An analysis of Lorentz's [1892b] and [1895] will throw light upon the erroneous statements in (Z_a), (Z_b), (Z_c), (Z_d) and elsewhere in Zahar's paper. First I shall outline *Lorentz's* methodology in [1892b] and [1895], and then compare this to Zahar's Lorentz—the one who supposedly did his physics according to the methodology of research programmes.

Lorentz began [1892b] by discussing his analysis given in detail in his [1886] of Stokes' ether theory and Michelson's experiment. He then discussed the Michelson and Morley experiment of 1887. *Then*, by means of a purely geometrical argument, *i.e.* one based upon the Newtonian addition law for velocities, Lorentz showed that their null result could be accounted for to second order in v/c by positing that the length of the interferometer arm in the direction of motion be shortened by an amount $1 - v^2/(2c^2)$. This subsequently became known as the *L.F.C.* Thus far, Lorentz has made no mention of what Zahar refers to as the 'Molecular Forces Hypothesis' (see (Z_a) and (Z_b)), hereafter referred to as *M.F.H.* Clearly, the *L.F.C.*, as first proposed by Lorentz in [1892b], was a physics of desperation—he says as much:[5]

(L_a): This experiment [Michelson and Morley] has been puzzling me for a long time, and in the end I have been able to think of only one means of reconciling its result with Fresnel's theory.

---

[1] Lorentz [1892a], p. 304.   [2] Lorentz [1895], p. 50.   [3] *Ibid.*, p. 85.
[4] Zahar [1973], p. 112.
[5] Lorentz [1892b], p. 221. See also the letter of Lorentz to Lord Rayleigh quoted above.

If one wishes to place this calculation within a hypothetico-deductive scheme, then one is forced to say that Lorentz deduced the *L.F.C.* from a law of mechanics.

Lorentz continues in [1892b] to give what is clearly a plausibility argument for the *L.F.C.*:[1]

> (L_b): Now, some such change in the length of the arms in Michelson's first experiment and in the dimensions of the slab in the second one is so far as I can see, not inconceivable. What determines the size and shape of a solid body? Evidently the intensity of the molecular forces; any cause which would alter the latter would also influence the shape and dimensions. Nowadays we may safely assume that electric and magnetic forces act by means of the intervention of the ether. It is not far-fetched to suppose the same to be true of the molecular forces. But then it may make all the difference whether the line joining two material particles shifting together through the ether, lies parallel or crosswise to the direction of that shift. It is easily seen that an influence of the order of p/V is not to be expected, but an influence of the order of p^2/V^2 is not excluded and that is precisely what we need.

Lorentz emphasises that the hypothesis of the *L.F.C.* is 'not inconceivable'. He then posits the hypothesis H', which is that 'it is not far-fetched to suppose' that the molecular forces which after all determine the shape of a body 'act by intervention of the ether'. Thus, at this point, Lorentz's only assertion about molecular forces is that they do not propagate instantaneously.

Lorentz concludes this passage by emphasising the necessity of proving that molecular forces are affected by motion through the ether in such a way that they influence the shape of an object in motion by at least a term of second order in (v/c). He then asserts that H' cannot be tested:[2]

> (L_c): Since the nature of the molecular forces is entirely unknown to us, it is impossible to test the hypothesis.

What is needed is the very bold hypothesis which we referred to as H_2—namely, that the 'influences of the motion of ponderable matter on electric and magnetic forces' is the same for molecular forces.[3] Thus, Lorentz's next step is to calculate the Lorentz force in S_r. Lorentz did this soon after the publication of [1892a]. His method was to make use of the fictitious system Q' so as to transform a problem of electrodynamics into one of electrostatics, the latter class of problems are easier to solve. As evidenced in [1892b], he obtained a result accurate only to second order in (v/c):

$$F'_x = F_x \tag{12}$$

$$F'_y = \left(1 + \frac{v^2}{2c^2}\right)F_y \tag{13}$$

$$F'_z = \left(1 + \frac{v^2}{2c^2}\right)F_z \tag{14}$$

where primed quantities refer to the fictitious system Q'. According to H_2, molecular forces transform like the Lorentz force, and, therefore, relative to S

---

[1] Lorentz [1892b], p. 221.   [2] *Loc. cit.*   [3] *Ibid.*, pp. 221–2.

propagate through the ether with the speed of light in vacuum. This is Zahar's *M.F.H.* Equations (*12*)–(*14*) indicate that a system in equilibrium under molecular forces in S_r should also be in equilibrium in Q', which is taken as at rest in the ether. Lorentz then argues that this can be so only if dimensions in S_r in the direction of motion are shortened by the amount $(1-v^2/(2c^2))$. Lorentz obtained this result by writing (*1*)–(*3*) in the approximate form

$$x_r = \left(1 - \frac{v^2}{2c^2}\right)x' \tag{15}$$

$$y_r = y' \tag{16}$$

$$z_r = z' \tag{17}$$

in order to compare dimensions in S_r and Q'. Thus, putting forth the *M.F.H.* allowed Lorentz to give theoretical support for the *L.F.C.* As if to emphasize this point, Lorentz goes on to assert:[1]

> (L_d): One may not of course attach much importance to this result; the application to molecular forces of what was found to hold for electric forces is too venturesome for that.

Lorentz continues by asserting that the *L.F.C. might not be the correct explanation* of the Michelson and Morley experiment:[2]

> (L_e): Besides, even if one would do so, the question would still remain whether the earth's motion shortens the dimensions in one direction, as assumed above, or lengthens those in directions perpendicular to the first, which would answer the purpose equally well.

In his [1895] Lorentz applies the electromagnetic theory that he had developed at length in his [1892a] to problems involving optical and electrical phenomena in moving bodies. Lorentz discussed two different transformations that he had developed in this treatise for two different sets of problems. The set of transformations:[3]

$$x' = \gamma x_r \tag{18}$$

$$y' = y_r \tag{19}$$

$$z' = z_r \tag{20}$$

$$t' = t_r \tag{21}$$

permitted Lorentz to prove that to order v/c electrostatic processes are unaffected by the earth's motion and also could be used to transform certain problems in electrodynamics to those in electrostatics. The primed coordinates refer to a fictitious system S', which bears some similarity to the Q' system from the [1892a]. S' is the electron's instantaneous rest system, in which the electron is dilated along the x'-axis by an amount $1/\sqrt{1-v^2/c^2}$. Using (*18*)–(*21*), Lorentz was able to relate the Lorentz forces in S' and S_r:

---

[1] *Ibid.*, p. 223.   [2] *Loc. cit.*
[3] Lorentz [1895], pp. 35–9. For a discussion of these equations see Miller [1973], esp. parts 3 and 4.

$$F_x = F'_x \tag{22}$$

$$F_y = \sqrt{1 - v^2/c^2} F'_y \tag{23}$$

$$F_z = \sqrt{1 - v^2/c^2} F'_z \tag{24}$$

which is a more exact version of (12)–(14).

The other system of transformation is:

$$x_r = x - vt \tag{25}$$

$$y_r = y \tag{26}$$

$$z_r = z \tag{27}$$

$$t_L = t - \frac{v}{c^2} x \tag{28}$$

where t_L is the local time and the spatial portion (25)–(27) pertains to the Galilean transformations. This system relates S_r and S. Lorentz postulated the mathematical concept of the local time coordinate to prove the covariance to first order in v/c of the Maxwell equations in charge free space or within a substance that is neutral, not magnetic and not a conductor.[1] Thus, equations (25)–(28) are the set of transformations that Lorentz used in the theorem of corresponding states. After having accounted systematically for the result of all first order optical experiments, Lorentz, in the final chapter of the [1895], discussed three experiments accurate to second order in v/c which could not be accounted for 'without further ado'. One of them is the Michelson and Morley experiment. Lorentz's methodology in dealing with the Michelson and Morley experiment parallels that given in his [1892b], but contains more details. An outline is sufficient for my purposes: After using the Newtonian addition law for velocities to discuss the Michelson and Morley experiment, Lorentz put forth the L.F.C. to explain their null result. However, as in [1892b], he goes on to assert that this account is not unique. For, if the dimensions of an object parallel to its direction of motion are changed in the ratio 1 to $1 + \delta$, and those normal to its motion in the ratio 1 to $1 + \epsilon$, then it follows that the Michelson and Morley result can be accounted for to second order in v/c if ϵ and δ satisfy the equation

$$\epsilon - \delta = \frac{v^2}{2c^2}. \tag{29}$$

Equation (29) is one equation for two unknowns; hence, either ϵ or δ would remain undetermined. The L.F.C. corresponds to $\epsilon = 0$. Then Lorentz tries to make 'this hypothesis' (one which he confesses to be 'surprising') more acceptable, and employs a characteristically modest tone that shows how tentative and dependent on a string of assumptions his suggestion really is.[2]

(L_f): Surprising as this hypothesis may appear at first sight, yet we shall have to admit that it is by no means far-fetched, as soon as we assume that molecular forces are also transmitted through the ether, like the electric and

---

[1] Lorentz [1895], '*Abschnitt* V'. See also Miller [1973], esp. part 4.
[2] *P.R.C.*, pp. 5–6.

magnetic forces of which we are able at the present time to make this assertion definitely. If they are so transmitted, the translation will very probably
affect the action between two molecules or atoms in a manner resembling
the attraction or repulsion between charged particles.

This is the *M.F.H.*, which Lorentz proposed *after* putting forth the *L.F.C.*
That Lorentz put forth the *M.F.H.* to assist him in making plausible the *L.F.C.*
is clear from his next statement in the [1895]:[1]

(L_g): From the theoretical side, therefore, there would be no objection to
the hypothesis.

Then, at the *end* of this section, Lorentz presents yet another plausibility
argument—full of honestly stated and interconnected assumptions—in support
of contraction only of the dimensions along the direction of motion. This final
section, quoted in its entirety, is:[2]

(L_h): It is worth noticing that we are led to just the same changes of
dimensions as have been presumed above if we, *firstly*, without taking
molecular movement into consideration, assume that in a solid body left to
itself the forces, attractions or repulsions, acting upon any molecule maintain one another in equilibrium, and, *secondly*—though to be sure, there is
no reason for doing so—if we apply to these molecular forces the law which
in another place we deduced for electrostatic actions. For if we now understand by S_1 and S_2 not, as formerly, two systems of charged particles, but
two systems of molecules—the second at rest and the first moving with a
velocity v in the direction of the axis of x—between the dimensions of which
the relationship subsists as previously stated; and if we assume that in both
systems the x components of the forces are the same, while the y and z components differ from one another by the factor $\sqrt{1-v^2/c^2}$, then it is clear that
the forces in S_1 will be in equilibrium whenever they are so in S_2. If therefore S_2 is the state of equilibrium of a solid body at rest, then the molecules
in S_1 have precisely those positions in which they can persist under the influence of translation. The displacement would naturally bring about this
disposition of the molecules of its own accord, and thus effect a shortening
in the direction of motion in the proportion of 1 to $\sqrt{1-v^2/c^2}$, in accordance
with the formulæ given in the above-mentioned paragraph. This leads to
the values

$$\delta = -\tfrac{1}{2}\frac{v^2}{c^2}, \quad \epsilon = 0$$

in agreement with (1).

In reality the molecules of a body are not at rest, but in every 'state of
equilibrium' there is a stationary movement. What influence this circumstance may have in the phenomenon which we have been considering is a
question which we do not here touch upon; in any case the experiments of
Michelson and Morley, in consequence of unavoidable errors of observation,
afford considerable latitude for the values of δ and ϵ.

This argument is similar to the one given in his [1892b]. S_1 and S_2 are what I

---

[1] *Ibid.*, p. 6.   [2] *Ibid.* pp. 6–7.

refer to as S_r and S', respectively. Using *(18)–(20)* and *(22)–(24)*, Lorentz argues that it is not unreasonable to propose that $\epsilon = 0$.[1]

In the final paragraph Lorentz once again emphasises that $\epsilon = 0$ may not be the correct choice; however, it is the only one which he can argue for on the basis of his electromagnetic theory, even though the argument makes use of a fictitious coordinate system. Lorentz still considers the *L.F.C.* as a physics of desperation, put forth to preserve his electromagnetic theory.

With this development, we can now easily show that Zahar's arguments for the non *ad hocness* of the *L.F.C.* are specious.

In (Z_a) Zahar asserts that 'Lorentz deduced the *L.F.C.* from a deeper theory', namely from what he calls the *M.F.H.* However, as we have just seen, Lorentz did no such thing. Moreover, the *M.F.H.* does *not*, contrary to Zahar, constitute a 'theory about molecular forces'.[2] For example, Lorentz's suggestion that molecular forces transform like electromagnetic forces does not even give us information about the functional form of molecular forces, as a theory about molecular forces would require.[3] Nor does Lorentz furnish such a theory elsewhere. See also (L_c), wherein Lorentz asserts that the 'nature of molecular forces is entirely unknown'. Thus, an objective rendering of Lorentz's [1892*b*] and [1895] makes manifest the incorrectness of Zahar's statement (Z_b), thereby also severely weakening his (Z_a), to which I now turn.

In (Z_a) Zahar, in order to support his claims about the *M.F.H.*, quotes at length from Lorentz's [1895]; however, this passage is from the *final* portion of Lorentz's lengthy work [1895]—compare (Z_a) with (L_h). As I emphasised, Lorentz in (L_h) first gave a plausibility argument for $\epsilon = 0$, and then concluded on a note of uncertainty as to the veracity of the *L.F.C.* Thus, Zahar is quite incorrect when he states:[4]

> *Lorentz Derived the L.F.C. from the M.F.H. . . . . the Michelson–Morley Experiment Lends Dramatic Support to the M.F.H.*

Moreover, the 'famous transformation' that Zahar mentions in (Z_a) is not what can be considered, even with hindsight, as the 'Lorentz transformation' of 1895. As I mentioned previously, there were two sets of transformations in the [1895]: equations *(18)–(21)*; and equations *(25)–(28)*. The former set was used in the passage from the [1895] quoted in (Z_a). The latter set is what can, in retrospect, be considered by us as (*c.* 1895) the 'Lorentz transformations'. They are the set Lorentz used to prove the theorem of corresponding states. Zahar is obviously unaware of this fact. Indeed, Zahar, in his discussions of Lorentz's work of 1892 and 1895, in particular of the 1895 theorem of corresponding states,[5] gives the impression that Lorentz used the set of transformations from S_r to Q' from his [1892*a*] in the *Versuch* [1895]—this is incorrect. Thus, also incorrect is the import of Zahar's statement that:[6]

> *the First Version of Lorentz's Theory of Corresponding States Arises out of the Realistic Interpretation of the Lorentz Transformation whose Origins were Purely Mathematical.*

[1] For further discussion see Miller [1973].   [2] Zahar [1973], p. 111.
[3] Another example: To assert merely that the gravitational force in Newton's theory of gravitation transforms like a vector gives no information about its functional form.
[4] Zahar [1973], p. 114; italics in original.
[5] Zahar [1973], pp. 111 ff.   [6] *Ibid.*, p. 111; italics in original.

In summary: It is clear from Lorentz's papers [1892b] and [1895] that his methodology was to posit the L.F.C. specifically to account for the Michelson and Morley experiment. Moreover, he obtained the L.F.C. from an argument based upon the Newtonian addition law for velocities. Then, Lorentz posited the M.F.H. for the purpose of constructing a plausibility argument in support of the L.F.C. that involved the use of certain concepts from his electromagnetic theory—namely, the transformation equations from S_r to Q' in [1892b], and from S_r to S' in [1895].[1] Lorentz stated explicitly in [1892b] and [1895] that the L.F.C. might not be the proper manner of accounting for the result of Michelson and Morley. In addition, Lorentz argued (in [1892b], p. 223, and [1895], p. 6) that the L.F.C. can very probably not be directly observed. Thus, to save his electromagnetic theory, Lorentz put himself into the position of arguing on the basis of a set of transformation equations to a fictitious coordinate system and a string of interconnected assumptions for the plausibility of an effect that probably could not be directly observed. Thus, Lorentz saw fit to put forth this 'surprising' hypothesis in a very modest manner.

These are the hard historical facts upon which Zahar must try to superimpose the methodology of research programmes. It will not do to distort history at one's convenience and then to criticise the analyses of other historians and philosophers of science with the aid of suitably redefined terms such as 'novelty' and 'ad hoc'.

Indeed, Zahar conveniently forgets to mention that, until 1904, the majority of physicists did consider the L.F.C. as ad hoc; in particular, Poincaré, whose criticism was of great concern to Lorentz,[2] and Lorentz himself. Moreover, Poincaré's reasons for considering the L.F.C. as ad hoc as it appeared in the *Versuch* [1895] (but less so as it was presented in Lorentz's [1904]) were quite logical: In the *Versuch* the L.F.C. was presented to account for one experiment, and, furthermore, it did not fit into the body of the treatise, which discussed a theory to account systematically only for first order effects. On the other hand, in Lorentz's [1904], the L.F.C. in conjunction with other hypotheses put forth in this paper could account for more experiments than that of Michelson and Morley, e.g. those of Trouton and Noble, Rayleigh, Brace and Kaufmann. Moreover, in Lorentz's [1904] the L.F.C. fits into the body of the theory presented, which claimed to be able to account for effects to all orders in v/c. Zahar may have omitted this well-known episode in the history of science because it was not consonant with fitting Lorentz's work into the methodology of research programmes.

Although Poincaré's judgement of the L.F.C. as ad hoc is one based upon logic—or as Lakatos would say, in terms of Poincaré's logical decision—judgements of ad hocness do not always follow from purely logical arguments. Holton

---

[1] In fact, Lorentz, in his review paper [1903]—to which Zahar does not refer—clearly separates the L.F.C. from the M.F.H. Lorentz, in Section 62 ('The Interference Experiment of Michelson', pp. 273–4) gives an argument based upon the Newtonian addition rule for velocities, asserting that the Michelson and Morley experiment could be accounted for if the dimensions of a body in motion are changed according to my equation (29). *Then*, in Section 64 ('Present Status of the Theory') Lorentz asserts that this hypothesis 'becomes somewhat more intelligible if we put forth that [molecular forces behave like the Lorentz force, *i.e.* the M.F.H.]' (p. 277).

[2] See Lorentz [1904], pp. 12–13. See also Miller [1973], esp. part 5.2.

argues for this point in his [1969]. He asserts that there are also instances where the judgement of a hypothesis as *ad hoc* is not made by narrowly conceived ideas of 'logic' alone. The weight of Holton's supporting arguments for these claims seem to be of great concern to Zahar as a proponent of the methodology of research programmes. This is evidenced in Zahar's paper, wherein Zahar permits Grünbaum to easily dispose of Popper's case for the *ad hocness* of the *L.F.C.*[1] and spends most of his space in an effort attempting to refute the examples cited in Holton [1969]. Whether one agrees with Holton's conjectures or not, a close study of his papers (especially his [1969]) reveals that Holton's methodology is first to present as objectively as is possible the historical facts relevant to the case under study, taking particular care *not* to distort the internal structure of the relevant literature, and only then to analyse or speculate—the reverse of Zahar's methodology.

As to Zahar's criticisms of Holton's [1969] themselves, they can be more quickly disposed of. The sentences from Holton [1969] to which Zahar directs what he claims to be factual criticism are:[2]

(H_a): This saving *Hülfshypothese* [the *L.F.C.*] is introduced completely *ad hoc* . . . *No explicit comment is made which connects this assumed shrinkage with the Lorentz transformations in their still primitive form, as published earlier in the book* . . .

(H_b): The contraction hypothesis when it was made was clearly and quite blatantly *ad hoc*—or, if one prefers to use the *patois* of the laboratory, *ingeniously cooked up for the narrow purpose which it was to serve* . . .

(H_c): The important point to note is that '*ad hoc*' is not an absolute but a *relativistic term*. Postulate 1 and 2 [Einstein's two postulates in his [1905]] may be said to have been introduced *ad hoc* with respect to the Relativity Theory of 1905 as a whole . . . But these postulates were *not ad hoc with respect to the Michelson experiment*, for they were not specifically imagined in order to account for its results . . .

Previously, I discussed how Lorentz, in the [1895], put forth a plausibility argument for the *L.F.C.*, using the transformations from the real inertial system S_r to the fictitious system S'. I then pointed out Zahar's misconception of this point. I also noted that the transformation from S_r to S' cannot be considered as the 1895 version of what were subsequently called the Lorentz transformations. Thus, Holton's (H_a) is correct.

Holton's (H_b) is a restatement of previously reported opinion c. 1900 concerning the *L.F.C.*, *e.g.* that of Poincaré. It is a historical fact which cannot be made to vanish simply by redefining the notion of *ad hoc*.

Zahar's claim to refutation of (H_c) is in the following footnote:[3]

I cannot make head or tail of Holton's sentence 'Postulates 1 and 2 may be said to have been introduced *ad hoc* with respect to the Relativity Theory

---

[1] Zahar [1973], p. 104. More precisely, according to Zahar Grünbaum showed that the *L.F.C.* is not *ad hoc$_1$*, as was claimed by Popper. However, if one wishes to analyze the status of the *L.F.C. c.* 1900, a goal to which Zahar aspires, then Grünbaum's argument is quite beside the point. The reason is that Grünbaum's argument turns upon the Kennedy–Thorndike experiment of 1932.

[2] Zahar [1973], pp. 105–6; italics by Zahar. [3] Footnote 1 in *ibid.*, p. 106.

of 1905 as a whole'. How can Einstein's two postulates, which constitute Special Relativity Theory or at any rate are part of Relativity Theory, be *ad hoc* with respect to Relativity Theory?

Ad hoc with respect to *x* means not derivable from *x*. The (H_e) sentence means that Postulates I and II are not derived from relativity theory, but are postulated prior to (and so become a part of) relativity theory. Evidence for this assertion is as follows: (*i*) Autobiographical remarks by Einstein; (*ii*) raising a 'Vermutung' to a postulate; (*iii*) argument that goes as follows—a careful reading of the scientific literature relevant to electron physics in 1905–10 reveals that Einstein's theory was widely considered in the community of physicists to be a generalisation of Lorentz's theory of the electron presented in Lorentz's [1904]. Indeed, after 1905 the nomenclature Lorentz–Einstein theory came into being. The prediction of this theory that was considered of prime importance was for the so-called transverse mass

$$m_T = \frac{m}{\sqrt{1 - v^2/c^2}}.$$

(As is well known, Einstein, due to an unfortunate choice for the definition of force in his [1905], arrived at a different expression for m_T. This oversight was corrected by Planck in 1906. Thus, Einstein's and Lorentz's predictions for m_T are symbolically identical.) Holton's point in (H_e) is as follows: From the result that $m_T = m/\sqrt{1 - v^2/c^2}$ one is not necessarily led to Einstein's two postulates of relativity, since m_T could also be deduced from Lorentz's theory of the electron of 1904. This leads me to another of Zahar's incorrect statements:[1]

(Z_e): The philosophical significance of the Theory of Corresponding States is that it could, as Poincaré showed, easily be turned into a theory observationally equivalent to Special Relativity.

Zahar also makes the claim[2] (repeats an old mistake) that Einstein's and Lorentz's theories are observationally equivalent. Special relativity is not observationally equivalent to Lorentz's theory of 1904. The reason is that a theory of light containing Lorentz's ether cannot account *exactly* for the optical Doppler effect, nor for observations of stellar aberration.[3] Neither are the two theories observationally equivalent concerning the *L.F.C.* The reason is that Lorentz's contraction is basically different from that of Einstein. In fact, Lorentz, by applying the classical notion of congruence to the behaviour of rods as implied by his theory, convinced himself that the *L.F.C.* could not be observed.[4] Lorentz did not realise the importance of the comparison of lengths using light signals. Einstein and Poincaré simultaneously and independently discovered this, although their reasoning on this point differed. Even if Lorentz had realized this point before the others did, the two contractions would still not be observationally equivalent. The reason is that in Lorentz's theory the class of ether-fixed

[1] Zahar [1973], p. 116.
[2] *Ibid.*, p. 123.
[3] This point will be discussed fully in a forthcoming work by the author. For some brief remarks the reader should consult Møller [1952], esp. pp. 8–10; Drude [1959], esp. the first footnote on p. 475.
[4] Lorentz [1892b], p. 223 and [1895], p. 6 of *P.R.C.*

reference systems is not equivalent to the class of inertial systems, whereas in Einstein's theory of relativity of 1905 this asymmetry does not exist. This lack of symmetry in Lorentz's theory between different sets of space-time coordinates results in a lack of symmetry in the observation of the contraction. Furthermore, the inclusion by Lorentz of true lengths and times, quantities which remain unknown to the observer in an inertial system, renders unclear the meaning of the contraction of a rod at rest relative to an inertial system S_r^1 by an observer in an inertial system S_r^2.

I should now like to comment upon Zahar's concluding section of his Part I, entitled '*The Rationality of Lorentz's Pursuing his own Programme after 1905*'.[1]

Contrary to what Zahar asserts,[2] there was no 'Relativity Programme' in the scientific community from 1905 to about 1910, because Einstein's work had not yet been disentangled from Lorentz's—there existed for most scientists only one theory, the Lorentz–Einstein theory. Moreover, the Lorentz–Einstein theory was part of a research effort that can be called the perfection of the electromagnetic world-picture, formally announced in 1900 by Wilhelm Wien.[3] The intent of this research effort was to formulate an electromagnetic field-theoretical description of the electron and, subsequently, of all matter. It was a reductionistic programme that attempted to reduce the laws of mechanics to those of electromagnetism, *i.e.* to Lorentz's fundamental equations.[4] Thus, contrary to Zahar's characterisation of Lorentz's theory, by 1904 Newton's third law of motion was no longer considered to be an integral part of this theory. I believe that Zahar, by referring to Lorentz's theory as the '*Theory of Corresponding States*',[5] has missed this point. An immediate result of Zahar's oversight is that Lorentz's theory of the electron is made to appear to be more universal than it actually was.[6]

In summary: It was quite 'rational' for Lorentz to continue to adhere to his theory of the electron until 1910. However, Lorentz maintained this viewpoint to the end of his life (1928). Can this fact be explained entirely on 'rational' grounds? Or, does it really serve the philosophy of science so narrowly to construe the idea of 'rational' or 'logical' scientific thinking by actual creative research

---

[1] Zahar [1973], pp. 122–23; italics in original.   [2] *Ibid.*, p. 122.

[3] Wien [1900]. See Miller [1973], esp. part 2, for further details.

[4] See Miller [1973] for further discussion. One of the reasons for the emergence of the electromagnetic world-picture was a presumed incompatibility between the laws of mechanics and those of electromagnetism: The laws of electromagnetism, *i.e.* those of Lorentz's theory of [1892a] and [1895], transformed according to the modified Galilean set of transformations (25)–(28). On the other hand, the laws of mechanics transformed according to the usual Galilean transformations, *i.e.* (25)–(27) with $t_r = t$ replacing (28). The former theory with its set of transformation equations could account systematically for the results of optical and electrical experiments to order v/c. The latter theory could not. In view of the successes of Lorentz's theory *c.* 1900, *vis-à-vis* the lack of success of mechanical models for electromagnetism, many physicists believed that the best way to resolve the tension between the laws of electromagnetism and those of mechanics was to reduce mechanics to electromagnetism. Thus, in light of these facts, we must consider Zahar's statement that 'Michelson's experiment did not *refute* the conjunction of Newton's laws and Maxwell's equations' ([1973], p. 115; italics in original) to be true only up to 1900.

[5] Zahar [1973], p. 116; italics in original.

[6] Incidentally, there were at least two other major theories of the electron to compete with the Lorentz–Einstein theory: Langevin-Bucherer and that of Abraham. See Miller [1973] for further discussion.

scientists that any deviation from it which occurred historically must be falsified or explained away?

I would finally comment on certain of Zahar's assertions concerning the work of Poincaré:[1]

> But despite this, one hesitates to count Poincaré among classical physicists such as Maxwell and Lorentz. Poincaré was the first scientist to recognise the group character of the transformation equations and probably also the first clearly to enunciate a physical principle of relativity. (Einstein is supposed to have carefully read 'Science and Hypothesis' before 1905.) Whatever the case may be, Poincaré showed, using Lorentz's own approach, how the Theory of Corresponding States could be made observationally equivalent to Special Relativity.

Zahar's statement that 'Poincaré was the first scientist to recognise the group character of the' Lorentz transformations can be contested. The reason is that this statement is also put forth by Einstein in his [1905].[2] Thus, we have another case in the history of science of simultaneous but independent discovery.

Zahar's statement that Poincaré was the 'first clearly to enunciate a physical principle of relativity. (Einstein is supposed to have carefully read *Science and Hypothesis* before 1905)' implies that Poincaré discussed in his book *Science and Hypothesis*, published in 1902, the principle of relativity as posited again in 1904. This is not true. In *Science and Hypothesis* Poincaré discussed the 'principle of relative motion'[3] in *classical mechanics*. This principle is a restatement of the Newtonian principle of relativity (*i.e.* Newton's Corollary V). In the final chapter, 'Electrodynamics', Poincaré mentioned the term 'principle of relativity' (in common use *c.* 1902), but in reference to the laws of geometry.[4] Furthermore, it has been repeatedly shown in recent publications that Poincaré's principle of relativity was basically different from Einstein's. Poincaré's contained Lorentz's ether; Einstein's did not. Poincaré's was to be one of the principles of a reductionistic programme; Einstein's embraced all branches of physics, and represented one of the principles of a viewpoint that did not attempt to reduce one branch of physics to another.

Thus I fear that Zahar is on the wrong track—the reason is that there is no such thing as what he refers to as Lorentz's T_3: for Lorentz always considered the local time as merely a mathematical coordinate transformation.

ARTHUR I. MILLER

Lowell Technological Institute, Lowell, Massachusetts

REFERENCES

DRUDE, P. [1959]: *The Theory of Optics*, translated from the German edition of 1900.
EINSTEIN, A. [1905]: 'Zur Elektrodynamik bewegter Körper,' *Ann. d. Phys.* **17**, 891–921, in *The Principle of Relativity*, translated by W. Perrett and G. B. Jeffery. This reprint volume is referred to as *P.R.C.* All references are to *P.R.C.*
HOLTON, G. [1969]: 'Einstein, Michelson and the "Crucial" Experiment,' *Isis*, **60**, pp. 133–97.

[1] Zahar [1973], p. 120.   [2] Einstein [1905], p. 51.
[3] Poincaré [1902], p. 111. See Miller [1973], esp. part 5, for a discussion of Poincaré [1902].
[4] *Ibid.*, p. 244.

LAKATOS, I. [1970]: 'Falsification and the Methodology of Scientific Research Programmes', in I. Lakatos and A. E. Musgrave (*eds.*): *Criticism and the Growth of Knowledge*, pp. 91–195.

LORENTZ, H. A. [1886]: 'De l'influence du mouvement de la terre sur les phénomènes lumineux', *Versl. Kon. Akad. Wetensch. Amsterdam*, **2**, p. 297 (1886); in *Collected Papers*, **4**, p. 153.

Unless indicated otherwise, all references to Lorentz's work will be to the *Collected Papers*.

LORENTZ, H. A. [1892a]: 'La théorie électromagnétique de Maxwell et son application aux corps mouvants,' *Arch. néerl.* **25**, p. 363 (1892); in *Collected Papers*, **2**, p. 164.

LORENTZ, H. A. [1892b]: 'The relative motion of the earth and the ether', *Versl. Kon. Akad. Wetensch. Amsterdam*, **1**, p. 74 (1892), in *Collected Papers*, **4**, p. 219.

LORENTZ, H. A. [1895]: *Versuch einer Theorie der elektrischen und optischen Erscheinungen in betwegten Körpern* (1st *ed.* 1895; 2nd *ed.* 1906). All references will be to 1906.

LORENTZ, H. A. [1903]: 'Weiterbildung der Maxwellschen Theorie. Elektronentheorie', *Encykl. math. Wiss.*, **2**, Art. 14, pp. 145–280 (1904; received December 1903).

LORENTZ, H. A. [1904]: 'Electromagnetic phenomena in a system moving with any velocity smaller than that of light', *Proc. Roy. Acad. Amsterdam*, **6**, p. 809 (1904), in *P.R.C.*, pp. 11–34.

MILLER, A. I. [1973]: 'A Study of Henri Poincaré's "Sur la Dynamique de l'Electron" ', *Archive for History of Exact Sciences*, **10**, pp. 207–328.

MØLLER, C. [1952]: *The Theory of Relativity*.

POINCARÉ, H. [1902]: *Science and Hypothesis*.

SCHAFFNER, K. [1972]: *Nineteenth-Century Ether Theories*.

WIEN, W. [1900]: 'Über die Möglichkeit einer elektromagnetischen Begründung der Mechanik', *Recueil de travaux offerts par les auteurs à H. A. Lorentz*, pp. 96–107.

ZAHAR, E. [1973]: 'Why did Einstein's Programme Supersede Lorentz's? (I),' *British Journal for the Philosophy of Science*, **24**, pp. 95–123.

Postscript:
Further Comments on what Einstein, Lorentz, Planck, and Poincaré Did — circa 1905

Zahar's (1978) rejoinder to Essay 5 is based on historical and scientific points that are substantively incorrect. I address this Postscript to these points, which also affords me the opportunity to enlarge on certain aspects of the reception of special relativity that are at times misunderstood.

1. Zahar prefaces his reply to me by discussing (p. 50):[1]

> a difficulty which I did not properly tackle in my [1973], namely: given that [relativity theory] had more to offer than its rival [Lorentz's theory of the electron] and given that Lorentz was an eminently rational man, why was he not quickly converted to Relativity (he was of course finally converted around 1914)?

Contrary to Zahar, in 1905 and for some years after, Einstein's first paper on the electrodynamics of moving bodies was received by virtually every interested physicist as offering important generalizations of Lorentz's theory of the electron, and not as a program alternate to Lorentz's theory. The scientific literature circa 1905 abounds with evidence that supports this point. For example, there are the following quotations, and more are to be found in my booklength historical study of electromagnetic theory and relativity theory in the period (1890-1911) (Miller [1981]): Europe's premier experimentalist on the characteristics of high-velocity electrons, Walter Kaufmann of Göttingen, wrote ([1905], p. 954) in what was most likely the first citation in print of Einstein's (1905) of a "recent publication of Mr. A. Einstein on the theory of electrodynamics which leads to results that are formally identical to those of Lorentz's theory." Kaufmann went on to discuss the "Lorentz-Einstein" equation for the electron's transverse mass (m_T). To the Berlin Academy on 23 March 1906, Max Planck ([1906a], p. 136) said: "Recently H.A. Lorentz and in a still more general form A. Einstein introduce the principle of relativity," where Planck conflated the principles of relativity of Lorentz-Poincaré and of Einstein which, although worded

---

[1] A.H. Bucherer, Messungen an Becquerelstrahlen. Die experimentelle Bestätigung der Lorentz-Einstein'schen Theorie, *Phys. Zeitschr. 9* (1908), p. 755; *Ber. d. Deutschen Phys. Ges. 6* (1908), p. 688.

similarly, were based on two dissimilar conceptual frameworks.[2] In his first venture into the Lorentz-Einstein theory, Paul Ehrenfest ([1907], p. 204) wrote that "The Lorentz relativity-electrodynamics in the formulation which Mr. Einstein has published it, is commonly regarded as a closed system." And Alfred H. Bucherer (1908) entitled the paper in which he presented empirical proof for Lorentz's deformable electron over Abraham's rigid electron, "Measurements on Becquerel Rays: Experimental Confirmation of the Lorentz-Einstein Theory." In summary, the consensus of the physics community was that Einstein's first paper on electrodynamics was a solid contribution to Lorentz's theory of the electron. It was primarily Ehrenfest's foundational analyses of 1909–1910 (after Einstein [1907a] had set Ehrenfest straight) of a rigid body that by about 1911 demarcated between the views of Lorentz and Einstein (see Miller, 1981, esp. Chapter 7).

Moreover, contrary to Zahar, Lorentz was never "finally converted" to Einstein's notion of the relativity of space and time. Zahar is driven to make this claim because it is consistent with the MSRP which characterizes the history of science as a parade of competing research programs. Zahar's supporting evidence for Lorentz's supposed conversion is (pp. 52-53):

> Lorentz thought he had good empirical reasons for *not* accepting Maxwell's equations as fully covariant. In his (1909) he mentioned Kaufmann's results as constituting serious evidence against the Relativity Principle. In the same book he reproduced his 1904 results without taking account of the corrections which Poincaré published in his (1905). These corrections would have made Lorentz's electrodynamics Lorentz-covariant and hence indistinguishable from Einstein's. As soon as Lorentz found good reasons — such as the result of Bucherer's experiment — for accepting the covariance of Maxwell's equations, he realized that the ether had lost all heuristic value and consequently joined the Relativistic camp. This conversion to Einstein's approach is clearly expressed in his (1914) paper on gravitation and in the footnotes of the 1915 edition of *The Theory of Electrons*. Thus Lorentz's conversion took place before the great *empirical* successes of General Relativity, e.g., before the explanation of the precession of Mercury's perihelion; this clearly indicates that heuristic considerations played an important part in Lorentz's thinking at the time (italics in original).

There are several problems with this passage. One is that, contrary to Zahar, covariance had been the goal of Lorentz's research in electrodynamics since 1892. His approach to covariance was constructive, and not axiomatic as was Einstein's in 1905. During the period 1892 to 1904 Lorentz postulated a series of coordinate (spatial and temporal) and electromagnetic field transformations for the purpose of rendering the Maxwell-Lorentz equa-

---

[2]M. Planck, *Zur Dynamik bewegter Systeme, Sitzungsber.* Berlin, 1907, p. 542; *Ann. Phys.* *26* (1908), p. 1. H. Minkowski, Die Grundgleichungen für die elektromagnetischen Vorgänge in bewegten Körpern, *Gött. Nachr. Math.-phys. Kl.,* 1908, p. 53.

tions written in an inertial reference system into the form they would have in a reference system at rest in the ether. In response to empirical data Lorentz had to supplement the various postulated coordinate and field transformations with other postulates such as the contraction of moving bodies (LFC). Lorentz (1904) achieved covariance for the case of an inertial system S_r that was the electron's instantaneous rest system. Poincaré (1905, 1906) generalized this result to the case where S_r need not be the instantaneous rest system, which required Poincaré to correct certain technical errors in Lorentz's (1904). But Lorentz's restricted covariance sufficed for deducing the electron's mass, and this may have been Lorentz's reason for not citing Poincaré's corrections in the 1909 edition of *Theory of Electrons*. Nevertheless, Lorentz (1909) displayed the Lorentz transformations in the symmetrical form introduced by Poincaré (1905, 1906), and went on to discuss the Poincaré stress which was a key contribution of Poincaré to Lorentz's theory of the electron.

Moreover, Zahar overlooked that in the 1909 edition of *Theory of Electrons,* Lorentz discussed the approach to covariance in "Einstein's theory" (p. 230) — "Einstein simply postulates what we have deduced..." — and he concluded: "[Einstein's] results concerning electromagnetic and optical phenomena (leading to the same contradiction with Kaufmann's results that was pointed to in §179) agree in the main with those which we have obtained in the preceding pages...". Thus, contrary to Zahar, Lorentz's statements in his (1909) constitute proof that far from being "converted," he considered his own theory of the electron to be indistinguishable — in terms of its results — from what he referred to as "Einstein's theory." In order to buttress this point I note that in the 1909 edition of *Theory of Electrons,* Lorentz referred to "Einstein's theory" (pp. 223, 229, 230) that was based directly on the "principle of relativity." Lest we conclude too hastily from Lorentz's text that in 1909 Lorentz considered Einstein's theory to differ fundamentally from his own theory of the electron, I offer the following quotation (Lorentz [1909], p. 230), "Yet, I think, something may also be claimed in favour of the form in which I have presented the theory." Consequently, to Lorentz there was only one theory which Einstein had presented in a different "form."

Lorentz went on to emphasize the "fascinating boldness of its [i.e., Einstein's theory] starting point," referring to Einstein's having raised to an axiom the theorem of corresponding states, in contrast with Lorentz's constructive approach. For Lorentz (1909) had previously emphasized (p. 230) that Einstein "may certainly take credit for making us see in the negative result of experiments like those of Michelson, Rayleigh and Brace, not a fortuitous compensation of opposing effects, but the manifestation of a general and fundamental principle"; Kaufmann (1906) had described Einstein's approach in just these terms.

previously, of an ether supporting the electromagnetic and gravitational fields?" Lorentz went on to write that (p. 145), "It seems to me that if we find satisfaction in it, there is nothing to prevent us from considering the two fields with everything that characterizes them (electric and magnetic forces, characteristic magnitudes, momenta, energy, tensions, currents of energy) as arising from the modifications produced in the interior state of an ether."

Similarly in 1922 at the California Institute of Technology, after what Zahar notes as the "great *empirical* successes of General Relativity," Lorentz held to his notion of the complete equivalence of the special theory of relativity and the Lorentz theory of the electron. Lorentz wrote ([1922], p. 221): "my opinion of time is so definite that I clearly cannot distinguish in my picture what is simultaneous and what is not."

While Lorentz continued to support his own theory of the electron rather than convert to Einstein's special theory of relativity, he made basic contributions to the formalization of Einstein's general theory of relativity. This situation could be said to be quite "rational" because Lorentz considered the curved space-time of the general theory of relativity as playing the role of an ether that supports light in transit (see Miller [1981]). If one's interest is in the history of science broadly defined, then one can inquire into the origin of the predilection of many scientists, for example, Newton, Maxwell, Poincaré, Lorentz, and Einstein, into theories that are based on contiguous action rather than action-at-a-distance. Then one would move into the realm of epistemology broadly defined (see Miller [1984]). The MSRP excludes the possibility that Lorentz continued to prefer his own theory of the electron after 1911 because, as it so states, in the history of science there are research programs competing and eventually one of them wins and requires immediate and general consent. Consequently, for Zahar it would have been "irrational" for Lorentz to have continued to support his theory of the electron after 1911, and any of Lorentz's reasons for doing so were mere "psychological facts" that are "irrelevant to an objective assessment of Lorentz's" work. To make matters worse, Zahar claims that Lorentz abandoned his theory after Bucherer's 1908 experiment; however, as I documented earlier, Lorentz considered Bucherer's experiment to have vindicated the Lorentz theory of the electron.

I next turn to Zahar's version of the effect of the Kaufmann experiments on Lorentz, Planck, and Einstein.

2. Zahar writes ([1978], p. 51):

> We know that Kaufmann's results, which seemed to contradict Special Relativity, were defused by Planck. Why Einstein immediately accepted Planck's arguments while Lorentz hesitated may be due to external factors. Lorentz was older and probably more conservative in his outlook than Einstein; he may have felt that the ether

had rendered so many services in the past that it should not be too light-heartedly disposed of. This may have led Lorentz to take Kaufmann's results at face value; or he may simply have omitted to read Planck's account which was anyway involved and difficult to follow. In the methodology of scientific research programmes (henceforth referred to as MSRP) there is room for such external considerations.

Kaufmann ([1906], p. 495) wrote that his *"measurement results are not compatible with the Lorentz-Einstein fundamental assumption"* (italics in original), that is, the principle of relative motion (as Kaufmann referred to the principle of relativity). The quantity that Kaufmann was testing empirically was the electron's transverse mass (m_T), and he compared his data with the m_T's of Abraham, Bucherer, and Lorentz. Kaufmann continued by calling for additional experiments to decide between the electron theories of Abraham and Bucherer and to seek the (p. 535) *"absolute resting ether"* (italics in original). Thus, Kaufmann's data refuted what he referred to as the "Lorentz-Einstein" theory, and not a theory of relativity of which no one besides Einstein was yet aware, and which furthermore did not yet have even a name. The name "Relativtheorie" was coined in 1906 by Max Planck for the Lorentz-Einstein theory.

At the Meeting of German Scientists and Physicians in Stuttgart, September 1906, Planck (1906b) scrutinized Kaufmann's experiment and data analysis. He found nothing seriously amiss, but concluded that in his "opinion [Kaufmann's data are] not a definitive verification of [Abraham's theory of the rigid electron] and a refutation of [Lorentz's]" (p.757). In the discussion session that followed Planck's lecture, Planck emerged a poor second to Kaufmann and Abraham. Planck could say only that he remained "sympathetic" toward what he called the "Relativtheorie" in order to distinguish the Lorentz-Einstein theory from Abraham's theory, which he called the "Kugeltheorie." Einstein ([1907b], p. 439) wrote: "That the calculations of Mr. Kaufmann are free of error follows from the fact that by employing another method of calculation, Mr. Planck arrived at results which are in complete agreement with those of Mr. Kaufmann." Therefore, contrary to Zahar, Kaufmann's results were not "defused by Planck."[4]

It is rather far-fetched for Zahar to write that Planck's (1906b) analysis may have been too "involved and difficult [for Lorentz] to follow." Surely Lorentz could follow Planck's elementary solution of the Lagrange equations for an electron entering a region of external constant electric and magnetic fields.

I next turn to what Zahar concedes to be "external factors."

3. Essay 5 contains quotations from Lorentz's (1892b), (1895) and a letter to Lord Rayleigh which demonstrate that Lorentz was puzzled over the result of the Michelson-Morley experiment and dissatisfied with having to resort to a physics of desperation for explaining it (i.e., the LFC). Whereas Zahar

(p. 54) emphasizes that "because the historian does not know all the facts... reconstructions are therefore not a luxury but a sheer necessity," he deems my reconstruction from Lorentz's original papers and letters to be based on "psychological facts which may be interesting in themselves but are totally irrelevant to an objective assessment of Lorentz's achievement." The reason is that Zahar's notion of "objective assessment" is the "rational" (logical) approach dictated by the MSRP (p. 52): "An allowable reconstruction is said to be rational if it minimizes the recourse to external factors for which there exists no independent evidence" (p. 54). Yet we recall that (p. 52) Zahar had permitted as an "external" consideration that Lorentz may simply have "omitted to read Planck's (1906b)...." To the best of my knowledge there "exists no independent evidence" that Lorentz "omitted" to read Planck's widely cited (1906b). This is another example of Zahar's violating his own artificial and constrictive tenets.

I gave another example in Essay 5 where it turned out that Zahar's redefinition of *ad hoc* was itself *ad hoc*. Zahar (1978) attempts to counter by claiming that "Miller's account [of the LFC] is based on a confusion between descriptive propositions and normative criteria [i.e., Zahar's]." At this juncture, however, Zahar finds it necessary to redefine his (1973) version of *ad hocness* as well as sharpening his normative criteria. He acknowledges help from Adolf Grünbaum who, Zahar reminds us, concluded that prior to 1932 the LFC was only "psychologically ad hoc" because the 1932 Kennedy-Thorndike experiment provided an independent test of the LFC.[5] At any rate, the Zahar-Grünbaum excursion into logic is uninteresting for the serious historian because straightaway Zahar writes (p. 52): "It is certainly not my intention to make historical facts vanish. But let us be clear that the putative fact in question is that, *in Poincaré's opinion,* the L.F.C. is *ad hoc.* This in no way implies that the L.F.C. must be *ad hoc* according to every normative criterion of *ad hocness*" (italics in original). Zahar continues by cooking up a new normative criterion specifically for the purpose of rendering the LFC non-*ad hoc*. But at least Zahar learned one lesson from Essay 5 — namely, that one criterion for a hypothesis not to be ad hoc is that it covers more than one experimental or theoretical puzzle (see also Essay 1 for discussion of Poincaré's "complementary force" that he proposed for ensuring the validity of the principle of action and reaction). Thus in order to remove the aura of the *ad hoc* from the new normative criterion, Zahar states that it permits the MSRP to demonstrate that "Einstein's explanations of Michelson's results and of Mercury's perihelion, Copernicus's explanation of stations and retrogressions, and Bohr's explanation of the Balmer series are all non-*ad hoc*" (p. 53). This is neither the occasion nor the place to digress on the fundamental distinction between explaining and accounting — for example, in his 1905 paper, Einstein did not propose to explain the result of the Michelson-Morley experiment; rather, according to the two principles stated in his (1905), the negative result of every ether-drift experiment, extant and to be performed, was a foregone conclusion.

Since the remainder of Zahar's note focuses on the LFC, I next review its historical status. It is important for Zahar to prove that the LFC was not *ad hoc.* If the LFC were *ad hoc,* then according to the MSRP the transition from Lorentz's theory in 1895 (which Zahar erroneously refers to as Lorentz's theory T_2) to the further developed version of 1904 (which Zahar erroneously calls T_3) would not have been a progressive problem shift. As shown in Essay 5 there were no T_2 and T_3; rather, Lorentz continuously developed the electromagnetic theory from its original formulation in 1892 through 1904.

According to Zahar, the Molecular Forces Hypothesis (MFH) implies the LFC. Essay 5 deals at great length with dispatching this contention, and Zahar has offered no new historical evidence to cast doubt on it. Briefly, the historical scenario obtained from the primary sources is that in section 90 of his (1895), in which Lorentz proposed the LFC specifically to explain the result of a single experiment. Lorentz deduced the LFC by applying the Newtonian addition law for velocities to the velocity of light (Zahar [1978], p. 57, admits to this point), even though the electromagnetic theory sought to obviate this law. Attempting to link the LFC to the electromagnetic theory, Lorentz used coordinate transformations that had been proposed in Section 23 for use in only electrodynamical problems [see Zahar's Eqs. (18)–(25) in my part 5 below]; then Lorentz postulated the MFH to render the LFC plausible. Therefore, it is simply incorrect for Zahar to write (p. 57) that a "non-*ad hoc* [derivation of the LFC] is in section 23 and the *ad hoc* one in section 90." Zahar continues on p. 57: "Only section 90 was included in the 'Principle of Relativity' which was later translated into English. Given the popularity of this book, it is hardly surprising that a whole myth arose about the *ad hoc* character of the Lorentz-Fitzgerald Hypothesis."

If there was a "myth" at all, Zahar invented it in 1973 because Lorentz, Poincaré, Abraham, and Einstein, among others, considered the LFC to be *ad hoc* for precisely the reasons that I stated above. In Essay 5 I quoted at length from Lorentz's writings, and Poincaré's most searing criticism was that "hypotheses are what we lack least" (1900). That persistent critic of the Lorentz theory of the electron who was highly regarded by Lorentz and Einstein, Max Abraham, wrote (1904) that Lorentz posited the MFH in order to render "plausible" the LFC. Zahar's (p. 56) quotation in extenso from Lorentz's (1892b) contains exactly what Abraham rightly referred to as Lorentz's having used the MFH as a plausibility argument for the LFC.[6] Regarding the Michelson-Morley experiment Einstein ([1907b], p. 412) wrote: "It is known that any contradiction between theory and experiment

Lorentz's (1909) contains substantive evidence negating Zahar's claim that Lorentz joined the "Relativistic camp" "as the result of Bucherer's experiment." For if Zahar's claim were true, then why did Lorentz not state his conversion in the 1909 edition of *Theory of Electrons*? Instead, Lorentz writes as follows:

> [Note] 87 (Page 230). Recent experiments by Bucherer[1] on the electric and magnetic deflexion of β-rays, made after a method that permits a greater accuracy than could be reached by Kaufmann (§179) have confirmed the formula (313) for the transverse electromagnetic mass, so that, in all probability, the only objection that could be raised against the hypothesis of the deformable electron and the principle of relativity has now been removed. In the mean time, this principle has already been the subject of several important theoretical investigations.[2]

Thus, according to Lorentz and Bucherer, Bucherer's data confirmed the theory of Lorentz and Einstein. I have reproduced Lorentz's two footnotes to his Note 87 in order to demonstrate that what Lorentz referred to as the "principle of relativity" is a generalization of his own theorem of corresponding states. Planck's paper of 1908 discussed the "principle of relativity presented by H.A. Lorentz and in a more general form by A. Einstein" (p. 6). Minkowski's paper was his major effort toward an electromagnetic world-picture based on Lorentz's theory of the electron.

Contrary to Zahar, Lorentz's supposed conversion to the special relativity theory appears nowhere in footnotes that Lorentz added in 1915 to *Theory of Electrons*. For example, Lorentz wrote in Note 72*, which he added in 1915 (p. 321): "If I had to write the last chapter now, I should certainly have given a more prominent place to Einstein's theory of relativity." Thus, in 1915 Lorentz wrote that in retrospect he should have discussed Einstein's "theory of relativity" in detail; I can find no passage where Lorentz wrote of his own "conversion to Einstein's approach." Furthermore, it was only in the 1916 edition of *Theory of Electrons* that Lorentz referred to Einstein's "theory of relativity."

For the purpose of forcing a square block into a round hole, the MSRP shaves much interesting history from developments in physics during the period 1905–1911. For example, why in the 1916 edition of *Theory of Electrons* did Lorentz refer to "Einstein's theory of relativity," whereas in the 1909 edition Lorentz had cited "Einstein's theory" and "Einstein's views." In 1915 a principal reason why Lorentz discussed two theories was due to Paul Ehrenfest's fundamental analysis of the notion of a rigid body that served to separate the Lorentz and Einstein views. Thus, in Note 72* of the 1916 edition Lorentz discussed at some length "Einstein's theory of relativity (§189) by which the theory of electromagnetic phenomena in moving systems gains a simplicity that I had not been able to attain." Lorentz's emphasis on "simplicity" is an indicator of his opinion that the two theories were equivalent and that Einstein's described electromagnetic phenomena in a simpler manner. In lectures given at Leiden University during 1910-1912, Lorentz expressed his opinion of the equivalence of the two theories more directly (Lorentz [1910-1912], p. 210):

> We thus have the choice between two different plans: we can adhere to the concept of an aether or else we can assume a true simultaneity. If one keeps strictly to the relativistic view that all systems are equivalent, one must give up the substantiality of the aether as well as the concept of a true time. The choice of the standpoint depends thus on very fundamental considerations, especially about the time.

> Of course, the description of natural phenomena and the testing of what the theory of relativity has to say about them can be carried out independently of what one thinks of the aether and the time. From a physical point of view these questions can be left on one side, and especially the question of the true time can be handed over to the theory of knowledge.

Thus whereas in these lectures Lorentz demonstrated a sure grasp of the subtleties concerning the relativistic notion of time,[3] he continued to maintain that special relativity and his own theory of the electron were equivalent both mathematically and physically. From Lorentz's point of view the relativity of time was not Einstein's most important result; rather, according to Lorentz, an important advantage of Einstein's postulating the equivalence between a set of space and time coordinates in the inertial reference system S_r and the ether-fixed system S was that Einstein then could deduce the correct relationship between the charge density and convection current in S_r and S (see Lorentz [1916], Note 72*; [1910-1912], pp. 212-231).

Without any supporting quotations, Zahar, in the passage under analysis, cites Lorentz's (1914) as the place where his "conversion to Einstein's approach is clearly expressed." Yet in Lorentz's (1914) we find that the term "ether" occurs throughout his derivation of the relativistic space and time transformations. To begin with, when Lorentz wrote (p. 120) that "without doubt my theory ought to be abandoned; it is much more fruitful to try a *'théorie relativiste'*," he meant abandoning his 1900 attempt at a *"théorie électromagnétique de la gravitation"* (p. 118), and not to abandon his theory of the electron. Then, Lorentz continued (p. 120), "Before speaking of it [i.e., *"la théorie relativiste"*] I shall recall briefly the origin and significance of the principle of relativity itself." Lorentz prefaced his discussion of Einstein's definition of clock synchronization by two observers A and B who are equipped with clocks, with sentences such as: "Let us imagine that A and his laboratory are at rest relative to the ether and that B with all his apparatus moves relative to this milieu with a velocity v that is constant in direction and magnitude." After reviewing Einstein's progress toward a generalized theory of relativity, Lorentz asked (p. 143): "Can we maintain the principle of relativity and can we speak, as I have done many times

was removed formally by the assumption of H.A. Lorentz and Fitzgerald according to which moving bodies experience a definite contraction in the direction of their motion. However, this assumption, introduced *ad hoc,* appeared to be an artificial means to rescue the theory [i.e., Lorentz's theory as it stood in 1895]."

So much for myths after 1913 (when the first German edition of *Principle of Relativity* was published by Teubner), unless one adheres to the MSRP, which discusses history as it ought to have occurred.

4. The syllogism that Zahar (p. 53) uses to reveal the "paradoxical consequence" resulting from my "attempt to rescue Holton's notion of *ad hocness*" is perfectly valid — as concerns logic, that is. But Zahar's elementary exercise in logic (with which he was assisted by Grünbaum) has little to do with the way that theoretical physics is done. Indeed there is nothing at all new in Holton's notion of *ad hocness,* for he was restating the opinion held by Abraham, Einstein, Lorentz, Poincaré, and most other physicists, then and now.

5. Zahar writes (p. 58): "It is my conjecture that Miller *invents (21) in order to try to disconnect the discovery of the L.F.C. from that of the Lorentz transformations*" (italics in original). In order for me to reply to Zahar's "conjecture" I reproduce the following equations from his [1978]:

"(18) $$x' = \gamma x_r \ [= \gamma^{[x-vt]}]$$

(19) $$y' = y_r \ [= y]$$

(20) $$z' = z_r \ [= z]$$

(21) $$t' = t_r \ [= t]$$

where $\gamma = (1 - v^2/c^2)^{-1/2}$. And:

(25) $$x_r = x - vt$$

(26) $$y_r = y$$

(27) $$z_r = z$$

(28) $$t_r = t - \frac{v}{c^2} x$$

(18') $$x' = \gamma x_r = \gamma(x - vt)$$

(19') $$y' = y_r = y$$

(20') $$z' = z_r = z$$

(21') $$t' = t_r - \gamma^2 \frac{v}{c^2} x_r = \gamma^2 \left(t - \frac{v}{c^2} x \right) \ ."$$

Lorentz (1895) employed Eqs. (18)–(21) for discussing electrodynamical phenomena in an inertial reference system S_r [with coordinates (x_r, y_r, z_r, t_r)], as if they were electrostatic phenomena that occurred in the reference system that has no physical significance S' [with coordinates $(x', y'z', t')$], Since Lorentz took S_r to be the charged particle's instantaneous rest system then, as Zahar writes (p. 58) *"the field quantities will be independent of t,"* (italics in original). But contrary to Zahar there is a time coordinate in S' which Lorentz did not have to state explicitly, owing to its superfluousness to the problem under consideration. Inspection of Eqs. (18)–(20) reveals that the time coordinate in S' is the same as the one in S_r, that is, the absolute time t_r, and hence it is incorrect physics for Zahar to claim that "Miller invents (21)...."

Incidentally it is also incorrect physics, and an historical error as well, for Zahar to include in Eqs. (18)–(21) the bracketed terms which relate S' to the ether-fixed system S with coordinates (x, y, z, t). Lorentz's goal at this point in the (1895) was to transform from S_r to a reference system that had all the mathematical properties of one fixed in the ether. The transformation from S_r to S' accomplished this purpose because it enabled Lorentz to rewrite the inhomogeneous wave equation for the scalar potential as a Poisson equation (Essay 2). Lorentz was not discussing a transformation between S and S'.

Zahar writes (p. 58):

Equations (25)–(28), which differ from (18′)–(21′) by second order quantities, do not constitute, as Miller claims, a predecessor in 1895 of the Lorentz transformation. A much more general Lorentz transformation, namely equations (18′)–(21′), had already been discovered *in 1892!* (italics in original).

Let Q' be the primed coordinate system in Eqs. (18′)–(21′), in which the velocity of light is $c\sqrt{1 - (v/c)^2}$ (see Essay 5). Lorentz (1892a) specialized to terms of first-order accuracy in (v/c) in order to prove that to order (v/c), electromagnetic processes occurred on the moving earth as if the earth were at rest in the ether, in agreement with first-order ether-drift experiments (the *"théorème général"*).[7] In (1892a) Lorentz's proof of the *théorème général* focused on demonstrating the covariance of cumbersome expressions for the field generated by oscillator ions in terms of phenomenological equations for the ions' force constants. In (1895) Lorentz bifurcated an approximate version of the transformation Eqs. (18′)–(21′) into Eqs. (18)–(21) and (25)–(28). This split persisted until 1904 when the two sets of transformations coalesced into the familiar Lorentz transformations (with some refinements added by Poincaré in 1905). Owing to its model independence Lorentz's (1895) theorem of corresponding states was a sharper version of the 1892 *théorème général*. Based on Eqs. (25)–(28), and appropriate transformation equations for electromagnetic field quantities, Lorentz's theorem of corresponding states asserted that to order (v/c) the Maxwell-Lorentz equations in S_r that described electromagnetic phenomena generated by matter in bulk that was nonmagnetic, nondielectric, and nonconducting, were the same as in the ether-fixed system S. The theorem of corresponding states referred systematically only to phenomena that concerned the optics of moving bodies. Lorentz applied the Eqs. (18)–(21) only to electrodynamical problems. The fact that Lorentz used two different sets of space and time transformations for two sets of phenomena which the electromagnetic theory purported to have unified (electromagnetism and optics) is indicative of the incompleteness of Lorentz's theory in 1895. Conversely, Lorentz (1892a) found that Eqs. (18')–(21') were inadequate for discussing electrodynamical or optical phenomena since they disagreed even with first-order ether-drift experiments. Consequently, I am at a loss to understand how Zahar can claim that Eqs. (18')–(21') were a "much more general Lorentz transformation" than Eqs. (25)–(28). Since the (1895) theorem of corresponding states was more widely applicable than the process for reducing electrodynamical to electrostatic problems, I wrote in Essay 5 that "in retrospect [Eqs. (25)–(28)] can be considered by us as (c. 1895) the 'Lorentz transformations'."

In conclusion, we have seen that the MSRP fails the test of historicity that was proposed by Lakatos (1970) himself: *"One tries to compare this rational reconstruction with actual history and to criticize one's rational reconstruction for lack of historicity"* (italics in original). No useful purpose is served by attempting to relegate the history of science to a branch of applied philosophy.

Notes

1. By an "eminently rational man" Zahar means that unknowingly Lorentz thought according to the guidelines that Imre Lakatos claimed to have existed a priori in the mind of every serious scientist, and which he alone was able to divine and offer to us as the MSRP. Thus, Zahar's terminology "rational" and "objective" must be understood within the context of the MSRP.
2. For example, a typical description of the content of Einstein's (1905) was that using the axiomatic method, with certain phenomenological assumptions on how clocks are synchronized using light signals, Einstein could deduce the Lorentz transformations, the Lorentz contraction, and without any assumptions on the motion of the electron, the electron's mass as well (for example, see Kaufmann (1906), Bucherer (1908), and descriptions of Einstein's (1905) in abstracting journals (Miller [1981])).
3. For example, Lorentz pointed out the paradoxical results that "when viewed superficially" the reciprocity of time could lead (see Miller, 1981, Chapter 7 for details).
4. The low-velocity electron deflection experiments of Adolf Bucherer

began to "defuse" Kaufmann's results. I have analyzed the complex state of affairs surrounding the demise of Kaufmann's results in my (1981), Chapter 12.

5. This is what Grünbaum (1977) refers to as the "timelessness of logical entailment" (in reply, see Miller [1977]). In fact, Grünbaum's application of logic to analysis in the history of science is detrimental to Zahar's emphasis on Bucherer's (1908) experiment. What are we to make of the fact that in 1938 Zahn and Spees (1938) found that Bucherer's experiment was flawed and could not distinguish between the transverse masses proposed by Abraham and Lorentz?

6. Moreover, Zahar is wrong to claim about Lorentz in 1892 (p. 57): "Every ether theorist like Lorentz, who was anyway inclined to reduce everything to electromagnetism, would naturally adopt the opposite assumption, namely that *all* forces transform alike (italics in original)." Wilhelm Wien did not suggest research toward an electromagnetic world-picture until 1900, and Lorentz was at first somewhat cautious about participating.

7. Lorentz's (1892a) did not discuss the Michelson-Morley experiment.

Bibliography

Abraham, M. (1903): "Prinzipien der Dynamik des Elektrons," *Annalen der Physik, 10,* pp. 105-179.

Abraham, M. (1904): "Die Grundhypothesen der Elektronentheorie," *Physikalische Zeitschrift, 5,* pp. 576-579.

Bucherer, A.H. (1908): "Messungen an Becquerelstrahlen. Die experimentelle Bestätigung der Lorentz-Einsteinschen Theorie," *Physikalische Zeitschrift, 9,* pp. 755-762.

Einstein, A. (1905): "Zur Elektrodynamik bewegter Körper," *Annalen der Physik, 17,* pp. 891-921. A new translation is in Miller (1981).

Einstein, A. (1907a): "Bemerkung zur Notiz des Herrn P. Ehrenfest: Translation deformierbarer Elektronen und der Flächensatz," *Annalen der Physik, 23,* pp. 206-208.

Einstein, A. (1907b): "Relativitätsprinzip und die aus demselben gezogenen Folgerungen," *Jahrbuch der Radioaktivität, 4,* pp. 411-462.

Ehrenfest, P. (1907): "Die Translation deformierbarer Elektronen und der Flächensatz," *Annalen der Physik, 23,* pp. 204-205.

Grünbaum, A. (1977): "Remarks on Miller's Review of Philosophical Problems of Space and Time," *ISIS, 68,* pp. 447-448.

Kaufmann, W. (1905): "Über die Konstitution des Elektrons," *Berlin Berichte, 45,* pp. 949-956.

Kaufmann, W. (1906): "Über die Konstitution des Elektrons," *Annalen der Physik, 19,* pp. 487-553.

Lakatos, I. (1970): "Falsification and the Methodology of Scientific Research

Programmes," in I. Lakatos and A. Musgrave (eds.): *Criticism and the Growth of Knowledge,* pp. 91-195.

Lorentz, H.A. (1892a): "La théorie électromagnétique de Maxwell et son application aux corps mouvants," *Arch. néerl. 25,* p. 363 (1892); in *Collected Papers, 2,* p. 164.

Lorentz, H.A. (1892b): "The relative motion of the earth and the ether," *Versl. Kon. Akad. Wetensch. Amsterdam, 1,* p. 74 (1892); in *Collected Papers, 4,* p. 219.

Lorentz, H.A. (1895): *Versuch einer Theorie der elektrischen und optischen Erscheinungen in betwegten Körpern* (1st. ed. 1895; 2nd ed. 1906).

Lorentz, H.A. (1904): "Electromagnetic phenomena in a system moving with any velocity smaller than that of light," *Proc. Roy. Acad. Amsterdam, 6,* p. 809 (1904), pp. 11-34, reprinted in part in *Principle of Relativity.*

Lorentz, H.A. (1909): *Theory of Electrons.*

Lorentz, H.A. (1910-1912): *Lectures on Theoretical Physics, vol. III,* translated by L. Silberstein and A.P.H. Trivelli.

Lorentz, H.A. (1914): "La Gravitation," *Scientia, 14;* in *Collected Papers, 7,* pp. 116-146.

Lorentz, H.A. (1916): *Theory of Electrons* (second edition).

Miller, A.I. (1974): "On Lorentz's Methodology," *The British Journal for the Philosophy of Science, 25,* pp. 29-45.

Miller, A.I. (1977): "Reply" [to Grünbaum (1977)], *ISIS, 68,* pp. 449-450.

Miller, A.I. (1981): *Albert Einstein's Special Theory of Relativity: Emergence (1905) and Early Interpretation (1905-1911).* (Reading, MA: Addison-Wesley, Advanced Book Program, 1981).

Miller, A.I. (1984): *Imagery in Scientific Thought: Creating 20th-Century Physics,* (Cambridge, MA: Birkhäuser Boston, Inc., 1984; Cambridge, MA: MIT Press, 1986).

Planck, M. (1906a): "Das Prinzip der Relativität und die Grundgleichungen der Mechanik," *Verh. D. Phys. Ges., 2,* pp. 202-204.

Planck, M. (1906b): "Die Kaufmannschen Messungen der Ablenkbarkeit der β-Strahlen in ihrer Bedeutung für die Dynamik des Elektronen," *Physikalische Zeitschrift, 7,* pp. 753-761.

Poincaré, H. (1900): "Sur les rapports de la Physique expérimentale de la Physique mathématique," a version is in H. Poincaré, *Science and Hypothesis.*

Poincaré, H. (1904): "L'état actuel et l'avenir de la physique mathématique," *Bulletin des Sciences Mathématics, 28,* pp. 302-324.

Poincaré, H. (1905): "Sur la Dynamique de l'Electron," *Comptes Rendus de l'Académie des Sciences, 140,* pp. 1504-1508.

Poincaré, H. (1906): "Sur la Dynamique de l'Electron," *Rendiconti del Circolo Mathematico di Palermo, 21,* pp. 129-176.

Wien, W. (1900): "Über die Möglichkeit einer elektromagnetischen Begründung der Mechanik," *Recueil de travaux offerts par les auteurs à H.A. Lorentz,* pp. 96-107.

Zahar, E. (1973): "Why did Einstein's Programme Supersede Lorentz's? (I)," *British Journal for the Philosophy of Science, 24,* pp. 95-123.

Zahar, E. (1978): "Einstein's Debt to Lorentz: A Reply to Feyerabend and Miller," *The British Journal for the Philosophy of Science, 29,* pp. 49-60.

Zahn, A.A. and A.H. Spees (1938): "A critical analysis of the classical experiments on the variation of electron mass," *Physical Review, 53,* pp. 511-521.

Essay 6

Introduction

P. W. Bridgman and the Special Theory of Relativity

Percy W. Bridgman wrote *A Sophisticate's Primer* (1962) for the reader "who feels the need to stand back a little for a critical scrutiny of what he has really got" (*SP*, p. 3).* This was his personal trademark as physicist and philosopher and it is present everywhere in Bridgman's précis of special relativity theory.

The reissue of *Sophisticate's Primer* is particularly welcome for it exhibits a quest for clarity in foundations in the style of Ernst Mach and Henri Poincaré, in whose philo-

---

* N.B. Papers and books are here cited as follows. Bridgman (1927) refers to Bridgman's book *The Logic of Modern Physics* in the bibliography. Cross-references to *Sophisticate's Primer* are indicated with the code (*SP*, p. ···). Bridgman's extant manuscripts are on deposit at the Harvard University Archives, and here cited by date, for example (MS 2 August 1959, p. ···). Some of the manuscripts were written over a period of days, with sequential paging, with the actual dates indicated. Full citations are given in the bibliography. I am grateful to the Harvard University Archives for permission to quote from these materials, and to Bridgman's daughter, Mrs. Jane Koopman, for permitting me to quote from the draft manuscripts of *Sophisticate's Primer*, which will be deposited in the Harvard Archives.

Reprinted from P.W. Bridgman, *Sophisticate's Primer of Relativity,* Second Edition. Middletown, CT: Wesleyan University Press, 1983.

sophical lineage Bridgman had placed himself in his first book on the philosophy of science, *The Logic of Modern Physics* (1927). Bridgman's profound and far-reaching analysis of space, time, and causality in *Sophisticate's Primer* is presented on a level that is basic enough to be appreciated by a wide spectrum of readers. It is a fine example of an analysis of foundations tempered by a philosophical view that continues to be relevant—the operational point of view. As the philosopher of science Adolf Grünbaum wrote twenty years ago in the Prologue to the first edition of *Sophisticate's Primer*, Bridgman's reflections on fundamental questions in this book "merit the serious attention of a wide public as the mature products of a long-standing intellectual concern."

Of Bridgman's many incisive analyses in *Sophisticate's Primer* I shall focus on the ones concerning the properties of light, clock synchronization, and the concept of the observer. Comparison of Bridgman's unpublished manuscripts on relativity from the period 1922–60, on deposit at the Harvard University Archives, with his published philosophical writings between 1924 and 1959 enables us to study his development of these notions into the form they took in *Sophisticate's Primer*.

As a mature philosopher-scientist, Percy Williams Bridgman (1954) recalled his reaction in 1914 to the state of electrodynamics and special relativity theory: "The underlying conceptual situation in the whole area seemed very obscure to me and caused me much intellectual distress, which I tried to alleviate as best as I could." Bridgman's urge to clean up these subjects to his own satisfaction later resulted in his proposing the operational view in *Logic of Modern Physics*. At the time of Bridgman's philosophical awakening in 1914, the 32-year-old Harvard

Assistant Professor of Physics already had a solid background in philosophy which he had begun to study seriously while a High School Senior in Newton, Massachusetts.**

At Harvard University, where he was awarded the B.A. in 1904 and the Ph.D. in 1908, he took courses with such Cambridge luminaries as Josiah Royce and George Santayana, in which he was almost certainly introduced to the pragmatism of William James. *The Logic of Modern Physics* was published in 1927 when Bridgman, then Hollis professor of mathematics and natural philosophy, was deeply involved in the research in high-pressure physics for which he was eventually awarded the Nobel Prize.

While the seeds of Bridgman's operational view can be traced to his "distress" of 1914 and then to his work of 1916–22 toward clarifying dimensional analysis, it was his deliberations on Albert Einstein's special theory of relativity that really served to crystallize the operational view for him. This work Bridgman began in earnest in 1922.

Early Assessment: 1922–27

The Relativity Theory of Einstein is the result of, and is resulting in, an increased criticalness with regard to the fundamental concepts of physics. . . . The general goal of criticism should be to make impossible a repetition of the thing that Einstein has done; never again should a discovery of new experimental facts lead to a revision of physical concepts simply because the old concepts had been too naive. Our concepts and general scheme of interpretation should be so broad and so well considered that any new experimental

---

** For biographical material on Bridgman (1882–1961), see Kemble (1970).

facts, not inconsistent with previous knowledge, may at once find a place waiting for them in our scheme. A program of consideration as broad as this demands a critical examination not only of the concepts of space and time, but of all other physical concepts in our armory. I intend in the following to wander over this whole broad field of criticism . . . no concept is to be admitted which does not bring with it its complex of operations; in fact, unless there is the complex of operations the concept has no meaning.

P. W. Bridgman (MS September 1923)

As Bridgman wrote in his notes of September 1923, had the physicists of 1905 been more critical of fundamental concepts, Einstein's work would have been unnecessary.[1]

---

1. For an analysis of the state of physical theory from 1890 to 1911 as it affected the emergence and early reception of the special relativity theory, I refer the interested reader to Miller (1981), which contains extensive references to the secondary literature and a new translation of Einstein's 1905 special relativity paper entitled "On the Electrodynamics of Moving Bodies."

Some of the aspects of the 1905 relativity paper that attracted Bridgman follow. Einstein based his view of physics on two axioms. The first axiom is a version of the principle of relativity from mechanics, widened to include electromagnetic theory:

> The laws by which the states of physical systems undergo changes are independent of whether these changes of state are referred to one or the other of two coordinate systems moving relatively to each other in uniform translational motion.

Then, instead of attempting to explain the failure of ether-drift experiments, Einstein proposed a second axiom:

> Any ray of light moves in the 'resting' system with the definite velocity c, which is independent of whether the ray was emitted by a resting or by a moving body.

In Section 1 of the relativity paper Einstein developed the theory's

Bridgman intended this sharply-worded commentary—
which objects to such analyses of time as Poincaré's
(1898)—to serve as an introduction to the operational view
he was developing in his 1927 book *The Logic of Modern
Physics*. There, after repeating essentially the same message,
and after paying due respects to Einstein's critical analysis[2]

---

key notion—the definition of simultaneity in a single inertial reference
system. But first he defined an inertial reference system ("one in which
Newton's equations hold"), and then he proposed an in-principle measurement operation for position relative to a reference system by means
of rigid measuring rods. Einstein next stressed the importance of a clear
understanding of time because, after all, the coordinates of a moving
particle are functions of time. These steps enabled him to formulate a
version of physical theory that did not contain such apparently unmeasurable quantities as the earth's velocity relative to the ether, which had
been eliminated in ether-based theories of electromagnetism by postulation of compensating effects like the Lorentz contraction.

Einstein went on to define operationally the concept of an event as
an occurrence at a particular position, measured relative to an inertial
reference system and registered by a clock at that position.

2. Bridgman (1927) elaborated further on Einstein's critical analysis of space and time: "It is precisely here, in an improved understanding of our mental relations to nature, that the permanent contribution of
relativity is to be found. We should now make it our business to understand so thoroughly the character of our permanent mental relations to
nature that another change in attitude, such as that due to Einstein, shall
be forever impossible." Bridgman (in 1959a) noted how his "attitude had
changed most drastically" on this point: "To me now it seems incom-
prehensible that I should ever have thought it within my powers, or
within the powers of the human race for that matter, to analyze so thoroughly the functioning of our thinking apparatus that I could confidently
expect to exhaust the subject and eliminate the possibility of a bright new
idea against which I would be defenseless." Although in *Sophisticate's
Primer* (*SP*, p. 4) Bridgman repeated in part his early assessment of
Einstein's work on special relativity theory, his later statements (*SP*, pp.

at appropriate junctures, Bridgman wrote: "In general, we mean by any concept nothing more than a set of operations; *the concept is synonymous with the corresponding set of operations*" (italics in original). Although the notion of defining a physical quantity operationally had been discussed even before Einstein, by Mach (1960) and Poincaré (1902), for example, it was clearly Einstein's use of this notion in the relativity paper that most impressed Bridgman. In fact, in *The Logic of Modern Physics* Bridgman entitled the section in which he introduced the operational point of view, "Einstein's Contribution in Changing Our Attitude toward Concepts."

At first, however, Bridgman was equivocal on the validity of the special relativity theory. This attitude is evident throughout *The Logic of Modern Physics*. He expressed it forcefully on 27 December 1922 in a lecture to the AAAS in Boston, in which he emphasized as well the theory's lasting philosophical importance (MS 27 December 1922):

> The physicist may therefore well doubt whether the theory of Einstein in its present form will ultimately survive, but entirely apart from the ultimate truth of the theory, there can be no question that matters can never return to their condition before the formulation of the theory, and besides addition to our knowledge of facts, Einstein has made changes in our points of view which must have their permanent effect. It is of these changes of view with respect to space and time that I wish to speak. Any one familiar with speculation before and after the theory cannot fail to remark the increase in self consciousness and self criticism of our

----

157, 160) have the effect of further removing the aura of logical positivism from the operational view.

attitude toward measurements in general. Now what appears to be E's most important service, and the one whose effect is most likely to survive, is one which in its statement appears almost trivial, but which nevertheless had far-reaching effects. This is merely the insistence on the requirement that the quantities which we use in our equations (or concepts in general) have a meaning.

One point of concern to Bridgman was that there were no exact inertial systems in nature[3] and that, consequently, the "special theory of relativity may very probably be only a close approximation" (Bridgman, 1924). In addition, as he wrote in *Logic of Modern Physics*, there was the possibility that "new kinds of experience" would demonstrate the lack of exact validity of special relativity.

Besides the "rapidly increasing array of cold experimental facts," several matters of principle concerned Bridgman. First, although light was central to special relativity theory, the theory did not get to the heart of the problem of the nature and constitution of light; and this problem was exacerbated at the time by complexities due to developments in the quantum theory of radiation. Within the operational framework Bridgman (1927) viewed the problem of the nature of light as follows. Can we really consider light to be a "thing that travels, thing not necessarily connoting material thing"? Bridgman could attribute no operational definition to the notion that light exists at every position intermediate between source and sink because, from the "view of operations, light means nothing more

---

3. Einstein (1923) considered the concept of the inertial reference system to be a "logical weakness" of the special relativity theory. On the other hand, the 1915 general relativity theory included observers in accelerated reference systems.

than *things lighted* [and] light as a thing traveling must be recognized to be pure invention" based on a combination of sense perceptions and "ordinary mechanical experience." One could suppose, Bridgman continued, that the propagation of a light wave could be detected by a suitably placed series of screens, with each screen so constructed that it destroyed only an infinitesimal part of the beam. This arrangement "evaporates," however, because contemporaneous quantum theory conceives light not as an irreducible quantity, but as an "exceedingly complicated thing." For example, a much-debated conundrum of the 1920s was how light quanta could explain interference.*

Nor does it help matters to consider light in transit as a transfer of energy. Conservation of energy entails time, because one must "integrate over all space the local energy at a definite instant of time." But in order to spread the notion of time over space, special relativity assumes light to be a thing traveling; consequently, a vicious logical circle ensues to which Bridgman offers no resolution.

Since Einstein had, for "convenience and simplicity" in dealing with optical problems, chosen to consider light to be a thing traveling, he could use the customary concept of velocity, in which an observer at rest in a reference system determines the distance covered by a material thing in a measured time interval on distance and time scales at rest along the system's coordinate axes. But in special relativity theory the notion of light propagating led to situations that were not "at all satisfying logically": light is neither a wave

---

* Miller, (1978) Visualization Lost and Regained: The Genesis of the Quantum Theory in the Period 1913–1927, and (1982) Redefining *Anschaulichkeit*.

disturbance in a medium, for the medium is not observed to have any effect on the light, nor is light a material "projectile," for then its velocity would not be independent of its source's motion, as in Walther Ritz's theory.

Renouncing the concept of light as a thing traveling, continued Bridgman, enables one to use an alternative notion of velocity, namely, a self-measured velocity which is a "hybrid sort of thing" in that it combines the time on a moving observer's clock with distances marked off along a resting coordinate system. An example would be the velocity determined by an automobile rider through comparison of his clock with markers on the road.[4] Despite the peculiarity of mixing coordinates, the self-measured velocity is independent of how time is spread through space because it is a single-clock velocity (see also *SP*, p. 106). From the Lorentz transformations it follows that the self-measured velocity of light is infinite, and so it is not a physical velocity. Bridgman found nothing amiss in this result because it agreed with assuming that light is not a thing traveling and hence need no longer be thought of as having the customary "physical velocity."[5] He proposed no alternative theory of space and time to include an infinite self-measured velocity of light. But he emphasized that this tack could lead to a theory that would be independent of the structure of light, though the "simplicity and mathematical tractability"

---

4. In the (1927) book Bridgman did not use the term "self-measured" velocity (see *SP*, p. 26).

5. The self-measured velocity of light is what Bridgman referred to in *Sophisticate's Primer* as a "here and now" velocity, thereby eliminating any notion of light as a thing traveling. However, in *Sophisticate's Primer* Bridgman did not consider the self-measured velocity of light be to infinite (*SP*, p. 138).

indigenous to special relativity theory would very likely be lost. Bridgman considered the loss of these qualtities to be worthwhile if it led to a theory closer to the requirements of the operational view. In fact, he continued, the "whole problem of the nature of light is now giving the most acute difficulty."

Bridgman's conclusion was that whereas Einstein's special relativity theory succeeded in dealing with optical phenomena with a "simple mathematical formula . . . the explanatory aspect is completely absent from Einstein's work." But analysis of special relativity theory and Einstein's own description of it reveal that the theory makes no assumptions about the constitution of matter or light and attempts only to account for phenomena, i.e., it is a theory of principle. Furthermore, special relativity theory can discuss the propagational aspects of light quanta. Bridgman was dissatisfied with that sort of theory primarily because of difficulties concerning the nature and constitution of light. In his opinion "it would seem that we ought at least to start over again from the beginning and devise concepts for the treatment of optical phenomena which come closer to physical reality."

Toward Sophisticate's Primer: 1935–60

Comparing special relativity theory to general relativity theory in 1936, Bridgman emphasized the special theory's firmer experimental basis and its "fewer philosophical considerations"—that is, considerations of a nonoperational sort. As is discussed later, Bridgman never accepted general relativity theory. Yet he continued to be vexed by the problem of describing the propagation of light. He wrote (in

1936) that the "fundamental arguments of the special theory demand a light signal, thought of as expanding spherically through space and apprehended by many observers, and therefore consisting of many photons." Thus he accepted the exact validity of special relativity theory only for "comparatively large scale phenomena."

Bridgman thought highly of the new quantum mechanics because of what he took to be its firm operational basis.[6] On the other hand, he considered contemporaneous problems of relativistic quantum mechanics to be strong indicators of the inapplicability of special relativity theory to the microscopic domain. He suggested that nonrelativistic quantum mechanics offered a clue toward resolving the problem of the nature of light propagation because it showed it "unprofitable to attempt to visualize" such phenomena as the "photon as a thing that travels . . . we seem to be at the point where we must learn to get rid of this adventitious aid in thinking."

Over twenty years later Bridgman's attitude toward special relativity theory had hardly changed, despite breathtaking advances in quantum electrodynamics that had resulted in a consistent blending of special relativity theory and quantum theory. Thus in *Sophisticate's Primer* he wrote that "conventional relativity theory is not concerned with the second step toward the microscopic—that is, it does not concern itself with photons" (*SP*, p. 139, see also pp. 162–163). Almost certainly Bridgman's 1961 position on applying special relativity theory to the microscopic domain was the one he held in 1927, when he wrote that on the

---

6. Bridgman (1927) applauded advances in quantum theory recently achieved by "Heisenberg-Born and Schrödinger," who had emphasized understanding concepts in terms of "physical operations."

atomic side one could hardly expect that the concepts of space and time would "have the same operational significance . . . that they have on the ordinary scale." In the strict operational sense he was right, because in relativistic quantum mechanics special relativity is applied formally as Lorentz covariance; one does not speak of observers with inertial reference systems equipped with clocks and rods who ride on submicroscopic elementary particles.

Thus in *Sophisticate's Primer*, Bridgman accepted special relativity theory to be valid *ab initio* in the macroscopic domain and he wrote:

> The "critical" attitude of this essay is critical only to the extent that it is necessary to find out what the theory essentially is. The theory as ordinarily understood is accepted without question as a working tool which, in its ostensible universe of discourse, gives the best control we have yet been able to acquire of phenomena [although this] does not mean that all problems of understanding have been solved. This essay is directed toward these problems of understanding (*SP*, pp. 4–5).

The interpretation of relativity theory as a "control" over nature is consistent with Bridgman's operational point of view.

A set of notes written in 1959 pinpoints two principal problems with special relativity theory that concerned Bridgman:

> The conventional formulations of relativity seem to me unsatisfactory in at least two respects. In the first place, the device of using two observers in two different frames of reference, and the talk about how a certain phenomenon would appear to the moving observer and what he would

say about it seems clumsy and raises the implication that the moving observer may be deceiving himself with respect to the "reality." Furthermore it does not correspond to what actually happens. I cannot get away from myself — what happens in the other frames of reference eventually has to be described and find its meaning in terms of what happens to me. I should not have to depend on what the moving observer tells me, but I, with my instruments, should be able to make the readings that correspond to what the conventional moving observer is supposed to do. I am to be the central clearing house, and all instrumental readings should be made in such a way as to be utilizable by the central clearing house. This does not obscure the fact, however, that I, the central observer, have to concern myself with moving sticks and moving clocks. In the second place, the conventional treatment of light is as a thing travelling — a light signal is thought of as an identifiable thing travelling which can be watched as it travels just as a moving wave on water can be watched. But light is not a *thing* travelling, and it should be handled in a way not to involve any unconscious consequences of thinking of it in this way. In order to do this, the whole question of the true nature of radiation, which has been bothering me for so long, will have to be considered (italics in original, MS 11 July 1959).

In the course of completing *Sophisticate's Primer*, Bridgman realized a means to respond operationally to the problem of interpreting how light propagates through space; this response also constituted part of his resolution to the first problem—namely, the reformulation of physics in terms of a single observer. Before we reach that point, let us accept for the moment Bridgman's conviction that there is "something progressive about light" (*SP*, p. 41), and turn to the nature of the velocity of light, and in particular whether it is isotropic and whether the one-way velocity of light can be measured. In order to set the stage for Bridgman's analysis let us digress briefly to survey previous work by Poincaré and Einstein.

In a remarkable paper of 1898 entitled, "La mesure du temps," Poincaré investigated the connections between the nature of the velocity of light and the notion of time. Poincaré demonstrated that astronomers measured the velocity of light received from stars by agreeing *ab initio* that space is isotropic for light propagation and that the velocity of light is the same in every direction. Although this assumption could not be tested experimentally, it conformed with the apparent symmetry of space, i.e., the "principle of sufficient reason," which in turn furnished support for it.[7] Consequently, astronomers agree also on the meaningfulness of an unambiguous value for the one-way velocity of light.

Poincaré's (1898) analysis of the epistemological problems inherent in defining a one-way light velocity was based on a variant of Hippolyte Fizeau's well known 1849 measurement of the velocity of light, which he accomplished as follows (Fig. 1): Let a ray of light be emitted from a point A at the time t_A as registered on a clock at rest at A; after traveling a known distance $\overline{AB}$ the light ray is reflected from a mirror at B and arrives back at A at time t'_A. Since only one clock is involved here no problems are created by the relation between phenomena that occur at widely separated

---

7. Poincaré's (1898) analysis was based on his view that the foundations of geometry allowed for no preferred directions in the space described by Newton's mechanics. For further discussion, see Miller (1981; in press).

spatial points, i.e., distant simultaneity. Thus, Fizeau measured the velocity of light to be

$$c = \frac{2\overline{AB}}{t'_A - t_A}.$$ (1)

Then he generalized this result to apply to the one-way light velocity from A to B and from B to A. To make this generalization, Fizeau had assumed implicitly the equality of the to-and-fro velocities and the isotropy of space for the propagation of light. In fact, he had measured only the to-and-fro velocity and then generalized this measurement to the equality of the velocities of light from A to B (c_{AB}) and from B to A (c_{BA}), i.e., $c_{AB} = c_{BA}$.

Poincaré addressed the question of whether c_{AB} or c_{BA} could be measured separately. For example, he wrote in "Measure of Time," suppose a clock were placed at B. Then the velocity of light c_{AB} is (Fig. 2)

$$c_{AB} = \frac{\overline{AB}}{t_B - t_A}.$$ (2)

A vicious circle results since the measurement of $(t_B - t_A)$ requires c_{AB}, and therefore "such a velocity could not be measured without measuring a time." As a result of this analysis he concluded that, to ensure the descriptive simplicity of Newton's laws of mechanics, he had to assume that the velocity of light was the same in every direction and that the one-way value was Fizeau's; and the proper definition of time is the one from Newtonian mechanics and is independent of the reference system's motion, in agreement with our sense perceptions.

Portions of Poincaré's 1898 analysis were rediscovered

by Einstein in 1905, who pushed them to startling conclusions that could be plumbed only within a philosophical framework that was to a large extent liberated from sense perceptions. After proposing in-principle operational definitions for distance and time, Einstein (1905) discussed a method for synchronizing clocks based on Fizeau's procedure for measuring the velocity of light that is a Gordian resolution of the situations in Figures 1 and 2 (Fig. 3). Einstein placed a clock at B, and then defined a "common 'time' for A and B" through the equation

$$t_B - t_A = t'_A - t_B \tag{3}$$

which means that

$$c_{AB} = c_{BA}. \tag{4}$$

But Einstein did not need to consider Eq. (3) or (4) a definition, since the equality of the to-and-fro velocities of light follows from the two axioms of relativity, which in turn imply the isotropy and homogeneity of space for light propagation. Without the two axioms of relativity, Einstein's placing of clocks at A and B would have removed neither the vicious logical circle exposed by Poincaré nor the necessity for Fizeau's implicit assumptions. Having broken the vicious logical circle of clock synchronization, Einstein went on to define the one-way velocity of light as $c_{AB} = c_{BA} = c$, where

$$c = \frac{\overline{AB}}{t_B - t_A}. \tag{5}$$

Bridgman's unpublished notes of 29 July 1959 reveal that he had devoted much effort to the design of an experiment that could yield the one-way light velocity, and he thought he had succeeded. Bridgman's analysis of the experiment in Figure 4 shows that an anisotropy is expected, although it would probably be extremely small. But, Bridgman continued, he would not be "scandalized by this result," because he could interpret the anisotropy as an indicator of how much the laboratory system differs from a true inertial system.

The problem of spreading time through space is linked with simultaneity and the definition of velocity. Bridgman's notes during 1959–60 reveal that he was much occupied with this problem, and he gave it extended treatment in *Sophisticate's Primer*. His interest in synchronizing clocks by methods that differed from Einstein's is expressed in some of his early unpublished notes. For example, in September 1923 he wrote that, in order to measure the to or fro velocity of light without becoming enmeshed in a vicious logical circle, "we must set our clocks at the two stations by other means than the optical means employed by Einstein" (p. 40).

Herbert E. Ives's series of papers during 1937–51 may well have been the catalyst for Bridgman's reconsidering this issue. Although Ives was a life-long critic of relativity, his criticisms were constructive and were offered at the highest levels of scholarship.[8] Since only the to-and-fro velocity of light has operational meaning, Ives considered that Einstein had no grounds for asserting that the to-and-fro velocities were equal to the round-trip velocity. Ives's basic objection was to the principle of the constancy of the velocity of light, which he found "not only 'ununder-

---

8. In a classic experiment of 1938 Ives and G. R. Stilwell confirmed empirically the phenomenon of time dilation, although they considered that their data supported the Larmor-Lorentz ether-based theory (see Miller, 1981).

standable,' [but] *not* supported by 'objective matters of fact' " (Ives, 1951). In order to avoid using light signals to synchronize clocks, Ives suggested an alternative method of infinitely slow clock transport which depended on the self-measured velocity of the transported clock (see *SP*, pp. 65–66). He succeeded in deducing somewhat complicated Lorentz-type transformations that were functions of the self-measured velocities of moving rods and clocks, but which reduced to the customary Lorentz transformations when those velocities became zero. Ives (1951) said that his Lorentz-type transformations were derived by the "operational principle championed by P. W. Bridgman." This comment evidently prompted Bridgman to undertake a detailed critique of Ives's procedures (MS 18 August 1959, p. 3). Although Bridgman was not in total agreement with Ives's derivation of the Lorentz-type transformations and his applications of the operational method, he became interested in the notion of infinitely slow clock transport (*SP*, pp. 64–66). In view of Bridgman's definition of operational meaning from the 1927 *Logic of Modern Physics*, it may seem surprising to find him in 1961 advocating a version of infinitely slow clock transport in which the self-measured velocity of a transported clock is extrapolated to zero. Ives did not go to that extreme but wrote: "[Permitting the self-measured velocities to go to zero] means that the setting clocks would not be moved, the distant clocks would have no epoch allocated, and the measurement could not be made" (Ives, 1951). Bridgman criticized this assertion in *Sophisticate's Primer* (*SP*, p. 66), and in notes written in 1959 in a manner that illuminates how he had further developed the operational view since 1927: "It seems to me [that Ives's] argument is fallacious. Operational methodology does not mean that the limiting process be carried out *physically*—there is no reason why the physical operation should not be carried out for several values of [the self-measured velocity] different from zero and then mathematical extrapolation be made to zero" (italics in original, MS 18 August 1959).[9] For in Bridgman's developed form of operationalism, "mathematical extrapolation" is a well-defined mental activity. Bridgman's *ab initio* acceptance of special relativity theory in *Sophisticate's Primer* led him to conclude that clocks set by infinitely slow clock transport would agree with those set by Einstein's method (*SP*, p. 66).

The Conventionality of Distant Simultaneity

Einstein's method of synchronizing clocks using light signals can be reformulated so as to emphasize the conventional, i.e., the nonempirical, status of the one-way velocity of light. This method of analyzing simultaneity had been investigated systematically by Hans Reichenbach (1958) and later elaborated on by Adolf Grünbaum (1973). In his notes and in *Sophisticate's Primer*, Bridgman devoted considerable thought to the so-called Reichenbach-Grünbaum thesis of the conventionality of distant simultaneity.

Reichenbach generalized Einstein's Eq. (3) to

$$t_B = t_A + \epsilon \, (t'_A - t_A), \qquad (6)$$

where in agreement with the principle of causality $0 < \epsilon < 1$. The choice of $\epsilon = \frac{1}{2}$ corresponds to Einstein's special relativity theory in which the space of inertial ref-

---

9. Bridgman had further developed his notion of operational definition in several works (e.g., 1938, 1959a), after the 1927 publication.

erence systems is isotropic and homogeneous for the propagation of light, in which the definition of clock synchronization is transitive and symmetric, and where the symmetrical Lorentz transformations hold. Other values of ϵ correspond to $c_{AB} \neq c_{BA}$ because from Eq. (5) it follows that

$$c_{AB} = c/2\epsilon \qquad (7)$$

and

$$c_{BA} = c/2(1 - \epsilon); \qquad (8)$$

as required by the operational meaning of the round-trip
light velocity

$$\frac{1}{c_{AB}} + \frac{1}{c_{BA}} = \frac{2}{c}. \qquad (9)$$

According to Reichenbach, Einstein chose $\epsilon = \frac{1}{2}$ for
descriptive simplicity. Since any value of ϵ may be chosen
within the open interval $0 < \epsilon < 1$, then certain predictions
of special relativity are dependent on the choice of ϵ — that
is, are conventional. Bridgman set out to investigate the
conventional aspects of special relativity in a manner that
went beyond anything hitherto written on this subject. In
the MS of 27 July 1959, pp. 3–4, Bridgman wrote:

> Given now a "stationary" single-clock system, we shall of
> course find that uniformly moving meter sticks are shortened
> and the "rates" of uniformly moving clocks are altered. A
> first question to be answered is whether the shortening and
> the change of rate are given by the Lorentz formulas for
> inertial systems? One would suspect yes. Let us examine this
> in detail.
> Imagine two "stationary" systems (1) and (2) with identical
> geometrical mesh systems, and with locally distributed
> clocks running at the same rate, but with different zero
> settings $x_1 = x_2$.[*]

Set the clocks by despatching a light signal from the origin and reflecting it back from a mirror at the point x, the clock at the origin in either (1) or (2) reading $2x/c$ when it returns. Assume the Einstein method of setting clocks in (1) and assume that in (2) the settings are consistent with a real difference of "go" and "come" velocity of light and that the clocks give this real difference.

Time of arrival of signal in (1) is x/c

" " " " " " (2) satisfies $t_2 = \epsilon(2x/c)$ [†]

(Ives, Grünbaum)

$$\text{Hence } t_2 - t_1 = \frac{x}{c}(2\epsilon - 1) = ax/c,$$

where $0 < a < 1$. This holds at all times. The difference of clock settings entails a difference of velocities measured in the two systems.

Imagine an object moving with velocity v_1 in (1) leaving the origin at $t_1 = t_2 = 0$, and arriving at $x = 1$ at t_1. Then

$$v_1 t_1 = 1$$

The same object in (2) satisfies

$$v_2 t_2 = 1$$

$$t_2 - t_1 = ax/c = 1/v_2 - 1/v_1 \qquad 1/v_2 = 1/v_1 + ax/c$$

Hence

$$v_2 = \frac{v_1}{1 + av_1/c} \qquad v_2 = \frac{v_1 c}{c + axv_1} : x = 1$$

Hence velocities differ in (1) and (2) by quantities of the first order.

What now about the shortening of meter sticks? We are concerned with two moving frames — (1) with its system of clock setting and (2) with its. In the (1) system we assume that the Lorentz transformation holds

$$x'_1 = \frac{x_1 - v_1 t_1}{\sqrt{1 - \beta_1^2}}$$

$$t'_1 = \frac{t_1 - \frac{v_1}{c^2} x_1}{\sqrt{1 - \beta_1^2}}$$

The contraction of the moving meter stick in (1) is the value of x_1 when $x'_1 = 1$ and $t_1 = 0$: $x_1 = \sqrt{1 - \beta_1^2}\, x'_1$, the conventional value. What now is the contraction in (2): This is the value of x_2 at $t_2 = 0$ corresponding to $x' = 1$. When $t_1 = 0$ at $x_1 = \sqrt{1 - \beta_1^2}$, $t_2 = ax/c = \frac{a}{c}\sqrt{1 - \beta_1^2}$. The end of the meter stick has overshot its mark. Where was it at the time $\frac{a}{c}\sqrt{1 - \beta_1^2}$ earlier? It has been moving with velocity v_2, hence the distance $a\frac{v_2}{c}\sqrt{1 - \beta_1^2}$ and x_2 when $t_2 = 0$ is $\sqrt{1 - \beta_1^2}\left(1 - \frac{av_2}{c}\right)$.

The important question now is whether the Lorentz transformation applies to (2). In other words is

$$\sqrt{1 - \beta_1^2}\left(1 - \frac{a\,v_2}{c}\right) = \sqrt{1 - \beta_2^2} = \sqrt{1 - \left(\frac{v_2}{c}\right)^2}$$

or

$$\sqrt{1 - \left(\frac{v_1}{c}\right)^2}\left(\frac{1}{1 + \frac{a\,v_1}{c}}\right) = \sqrt{1 - \left(\frac{v_1}{c}\right)^2\left(\frac{1}{1 + \frac{a\,v_1}{c}}\right)^2}$$

evidently not for

$$\sqrt{1 - \left(\frac{v_1}{c}\right)^2} \neq \sqrt{\left(1 + \frac{av_1}{c}\right)^2 - \left(\frac{v_1}{c}\right)^2}.$$

[*] Since Bridgman assumed "identical geometrical mesh systems," the coordinate grids in (1) and (2) are identical.

[†] The result for t_2 follows from Eq.(7).

Thus Bridgman found that the root of the conventionality of distant simultaneity resides in the difference in settings for the zero points (i.e., of synchronization) between
two clocks running at the same rates in their rest systems.
Furthermore, a consequence of this difference in synchronization is a conventionality of relative velocities, which in
turn leads to the result that the velocity of (1) relative to
(2) is not the negative of the velocity of (2) relative to (1);
basic symmetries of physics are dropping by the wayside.

Bridgman's conclusion on his final equation is:

> This result is at first disconcerting. The shortening of the
> meter stick is not given by the same formula when the
> method of setting clocks is changed. Does this mean that the
> "shortening of the meter stick" contains an element of con-
> vention in the canonical transformation? This does not feel
> right. Physically it looks as though it ought to be possible to
> give a meaning to "length of a moving meter stick" ir-
> respective of the method of setting clocks (MS 27 July
> 1959, p. 4).[10, 11]

---

10. In his notes (MS 1 August 1959, p. 10), Bridgman recalled
hearing about a method proposed by Einstein to determine length that
involved two meter sticks of equal rest lengths moving in opposite directions with the same velocity, a method that was independent of any element of conventionality. In *SP* (pp. 92–93) Bridgman elaborated on that
method. His critique is incorrect, however, because the velocities need
not be equal and calculation shows that the length of the moving rods
measured in the stationary system is less than either of their rest lengths.
Einstein (1911) proposed this method without offering any calculation.
See Winnie (1972) for details. Among other cases that are independent
of conventionality are the round-trip time dilation effect and the increased distance of travel for unstable elementary particles that are
traveling close to the velocity of light (see Winnie, 1970).

11. In MS 26 May 1960 (pp. 11–12) Bridgman speculated on the
possibility of framing physical theory in a nonconventional mathematics

A few pages later Bridgman wrote:

> There is something funny about the "conventionality" of
> the single directional velocity of light. Now different values
> of ϵ in the stationary system correspond to different methods
> of setting clocks. Each one of these methods should give a
> physically consistent description of all physical phenomena
> if the precise choice of ϵ is a convention, and therefore to
> each different stationary system with a different ϵ, there
> should correspond a Lorentz transformation. Perhaps the
> answer is that the Lorentz transformation is not the most
> general one satisfying the requirements of relativity. [Bridg-
> [my][*]
> man inserted this sentence on 10 August 1959.] But all our

that conformed to the operationalism view and so avoided from the start
the notion of one-way light velocity. He went on to write:

> No sooner is a system of coordinates established in which the time
> variable is spread through space than we *have* to talk about "one-
> way" velocity, willy nilly, for the time derivative of the space coor-
> dinate is such a one-way velocity.
>
> The considerations of the last paragraph offer, I believe, a
> logical way out of the dilemma in which we apparently find our-
> selves. I believe, however, that average physicists, among whom I
> count myself, will react in a somewhat different way. The physicist
> has found conventional mathematics applicable in such an over-
> whelming number of physical situations, that I think he will react
> to the analysis above by the expectation that if the propagation of
> light is not isotropic, the failure of isotropy will be found to be very
> small. I think he would be very much surprised if future experi-
> ment ever showed that under some conditions a value for ϵ differing
> from ½ by first order terms might be demanded by experiments in
> terrestrial laboratories, a possibility which the analysis of Reichen-
> bach leaves open (italics in original).

Bridgman provided no details for these private speculations, and
he crossed the pages with diagonal pencil marks. By April 1961, he had
concluded that the one-way light velocity is unmeasurable (see pp. xxxii–
xxxiv below).

[*] The [my] is Bridgman's own insertion.

analysis seems to show that this is not possible (MS 1 August 1959, p. 10).

At least one aspect of the conventionality thesis that he may have considered "funny" he went on to describe thus:

> If it is only the go-come velocity of light that has operational meaning there is going to be an important reaction on conventional arguments in cosmology, the light from distant parts of the universe being unidirectional (ibid.).

Poincaré had in 1898 also stressed the point that it would be impossible to do research in astronomy without assuming the isotropy and homogeneity of space for light propagation, and that the one-way velocity of light was c.[12]

Detailed investigation of the conventionality of length and time in which ϵ is different in different inertial systems entails deducing ϵ-dependent Lorentz-type transformations, which Bridgman's unpublished notes reveal he was unable to derive. Yet he was convinced that the Lorentz transformation was the most general transformation satisfying special relativity. Bridgman's placement of the insert into the passage of 1 August 1959, quoted from above, emphasizes that he posed and then found to be unworkable the notion that the "Lorentz transformation is not the most general one satisfying the requirements of relativity." Thus he was led to conclude that the value of ϵ had to be the same in every inertial reference system, and since special relativity was valid, then ϵ had to be ½. This was pure

---

12. But Poincaré's conclusion held strictly only for astronomy. For in ether-based electromagnetic theory the space of inertial reference systems was in-principle not isotropic and homogeneous for the propagation of light. However, the null results of ether-drift experiments asserted otherwise (see Miller, 1981, esp. Chapter 1).

speculation, however, since he lacked the mathematical proof.

In September of 1960 Bridgman returned to the experiment in Figure 4, and found a formula for the anisotropy of light which was consistent with no difference of measured transit times from A to B and from B to A.[13] This result was perplexing because, as he had written in the manuscript of 30 July 1959 (see Fig. 4): "If these two Δ's are the same then the difference between the go and come times is not measurable, and their equality is a convention." Thus, he had to conclude in 1960 (MS 4 September 1960):

> It would seem that we are here confronted with a genuine logical and semantical dilemma. Is there any meaning to the [v_{AB} and v_{BA}] which we have used in our analysis beyond a purely paper-and-pencil meaning? Furthermore, if the principle of relativity is true, then the propagation of light must be isotropic. But there may be a conspiracy of nature which prevents us from checking whether light propagation is isotropic, and therefore, in so far, prevents us from checking whether the principle of relativity is true. In other words, there may be a conspiracy of nature to prevent us from checking the truth of the principle of relativity. If this is the case, then the principle of relativity itself must be recognized to be a "convention."

But earlier in this manuscript he had developed the proof that would appear in the published version of *Sophisticate's*

---

13. He wrote (MS 4 September 1960, p. 8):

$$\frac{1}{v_\theta} = \frac{1}{c} + \frac{1}{2}\left(\frac{1}{v_{AO}} - \frac{1}{v_{BO}}\right)\cos\theta$$

where the angle θ is defined in Figure 4, and v_{AB} and v_{BA} are found by setting θ equal to $\frac{\pi}{2} + \phi$ and $\frac{\pi}{2} - \phi$, respectively. Consequently, the times Δ_{AB} and Δ_{BA} become equal.

Primer that $\epsilon = \frac{1}{2}$ always owing to the principle of relativity (*SP*, pp. 88–89). This may well have been the part of the "analysis" that Bridgman mentioned in the passage from the notes of 1 August 1959, p. 10, that I quoted from previously, that proves that $\epsilon = \frac{1}{2}$. For the moment I postpone commentary on Bridgman's published proof which is incorrect because it is based on a space and time transformation whose spatial part displays the reciprocity of velocities between two inertial reference systems—that is, Bridgman did not include the conventionality of relative velocites.[14]

From January through early March of 1961 Bridgman wrote the first drafts of *Sophisticate's Primer*, in which he used the experiment from the notes of 1959 to argue at some length that a one-way velocity of light was not operationally definable. First of all, Bridgman noted the formula for the anisotropy of light, given in footnote 13, in which the measured transit times from A to B and from B to A were equal. More basically he realized that he had overlooked the point that the velocities of light from B to O and from A to O are two-clock velocities, as, he emphasized, are all velocities, because by definition velocity is the distance traveled in a time interval measured on two clocks that are situated at the beginning and end of the traversed spatial interval. Consequently a vicious logical circle ensues. As a result Bridgman wrote in the February 1961 manuscript of *Sophisticate's Primer* that Einstein had concluded correctly that "the question, is light *really* propagated isotropically?" can be answered only by posing an

---

14. In an epilogue to the first edition of *Sophisticate's Primer*, Grünbaum noted that Bridgman's conclusion was incorrect, but did not note where the error lay.

axiom system, or by definition if you prefer, because the isotropy of space for light propagation is linked to how clocks are synchronized.

Then on 7 April 1961 Bridgman inserted a page not intended for publication into the revised manuscript of *Sophisticate's Primer*. The inserted page is an interesting historical document, and in fact Bridgman very likely addressed it to posterity, since it instructs us how to read some of his unpublished notes of 1959, as well as the earlier versions of *Sophisticate's Primer*. We recall also that Bridgman was aware of his progressive illness. Bridgman wrote:

> In reading over any old material it is to be kept in mind that my thinking has experienced [certain] major changes. [One of them occurred] early in 1961, that "one-way" velocity of light can have no physical significance by itself, but is essentially a two-clock concept and has meaning only when the method has been specified by which time is spread over space. This realization negatived my efforts to find some physical method of measuring one-way velocity . . . .

This intellectual upheaval rendered the problem of measuring one-way light velocity "illegitimate" (*SP*, p. 47), because it involves a vicious logical circle. Thus, in the published version of *Sophisticate's Primer* after a brief description of the experiment in Figure 4 (*SP*, p. 47), Bridgman dismissed this measurement with short shrift (see also, *SP*, p. 48.)

What about the published incorrect proof in *Sophisticate's Primer* that $\epsilon = \frac{1}{2}$? Could this proof have been the result of an error on Bridgman's part? Having studied Bridgman's extant manuscripts I would say yes, but not merely an error. Rather I should like to suggest that it was an act of desperation. For as we have seen from his notes, after much travail he had convinced himself of the experimental impossibility of measuring the one-way light velocity, and also of the necessity to assume that the one-way light velocity would always be c in order to do any research, for example, in astronomy; furthermore, he was convinced of the validity of the special relativity theory within a restricted domain of applicability. Thus he next had to divest a basic quantity of the theory—the velocity of light—of any conventional characteristics. In fact, if the one-way light velocity were a convention, then, as Bridgman wrote in the MS of 4 September 1960, the "principle of relativity itself must be recognized to be a 'convention.' " According to Bridgman, however, a convention is not a basic statement of a physical theory but is a "more or less incidental consequence of more deep-seated properties" (see Bridgman, 1927, p. 117). Here he was in a quandary about what to do, because he had to present a proof for the necessity and sufficiency of $\epsilon = \frac{1}{2}$, so that "Einstein had no choice" (SP, p. 90).[15] Considering himself to be constrained to toe the operational line of sticking to the facts and not speculating in print on a quantity as basic as velocity, he offered the proof in *Sophisticate's Primer*, which is the one from the MS of 4 September 1960. Recalling that Bridgman usually refined and developed his writings further, I feel sure that he would have revised this proof in a later edition of *Sophisticate's Primer*. And almost certainly

15. In the revised MS of *Sophisticate's Primer*, Bridgman crossed out the statement that Reichenbach's "ϵ would have to be a function of v" in order to have ϵ be other than $\frac{1}{2}$. This, in fact, is the case (see Winnie, 1970).

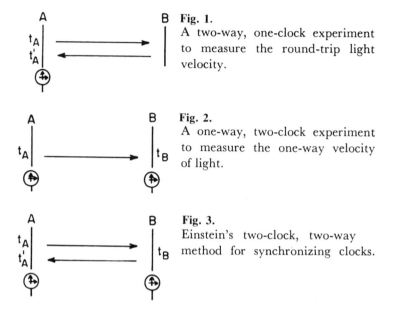

Fig. 1.
A two-way, one-clock experiment to measure the round-trip light velocity.

Fig. 2.
A one-way, two-clock experiment to measure the one-way velocity of light.

Fig. 3.
Einstein's two-clock, two-way method for synchronizing clocks.

he would have maintained that, whereas it is the case that there are alternate theoretical descriptions of phenomena, the individual scientist judges one among these theories to be of the greatest value because, as Einstein wrote in the 1905 relativity paper, it has removed "asymmetries which do not appear to be inherent in the phenomena."

From basic statements that are presumed to be independent of one-way velocity assumptions, John A. Winnie deduced ϵ-dependent Lorentz transformations that revert to the usual Lorentz transformations when $\epsilon = \frac{1}{2}$. (Winnie, 1970). Winnie rediscovered some of Bridgman's results on the conventionality of relative velocity and of length contraction—in particular, the conventional aspects of special relativity are rooted in real differences of clock synchronizations. Bridgman may well have glimpsed the sheer artificiality of the conventionality of distant simultaneity, which destroys symmetries of the sort that philosopher-scientists such as Einstein and Poincaré had in their own ways believed to be intrinsic to physical theories (see also Holton, 1973).[16]

The Observer

The first problem that Bridgman attacked in the manuscript of 11 July 1959 on the usual formulation of special relativity theory concerned the observer. As Bridgman put it, according to the operational view, "I am to be the central clearing house" (MS 11 July 1959, p. 1). This, in fact, is the way science is done when a single observer correlates meter readings whether he is performing a table-top experiment or correlating the registrations on instruments in different inertial reference systems through the application cf the Lorentz transformations. Bridgman described this mode of formulating special relativity theory with the Machian term "minimum point of view" (*SP*, p. 163). Einstein, on the other hand, had assumed a "psychological attitude" that entailed several independent observers. In order to support his view that relativity can be addressed only by a single observer who correlates meter readings, Bridgman in *Sophisticate's Primer* cited the Terrell effect (*SP*, p. 130).

---

16. For example, for a certain value of ϵ the length contraction of a rod oriented along K''s x-axis and observed in K can be eliminated for the relative velocity of K' relative to K in the positive x-direction, but not in the negative x-direction; this is the case also for time dilation. In these cases ϵ is not the same in K and K'. Winnie also deduced the conventionality of infinitely slow clock transport.

In almost every exposition of special relativity theory, wrote Bridgman, the device of one or more observers is used "uncritically" to discuss how a moving rod appears. In 1959 James Terrell, by taking into account that light does not reach the eye simultaneously from different parts of a moving object, used special relativity theory to calculate that an observer does not "see" a Lorentz contraction but rather sees other sorts of distortions. Consequently, Bridgman concluded that since expositions and applications of special relativity theory had been carried out for over fifty years without awareness of the Terrell effect, then the device of observers traveling with frames of reference and discussing what they see is superfluous, since it turns out

(*opposite page*) A page from Bridgman's MS "Inertial Systems" where he proposed a measurement for the one-way light velocity. There are clocks and mirrors at A and B and a single observer at 0 who is also equipped with a clock. A light beam goes from A to B. An observer at 0 registers the time difference between the flash of light at A and the flash of light at B that denotes the beam's arrival at B to be

$$\Delta_{AB} = \frac{AB}{v_{AB}} + \frac{BO}{v_{BO}} - \frac{AO}{v_{AO}},$$

where $AB = 1$, $AO = BO$, and v_{AB}, v_{BO}, and v_{AO}, are the velocities of light in the directions AB, BO, AO, respectively. Similarly, for a flash of light emitted from B

$$\Delta_{BA} = \frac{BA}{v_{BA}} + \frac{AO}{v_{AO}} - \frac{BO}{v_{BO}},$$

where v_{BA} is the velocity of light in the direction BA. Bridgman defined the angle θ to be the angle between the direction of light propagation and the direction from A to B.

Fig. 4. This suggets that we should go back and reexamine our assumption that operations are possible which will distinguish between the "go" and the "come" velocity of light

30/7/59

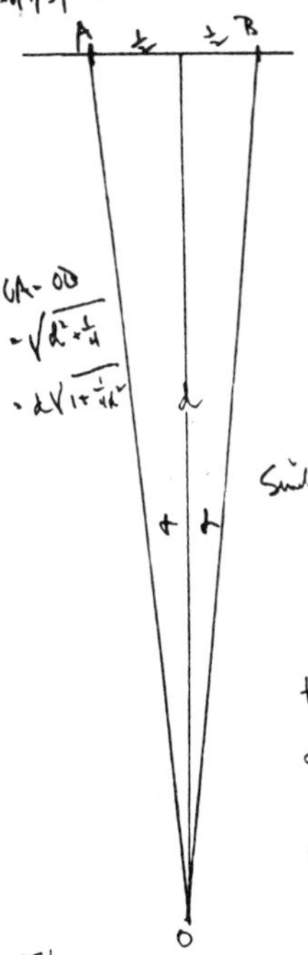

$OA = OB$
$= \sqrt{d^2 + \frac{1}{4}}$
$= d\sqrt{1 + \frac{1}{4d^2}}$

Given two points A & B unit distance apart, a distant observing station O with clock, & mirrors at A & B directed toward O to announce the arrival of lights at A & B.

Send first a beam of light from A to D. The observer at O times the difference between his observation of the arrival of lights at A & B. Assume the velocity of light depends on the direction. Then the difference of arrival time of the beams is

$$\Delta_{AB} = \frac{1}{v_{AD}} + \frac{d\sqrt{1+\frac{1}{4}d^2}}{v_{BO}} - \frac{d\sqrt{1+\frac{1}{4}d^2}}{v_{AO}}$$

Similarly

$$\Delta_{BA} = \frac{1}{v_{BA}} + \frac{d\sqrt{1+\frac{1}{4}d^2}}{v_{AO}} - \frac{d\sqrt{1+\frac{1}{4}d^2}}{v_{BO}}$$

If these two Δ's are the same then the difference between the go & come times is not measurable, and their equality is a convention.

If v is a function of direction, then if we assume that they are in the order indicated, there is at least partial compensation, and if v is the proper function of direction it is conceivable that the compensation may be complete so that $\Delta_{AB} \equiv \Delta_{BA}$. For

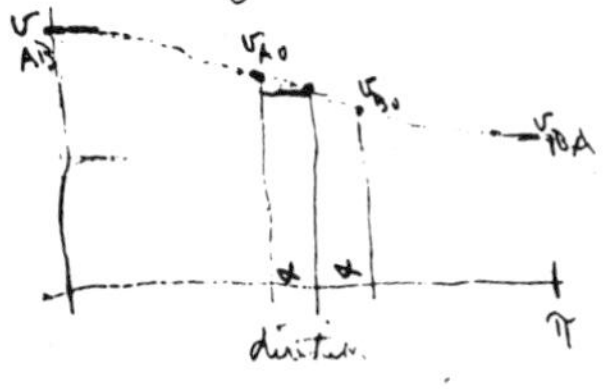

$$\Delta_{AB} = \frac{1}{v_{AB}} + d\sqrt{1+\frac{1}{4}d^2}\left(\frac{1}{v_{BO}} - \frac{1}{v_{AO}}\right)$$

$$\Delta_{BA} = \frac{1}{v_{BA}} - d\sqrt{1+\frac{1}{4}d^2}\left(\frac{1}{v_{BO}} - \frac{1}{v_{AO}}\right)$$

Let us now make a plausible assumption about the dependence of v on direction. Try

$$v_\theta = \frac{v_{AB} + v_{BA}}{2} + \frac{v_{AB} - v_{BA}}{2}\cos\theta$$

that an observer "may be deceiving" himself (*SP*, p. 130). Special relativity theory can be applied by a single observer at rest in the laboratory who correlates instrument readings from instruments either at rest or moving relative to him (*SP*, p. 151). For Bridgman these meter readings were the sole physical reality, and seeking these readings under every conceivable circumstance was "part of the ultimate task of the physicist" (*SP*, p. 152). Thus the electromagnetic field, and the propagation of light as well, are ensembles of meter readings, and a "field-in-the-absence-of-an-*instrument* has no meaning" (italics in original, *SP*, p. 157). This view of physical reality accorded Bridgman an instrumentalist definition of the electromagnetic field equations—they permit us to refine our meter readings (*SP*, p. 159). In agreement with Gustav Kirchhoff's widely quoted 1876 statement, Bridgman concluded that a "simple description, it seems to me, is the inevitable end of any analysis of which we are capable" (ibid.).

Consistent with his philosophy of science, Bridgman deemed that the role of the individual observer—the "I"—was essential to the enterprise of doing science. He wrote (1949) that science is "public" only to the extent that two or more observers can agree on a phenomenon. But, he emphasized, "it is a matter of simple observation that the private comes before the public." In my opinion, Bridgman's strong belief and life-long struggle to maintain the "I" in the scientific enterprise lay behind his never accepting general relativity theory. Most likely as a result of not having read Bridgman's earlier writings, Einstein (1949) missed this point when he dismissed Bridgman's at times strongly put and not always substantively based criticisms of general relativity theory with the comment that "to consider a logical system as physical theory it is not necessary to demand that all of its assertions can be independently interpreted and 'tested,' 'operationally.' " Here Einstein addressed himself only to the part of Bridgman's critique that concerned Einstein's transference of physical reality from space-time intervals to the metric tensor. Bridgman's deep scientific-philosophic concerns about general relativity were: (1) that it focused on the coordinate system rather than on events themselves because there were no standard measuring clocks and rods as in special relativity theory; and (2) that "certainly the feeling that one gets from reading many of the fundamental expositions, even those of Einstein himself, is that this business of getting away from a special frame of reference and observer to something not fettered to a special point of view is very important," i.e., the general relativity theory's emphasis on covariance (1936). This led Bridgman (1949) to conclude: "Perhaps the most sweeping characterization of Einstein's attitude of mind [i.e., realist philosophy] with regard to the general theory is that he believes it possible to get away from the special point of view of the individual observer and sublimate it into something universal, 'public,' and 'real.' " Bridgman's opinion of realism versus operationalism in the 1940s was very likely the one that appears in *Sophisticate's Primer* (pp. 154–155) and that lies at the crux of his concern regarding the general theory—i.e., that it excluded the observer. He had expressed this point most strongly in 1936:

> It cannot be too strongly emphasized that there is no getting away from preferred operations and a unique standpoint in physics; the unique physical operations in terms of which interval has its meaning afford one example, and

there are many others also. There is no escaping the fact that it is *I* who have the experiences that I am trying to coordinate into a physical theory, and that *I* must be the ultimate center of any account which I can give. I and my doings must be specially set apart and, in perhaps the only possible sense of the word, constitute an absolute. It seems to me that to attempt to minimize this fact constitutes an almost wilful refusal to accept the obvious structure of experience. In so far as the general spirit of relativity theory postulates an underlying "reality" from which this aspect of experience is cancelled out, it seems to me to be palpably false, and furthermore devoid of operational meaning.

Furthermore, in the light of this quotation we can understand Bridgman's additions to his notes of 1 August 1959—concerning all "our [my]" analyses (see p. xxx of this Introduction).

Bridgman's emphasis on the "I" was sometimes criticized as solipsistic. His calling attention to the Terrell effect in *Sophisticate's Primer* can be taken to be an explicit science-based defense of the consistency of maintaining the point of view of a single observer who reads instruments. With the forthrightness that had become his trademark, in the *The Way Things Are* (1959b) he defended the operational philosophy against the charge of solipsism. (Unfortunately the Terrell effect was as yet unknown when he wrote that book.) Here Bridgman wrote that "there *has* to be a plurality of sorts because there are many people, and there *has* to be a dichotomy of sorts" because there is, for example, your-science and my-science. This necessary plurality and dichotomy he took as grounds for not uncritically accepting any unitary world view, even the positivistically inclined Unity of Science movement that emphasized public science. The problem that Bridgman faced was how to fit this plurality and dichotomy into "my picture of the world," without being accused of solipsism. In *The Way Things Are* he provided a Gordian solution of his own. Bridgman dismissed as "innocuous" the unavoidable pluralistic aspects of "viewing everything from myself as center," because this description of nature gives us the "only unity we can use, the only unity we need, and the only unity possible in the light of the way things are."

Although one may not always agree with the operational point of view, I feel sure that students and teachers alike will pay Bridgman's final book on the philosophy of science that highest compliment of picking up a pencil in order to follow better his train of thought, and to write their own views in the margins as a result of their having been inspired to "stand back a little for critical scrutiny of what he has really got."

Acknowledgment

It is a pleasure to acknowledge Ms. Maila Walters' assistance with the Bridgman Archives at Harvard University.

[*Bibliography* begins on next page.]

Bibliography

Bridgman, Percy W.
  Manuscripts:
  27 December 1922   Preliminary for Space Time Symposium.
  September 1923   Critical Discussion of Relativity.
  11–18 July 1959   Reformulation of Special Relativity.
  24 July –   Inertial Systems.
    2 August 1959
  18 August 1959   Comment on Ives' Papers on Relativity.
  26 May 1960   Distance as a 'Convention.'
  4 September 1960   Two Footnotes to Special Relativity
                 Theory
  January–April 1961 Manuscripts of *Sophisticate's Primer*

  Published Materials:
  1924   A Suggestion as to the Approximate Character of the
        Principle of Relativity, *Science*, *59*, 16–17.
  1927   *The Logic of Modern Physics* (New York: Macmillan).
  1936   *The Nature of Physical Theory* (New York: Dover,
        1936). Based on three Vanuxem lectures delivered
        during December 1935 at Princeton University.
  1938   Operational Analysis, *Philosophy of Science*, *5*, 114

(1938); reprinted in P. W. Bridgman, *Reflections of a Physicist* (New York: Philosophical Library, 1955), pp. 1–26.

1949 Einstein's Theories and the Operational Point of View, in P. A. Schilpp (ed.), *Albert Einstein: Philosopher-Scientist* (Evanston, Ill.: Library of Living Philosophers, 1949), pp. 333–354.

1954 Remarks on the Present State of Operationalism, *Scientific Monthly*, 79, 224–226 (1954); reprinted in *Reflections of a Physicist*, pp. 160–166.

1959a The Logic of Modern Physics after Thirty Years, *Daedalus, 88, 518–526.*

1959b *The Way Things Are* (Cambridge, Mass.: Harvard University Press.

1962 *A Sophisticate's Primer of Relativity* (1st ed. Middletown, Conn.: Wesleyan University Press.

Einstein, Albert

1905 Zur Elektrodynamik bewegter Körper, *Annalen der Physik*, *17*, 891–921 (1905); translated in Miller (1981).

1911 Zum Ehrenfestschen Paradoxen, *Physikalische Zeitschrift*, *12*, 509–510.

1923 Fundamental Ideas and Problems of the Theory of Relativity. Lecture delivered on 11 July 1923 to the Nordic Assembly of Naturalists at Gothenburg, in acknowledgement of the Nobel Prize. Reprinted in *Nobel Lectures, Physics: 1901–1921* (New York: Elsevier, 1967), pp. 479–490.

1949 Reply to Criticisms, in P. A. Schilpp (ed.), *Albert Einstein: Philosopher-Scientist* (Evanston, Ill.: Library of Living Philosophers, 1949), pp. 665–688.

Grünbaum, Adolf

1962 Prologue and Epilogue to P. W. Bridgman, *A Sophisticate's Primer of Relativity* (1st ed., Middletown, Conn.: Wesleyan University Press.

1973 *Philosophical Problems of Space and Time. Boston Studies in the Philosophy of Science, 12* (Dordrecht: Reidel).

Holton, Gerald
1973 *Thematic Origins of Scientific Thought: Kepler to Einstein* (Cambridge, Mass.: Harvard University Press).

Ives, Herbert E.
1951 Revisions of the Lorentz Transformations, *Proceedings of the American Philosophical Society*, *95*, 125–131.

Kemble, Edward (with Francis Birch and Gerald Holton)
1970 Bridgman, Percy Williams, in C. C. Gillispie (ed.), *Dictionary for Scientific Biography*, *2*, 457–461.

Mach, Ernst
1960 *The Science of Mechanics: A Critical and Historical Account of its Development* (1st German edition 1883; LaSalle, Ill.: Open Court, 1960), translated by T. J. McCormack.

Miller, Arthur I.
1978 Visualization Lost and Regained: The Genesis of the Quantum Theory in the Period 1913–1927, in J. Wechsler (ed.), *On Aesthetics in Science* (Cambridge, Mass.: MIT Press), pp. 72–102.

1981 *Albert Einstein's Special Theory of Relativity: Emergence (1905) and Early Interpretation (1905–1911)* (Reading, Mass.: Addison-Wesley, 1981).

1982 Redefining *Anschaulichkeit*, in *Festschrift for Laszlo Tisza* (Cambridge, Mass.: MIT Press).

in press Poincaré and Einstein: A Comparative Study, in press A. I. Miller, *On the Nature of Scientific Discovery* (Cambridge, Mass.: Birkhäuser, in press).

Poincaré, Henri
1898 La mesure du temps, *Revue de Métaphysique et de Morale*, *6*, 371–384 (1898); reprinted in H. Poincaré, *Value of Science* (Paris: Flammarion, 1905; New York: Dover, 1958), pp. 26–36, translated by G. B. Halsted.

1902   *Science and Hypothesis* (Paris: Flammarion; New York: Dover, 1952), translator unknown.

Reichenbach, Hans
  1958   *The Philosophy of Space and Time* (Berlin: Gruyter, 1924; New York: Dover, 1958), translated by M. Reichenbach and J. Freund.

Winnie, John A.
  1970   Special Relativity without One-Way Velocity Assumptions, *Philosophy of Science*, I, *37*, 81–99; II, 223–238.
  1972   The Twin-Rod Thought Experiment, *American Journal of Physics*, *40*, 1091–1094.

Recent Praise for Arthur I. Miller

Imagery in Scientific Thought
Creating Twentieth Century Physics

"Miller's book differs from the many other studies of the origins of twentieth century physics in treating together three facets of the development of relativity and quantum theory . . . an internal history of the major ideas; an account of the epistemologies and the mental images of Boltzmann, Poincaŕe, Einstein, Bohr, and Heisenberg . . . a discussion of the adequacies of gestalt psychology and of Piaget's genetic epistemology to account for creative thinking.

Students of physics will find a clear analysis both of the fine structure of the genesis of Einstein's 1905 relativity paper (including convincing corrections to Max Werthei-mer's gestalt reconstruction) and of Heisenberg's developing understanding of quantum theory. They will also be helped to visualize quantum theoretical concepts such as exchange forces. Historians of science are challanged to apply developments in cognitive science to the invention of scientific ideas, especially to the role of visualization in creative thinking."

—Choice

"With impressive documentation, using much hitherto unavailable material, physicist-historian Miller discusses the often contrasting styles of the creators of modern physics . . . . Physicists, psychologists, and anyone with an interest in the question of what constitutes the 'creative scientific mind' will find this fascinating book rewarding reading."

— Felix M.H. Villars
—Massachusetts Institute of Technology

1984/ xvi, 355 pages / illus.
ISBN: 0-8176-3196-8
A *Pro Scientia Viva* Title